M. E. Fine

HEAT TREATMENT OF FERROUS ALLOYS

HEAT TREATMENT OF FERROUS ALLOYS

Charlie R. Brooks

Professor of Metallurgical Engineering
University of Tennessee

Hemisphere Publishing Corporation
Washington New York London

McGraw-Hill Book Company
New York St. Louis San Francisco Auckland Bogotá Düsseldorf
Johannesburg London Madrid Mexico Montreal New Delhi
Panama Paris São Paulo Singapore Sydney Tokyo Toronto

HEAT TREATMENT OF FERROUS ALLOYS

1 2 3 4 5 6 7 8 9 0 L I L I 7 8 3 2 1 0 9

This book was set in Press Roman and edited by Communication Crafts Ltd. LithoCrafters, Inc., was printer and binder.

Library of Congress Cataloging in Publication Data

Brooks, Charlie R. date
Heat treatment of ferrous alloys.

Bibliography: p.
Includes index.
1. Steel alloys–Heat treatment. 2. Cast iron–Heat treatment. 3. Physical metallurgy. I. Title.
TN751.B77 672'.3'6 78-16513
ISBN 0-07-008076-3

CONTENTS

PREFACE

In teaching physical metallurgy, it is important that the students obtain an understanding of the relationships between the principles of the behavior of metallic materials and the industrial processes actually used in the mechanical, thermal, and chemical treatment of these materials. I believe that is especially important to stress the relationships between phase equilibria, kinetics of phase transformations, and the development of microstructure, and the properties obtained. I have attempted to do this in this book, which deals with the physical metallurgy principles of the heat treatment of steels and cast irons. Emphasis is placed on the relationships between the phase diagrams, the kinetics of the solid-state changes, and the microstructure developed. The effect of the microstructure on mechanical properties is then emphasized. This approach is taken for several classes of alloys (e.g., tool steels, stainless steels), and the heat treatments of commercial alloys are used to relate these principles to industrial practices.

There are no problems listed for students' use. Instead, I have examined in some detail in a separate chapter the design of specific heat treatments. The only prerequisite to the use of this book is an introductory course in metallurgy or material science. The book will be usable in courses for students in materials and metallurgy, as well as in service courses for mechanical engineering students and those in related areas.

Although it is clear that SI units will eventually be used, at the present time the American Engineering System of units is still widely employed. Thus I have made no serious attempt to convert data to the SI system; some conversions factors are in Appendix II. The magnification of the microstructures is given by a marker whose real length is shown in micrometers (μ), units of 10^{-4} cm. The method of imaging the microstructures dictates to some extent their appearance, so I have indicated those which have been obtained by using an optical microscope by OM and those obtained by using a scanning electron microscope by SEM.

I especially express my appreciation to my colleague Professor E. E. Stansbury (under whom I studied) for allowing me to use some of his ideas in this book. Appreciation is expressed also to my colleagues Professors J. E. Spruiell and W. T. Becker for reading part of the manuscript and for making suggestions. The support of the Chemical, Metallurgical and Polymer Engineering Department in writing this book is appreciated.

Charlie R. Brooks

CHAPTER

ONE

THE HEAT TREATMENT OF STEELS: A REVIEW

Steels are a class of iron-carbon alloys, with other elements added, which comprise one of the most widely used materials, both as final products (e.g., automobile parts, electrical transformer parts) and in manufacturing equipment for processing (e.g., rolling mills for fabricating copper sheet, extrusion presses for processing polymers, reactors for carrying out chemical reactions). One of the main reasons for their wide use is the range of properties which can be induced by various heat-treating procedures. As an illustration of this, consider Fig. 1-1. Here the microstructures of various iron-carbon alloys, in various heat-treated conditions, are shown, along with some mechanical-property data. It is seen that an increase in the carbon content of iron from essentially zero to 0.8% will increase the yield strength by about a factor of 4 (from 15,000 psi to 65,000 psi) and lower the ductility (from 62% to 14%). As will be made clear in this chapter, this increase in hardness is due to an increase in the amount of a carbide (Fe_3C) when the carbon content is increased. Upon reaching 0.8% carbon, the microstructure consists of alternate plates of iron carbide (Fe_3C) and essentially pure iron (see Fig. 1-1). The carbide, being hard, strengthens the iron by its presence.

The influence of heat treatment on the properties is illustrated for the 0.8% carbon steel. If this steel is heated to about 1000°C for 1 hour, and then cooled slowly (e.g., in 24 hours) to 25°C, the microstructure is as shown in Fig. 1-1*d* and the steel has a yield strength of 65,000 psi. However, if the steel is heated to 1000°C for 1 hour, then cooled very rapidly to 25°C (water quenched), the microstructure is radically altered (Fig. 1-1*f*) and the yield strength is increased by about a factor of 5 over the slowly cooled condition (65,000 psi compared to 300,000 psi). Further, the ductility is decreased from 14% to about 1%; thus, although the water quench strengthened the steel, it is now rather brittle. However, if the water-quenched steel is reheated to

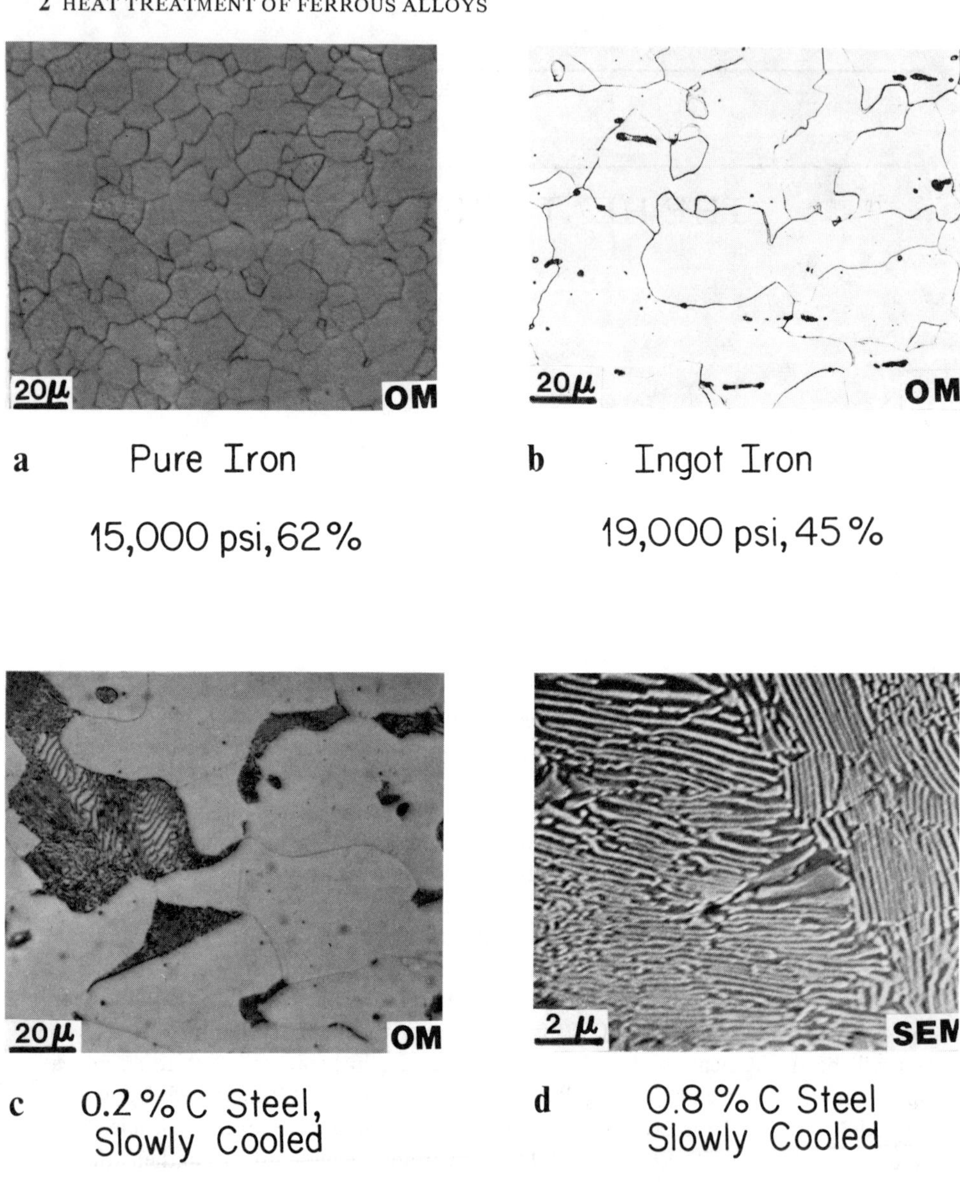

Figure 1-1 Microstructures of steels in various heat-treated conditions. The approximate yield strength and elongation at fracture are shown also (tested at 25°C) *(Adapted by courtesy of Professor E. E. Stansbury.)*

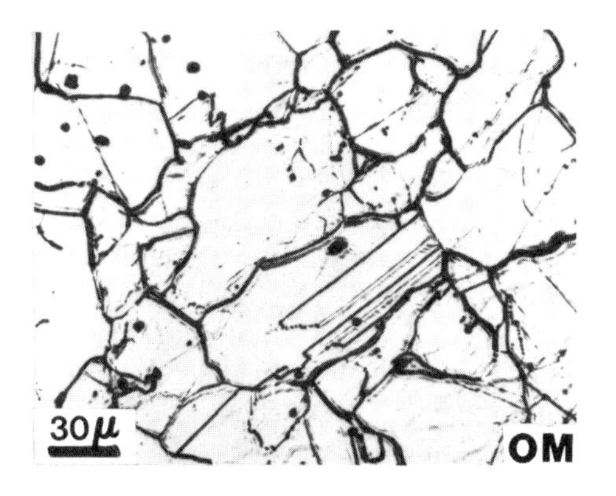

e 0.8% C Steel
Above 723 °C

f 0.8% C Steel,
Water Quenched

300,000 psi, about 1%

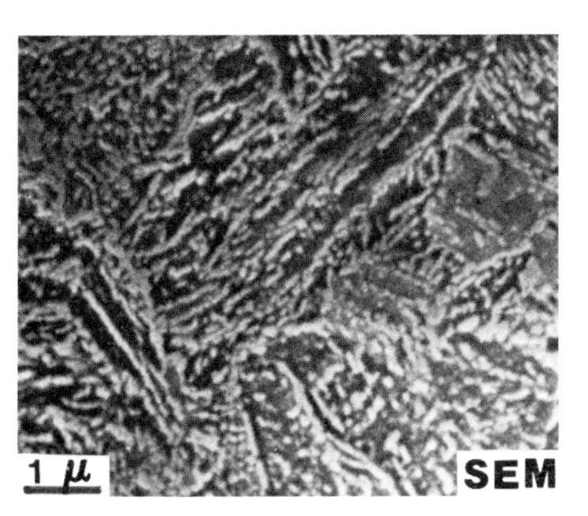

g 0.8% C Steel, Water
Quenched, reheated to
500 °C for 1 hour

140,000 psi, 7%

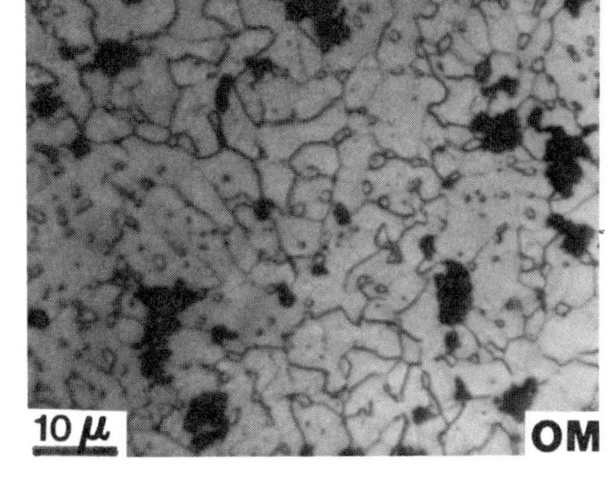

h 0.8% C Steel, Water
Quenched, reheated to
700°C for 100 hours

60,000 psi, 20%

Figure 1-1 *(continued)*

500°C for 1 hour, then cooled to 25°C, the microstructure is as shown in Fig. 1-1*g* and the elongation of the steel is increased to 7%, although the strength has decreased from 300,000 to 140,000 psi.

The point is that the wide use of steels arises from the range of properties which can be developed, either by altering the carbon content, or, for a given carbon content, by altering the heat treatment. This range of properties is associated with the type, size, and distribution of phases (e.g., the iron carbide), frequently on a scale approaching atomic dimensions.

This chapter *reviews* in some detail the principles of the heat treatment of the simpler steels as background to detailed treatment of the methods of choosing specific heat treatments and the principles of the heat treatment of complex steels.

1-1 IRON–CARBON PHASE EQUILIBRIUM

The wide range of properties developed in steels by utilizing different thermal treatments is associated with the decomposition of the high-temperature, face-centered cubic form of iron, austenite (or γ). Thus, attention can immediately be centered on that part of the iron-carbon diagram which concerns the phases present below about 1100°C.

The equilibrium diagram for the iron-carbon system is reproduced in Fig. 1-2. Below 1100°C, there are three phases which will be of concern (see Fig. 1-3), provided that the carbon content is in the range normally encountered (i.e., 0% to 2%). Above 723°C, the phase of interest is face-centered cubic (fcc) *austenite*, or γ. The solubility of carbon in this structure is relatively high, being a maximum of about 2% carbon at the eutectic temperature, 1130°C. The solubility of carbon in the low-temperature, body-centered cubic (bcc) form of iron (α, or *ferrite*) is quite low, being a maximum of about 0.025% carbon at the eutectoid temperature of 723°C. In the proportion of three iron atoms to one carbon atom, a compound is formed which is Fe_3C, *iron carbide*, or carbide. Its chemical composition is 6.67% carbon, corresponding to one carbon atom to three iron atoms, or 25 atom % carbon. Because of its negligible solubility for iron or carbon, Fe_3C is shown on the phase diagram (Figs. 1-2 and 1-3) as a single vertical line.

The following review is given as an aid to the understanding of phase diagrams. The line *abcf* in Fig. 1-4 gives the variation with temperature of the solubility of carbon in ferrite. Thus, at a given temperature, if progressively increasing numbers of carbon atoms are placed in the interstices between the iron atoms in the bcc ferrite lattice, a limit is reached beyond which the additional carbon atoms will not dissolve in the bcc ferrite but instead "react" with some of the iron atoms to form another phase. For example, at 780°C (see Fig. 1-4), the solubility limit is about 0.01% carbon (point *b*); additional carbon added to the ferrite beyond this value causes the precipitation of austenite, the austenite containing 0.4% carbon (point *d*, Fig. 1-4). At 650°C, the addition of carbon to ferrite beyond the solubility limit at this temperature (about 0.01% carbon) causes the precipitation of iron carbide.

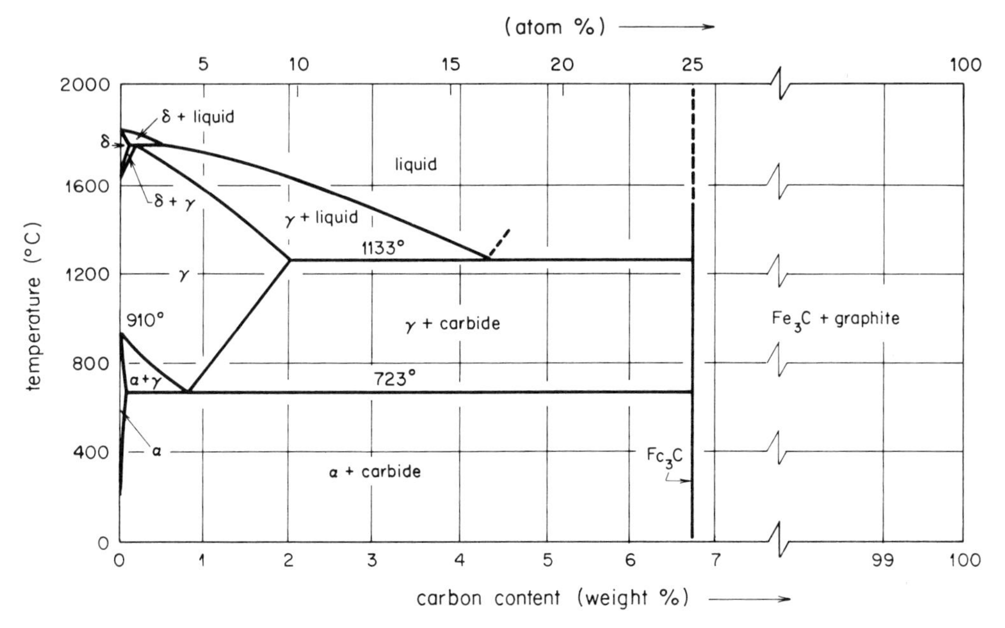

Figure 1-2 The iron-carbon phase diagram. *(Adapted from "Metals Handbook," 8th ed., vol. 8, American Society for Metals, Metals Park, Ohio, 1973.)*

It is seen, then, that the solubility of carbon in the fcc form of iron is much greater than in the bcc form. Although the fcc lattice, based on a rigid-sphere model, has less free (or unoccupied) volume, the factor of importance is the size of interstices available in which to place the carbon atoms. Calculations for the fcc and bcc structures show that the available interstices in the fcc austenite are much larger than in the bcc ferrite; thus austenite has a much greater solubility for carbon than does ferrite.

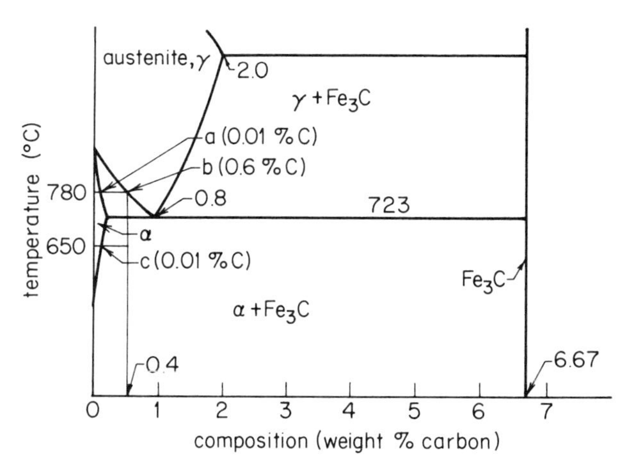

Figure 1-3 Portion of the iron-carbon phase diagram of interest in the heat treatment of steels.

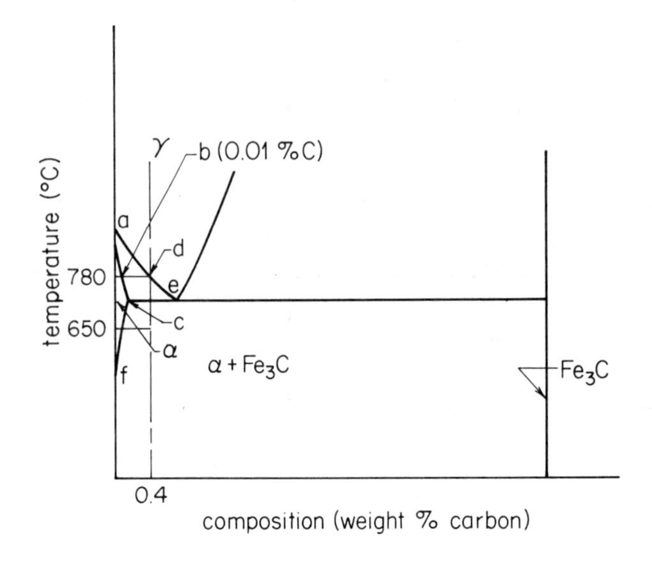

Figure 1-4 Portion of the iron-carbon phase diagram.

1-2 MICROSTRUCTURE AND PROPERTIES OF SLOWLY COOLED STEELS

Consider now the microstructural changes that occur upon cooling a 0.4% carbon steel from the austenite, say from 1000°C. It is assumed that the cooling rate is sufficiently slow that the phase diagram is followed. Such a cooling rate can be obtained by cooling to 25°C in, say, 24 hours. In the terminology of the heat treating of steels, this heat treatment is sometimes called *annealing.* Unfortunately, annealing is also used in another context concerning the heat treatment of cold-worked materials. In this chapter, however, annealing will be taken to mean slow cooling from the austenite region.

Referring to the phase diagram and to Fig. 1-5, the following description of austenite decomposition upon slow cooling can be developed. Upon reaching the solubility line *ae* at 780°C, the two-phase, α and γ, region is entered. Hence ferrite must begin to form in the austenite. The most reasonable location for the fcc iron structure to rearrange to form the bcc ferrite is at austenite grain boundaries, since in the localized region of the boundary the iron atoms are already somewhat randomly arranged and can move more freely than in the bulk of the grain. Thus it will be assumed that the ferrite nucleated in the austenite grain boundaries. Such a nucleation is illustrated in Fig. 1-5. Figure 1-5*a* shows a two-dimensional representation of two austenite grains, in a hexagonal array. A group of 12 atoms in the region of the boundary rearrange to form a square array, representing nucleation of a ferrite crystal. This is shown in Fig. 1-5*b*. This ferrite nuclei then grows by atoms in the austenite at the austenite-ferrite interface taking up the square arrangement. Figure 1-5*c* depicts the appearance of the ferrite grain when it has grown to contain 32 atoms. This is a schematic representation, and the local adjustments right at the grain boundaries are not shown.

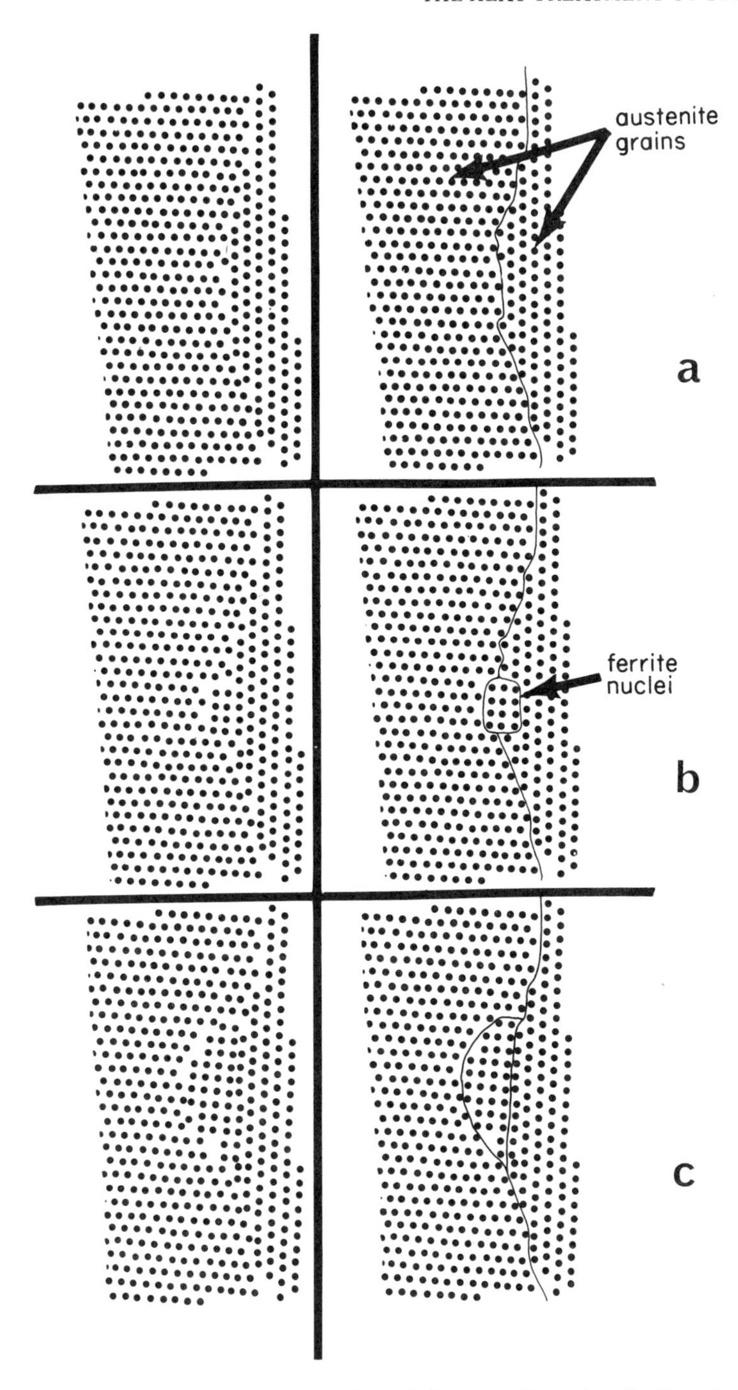

Figure 1-5 Schematic representation of the nucleation of a ferrite grain at an austenite grain boundary. The sequence of pictures is for increasing time from (a) to (c). The left-hand and right-hand pictures are the same region, except that a line is shown along the grain boundary.

Now consider what happens to the carbon atoms as the ferrite forms. If it is assumed that the grouping of 12 atoms in Fig. 1-5*b* which will become the ferrite (Fig. 1-5*c*) extends back into the paper about 4 atom layers, then the volume under consideration contains about 64 atoms. If the austenite contains 0.4 *weight %* carbon, which corresponds to about 2 *atom %* carbon, then these 64 iron atoms will have contained within their volume about 1.3 carbon atoms, *on the average.* Thus the volume shown in Fig. 1-5*b* should have either one or two carbon atoms in it. Of course, these might not be visible in the particular plane of cut shown in the figure, but a few are shown on this plane in Fig. 1-6. Now, realizing that the solubility of carbon in ferrite is quite low (we will assume it to be zero), when these 64 iron atoms are rearranged to the bcc ferrite (square array in Fig. 1-5*c*), then the carbon that *was* in this volume must now be in the austenite-ferrite interface. Thus the concentration of carbon atoms in the interface is higher than in the bulk austenite (at a distance away from the interface). This means that a concentration profile similar to that shown in Fig. 1-6 is generated, with carbon in the austenite diffusing back into the austenite away from the austenite-ferrite interface as the ferrite crystal grows. Since diffusion of carbon atoms requires a finite time, the rate at which the interface advances is limited.

If the temperature is held constant just below the phase boundary, say at 775°C (Fig. 1-7), then the ferrite crystal shown in Fig. 1-6 will continue to grow until the austenite has a uniform carbon content given by the austenite-ferrite phase boundary, 0.41% carbon (Fig. 1-7). If cooling is not interrupted, but continues slowly, the ferrite crystal continues to grow, and the chemical composition of the austenite at any temperature is given by the intersection of the isotherm with the line *ae* (Fig. 1-7). Thus just above the eutectoid temperature of 723°C, the austenite contains essentially 0.8% carbon, and the ferrite contains 0.025% carbon. At this temperature, which we will designate as 723+ °C, the amount of α and γ can be calculated. If we assume that we have 100 mass units of steel (say 100 g), then

$$100 = W_\alpha + W_\gamma$$

where W_α is the weight of α present, and W_γ is that of the γ. A carbon balance gives

Amount of carbon in steel
= amount of carbon in ferrite + amount of carbon in austenite

This expression can be written as

(Amount of steel)(% carbon in steel)
= (amount of α)(% carbon in α) + (amount of γ)(% carbon in γ)

or

$$(100)(0.4) = (W_\gamma)(0.8) + (W_\alpha)(0.025)$$

Substituting for W_α in the expression (from above) $W_\alpha = 100 - W_\gamma$,

$$40 = 0.8W_\gamma + 0.025(100 - W_\gamma)$$

Rearranging this expression, we get $W_\gamma/100$, or the fraction of the 100 g of steel which is austenite at this temperature of 732+ °C. This expression becomes

$$\frac{W_\gamma}{100} = \frac{0.4 - 0.025}{0.8 - 0.025}$$

which is almost 50%. Thus the steel contains 50% austenite and 50% ferrite.

Now consider what changes occur in the steel upon cooling from just above 723°C to just below 723°C. First, we examine what would happen upon removing the austenite regions from the ferrite regions, and cooling the austenite regions to just below the eutectoid temperature (say 723− °C). The austenite, just removed from the steel at 723+ °C, contains 0.8% carbon, and according to the phase diagram

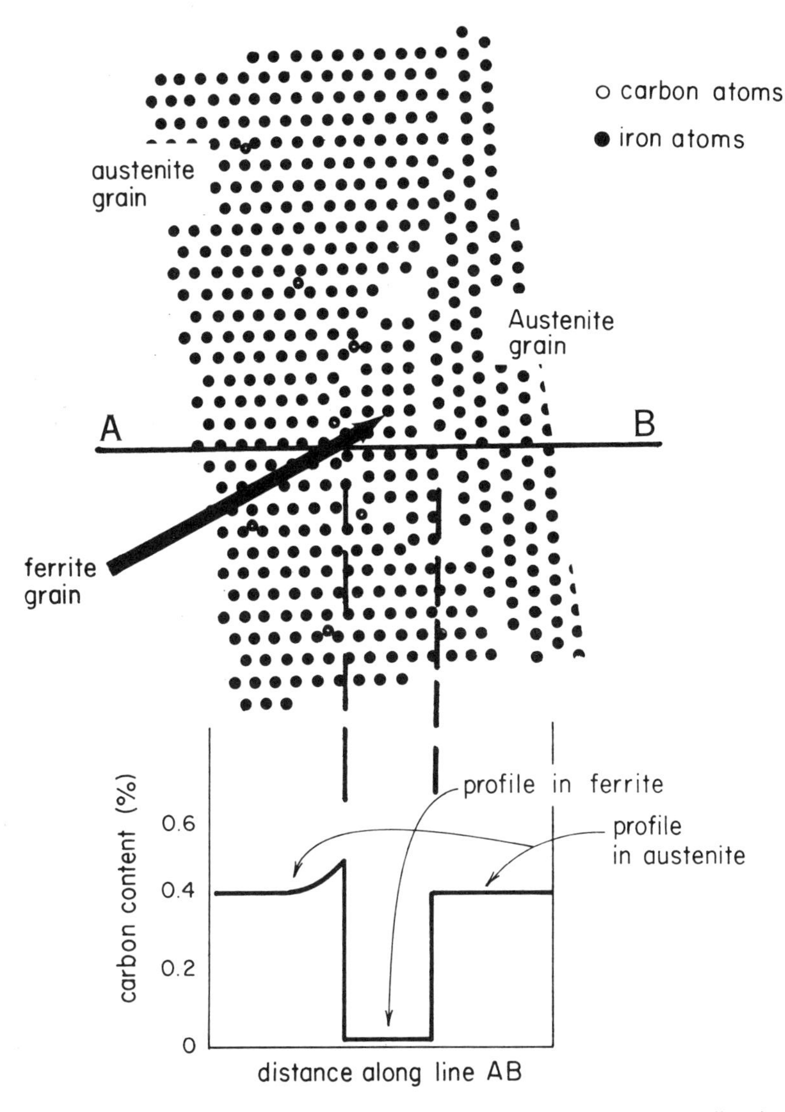

Figure 1-6 Carbon profile (on the average) on a plane perpendicular to the line AB.

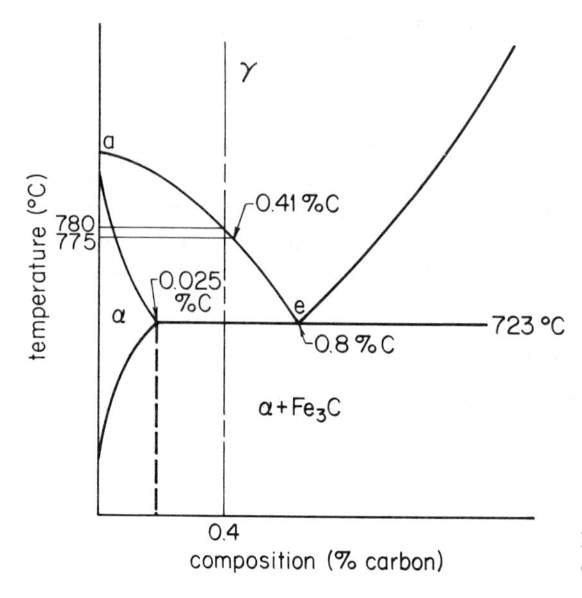

Figure 1-7 Portion of the iron-carbon phase diagram.

(Fig. 1-7), upon cooling such a steel from 723+ °C to 723− °C, the austenite will decompose to ferrite and carbide. Experimentally, it is found that the morphology of these two phases consists of alternate parallel plates, an arrangement which is referred to as *pearlite* (see Fig. 1-8).

Now we return to a consideration of the 0.4% carbon steel above. We examine what happens to the microstructure at 723+ °C, which consists of primary ferrite and austenite (containing 0.8% carbon), upon cooling to 723− °C. According to the description in the preceding paragraph, the austenite should form pearlite. The only question is, what is the effect on the eutectoid reaction of the presence of the proeutectoid, or primary ferrite? In some similar transformations, the presence of a proeutectoid phase can influence the morphology of the eutectoid structure (an example is the divorced eutectic structure). In the type of steels being discussed here, it is observed that the presence of the primary ferrite has no great influence on the morphology of the ferrite and carbide mixture formed from the decomposition of the austenite. That is, the austenite forms pearlite. Thus at 723− °C, the microstructure consists of about 50% pearlite and 50% primary ferrite. The microstructure would appear as in Fig. 1-8.

The final consideration is of the changes that occur upon cooling from 723− °C to 25°C. To see what is involved, consider an iron-carbon alloy containing just slightly less than 0.025% carbon and at 723°C. The phase diagram shows this alloy to be in the single-phase ferrite region. However, upon slow cooling, the alloy enters the two-phase ferrite and carbide region as the solubility line of carbon in ferrite is crossed (Fig. 1-7). Thus, the ferrite precipitates carbide upon cooling. The amount of carbide formed is small, but this slight increase in the amount of hard carbide at 25°C over that present at 723− °C should cause an increase in the hardness. However, the pre-

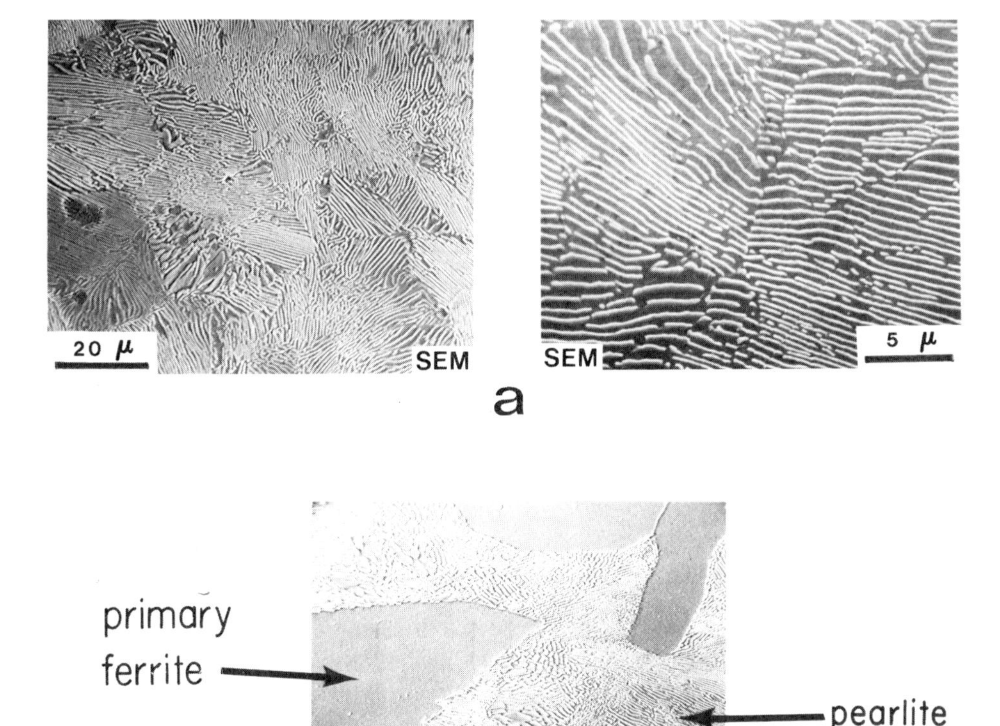

Figure 1-8 Micrographs of (a) pearlite in a 0.8% C steel and of (b) primary ferrite and pearlite in a 0.4% C steel. Both steel were slowly cooled from austenite (annealed).

cipitation of the carbide reduces the amount of carbon in solution in the ferrite, and hence reduces the solid-solution strengthening effect. The final hardness will depend upon the magnitude of these two contributions. Figure 1-9 shows that the effect is a very slight decrease in hardness.

Thus, upon slowly cooling a steel from 723°C to room temperature, any ferrite present will precipitate carbide. In the case of the 0.4% carbon steel discussed above, both the primary ferrite and the ferrite in the pearlite will precipitate some carbide upon cooling. However, the slight change in hardness will be so small compared to the hardness of the steel that the effect will not be easily measurable. Also, the amount of carbide precipitated is so small that it is difficult to detect in the microstructure. The

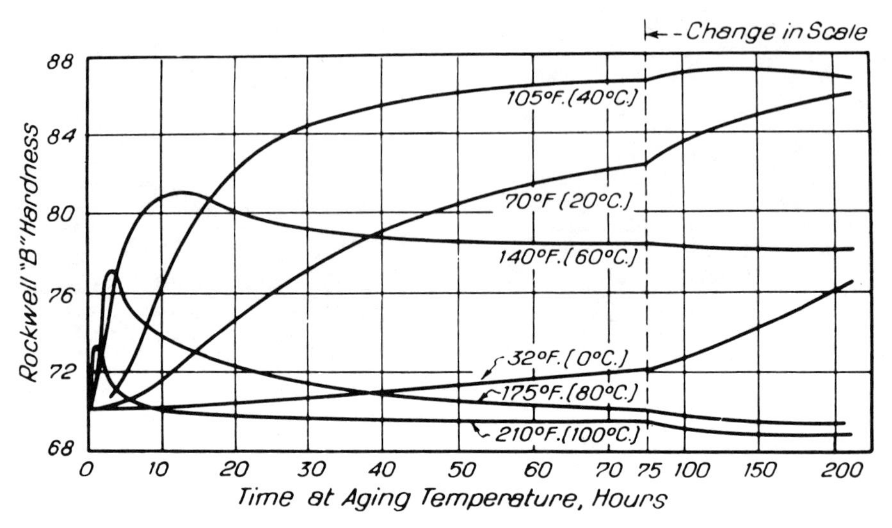

Figure 1-9 Effect of aging time and temperature on the hardness of a 0.06% carbon steel, initially cooled rapidly from 720 to 25°C prior to aging. Note that the hardness at 100°C after equilibrium is reached, and the alloy is overaged, is slightly less than the initial hardness. Thus the hardening contribution from the formation of the hard Fe_3C is diminished by the reduction of solution hardening by the precipitation of the carbide. *(From E. S. Davenport and E. C. Bain,* Trans. ASM, *vol. 23, p. 1047, 1935.)*

net effect, then, is that the analysis and calculations made slightly below the eutectoid temperature (i.e., 723− °C) can be considered to be valid at 25°C.

From the above discussion, it should be clear that in annealed steels increasing carbon content causes an increase in the amount of carbide present at room temperature. Thus, since the carbide is considerable harder than the ferrite, it should be expected that the hardness will increase as the carbon content increases. That is, it would appear that the carbide is increasing the hardness of the ferrite by dispersion or particle hardening. However, it is also noted that as the carbon content increases from zero to 0.8%, the amount of pearlite goes up from zero to 100%, the remainder being primary ferrite. Thus the structure consists of pearlite dispersed with ferrite. Since pearlite is harder than the ferrite (since pearlite itself is ferrite strengthened by the presence of 12 wt. % carbide), it is perhaps best to credit the increasing hardness of the steels with increasing carbon content to the increasing amount of pearlite. The dependency of strength on carbon content is shown in Fig. 1-10. Above 0.8% carbon in the steel, the amount of pearlite decreases; the rest of the microstructure is the hard primary carbide, but since it is there only in small amounts the hardness is about the same as that of pearlite.

Multiphase mixtures will tend to alter their shape in order to minimize the total surface energy. That is, the equilibrium configuration will be single crystals of the phases in contact, with a configuration that depends upon the relative interfacial surface energy. This means that the annealed steels just discussed are not really at equilibrium at room temperature, since the microstructure consists of many crystals of both the ferrite phase and the carbide phase. Thus, these ferrite-carbide mixtures

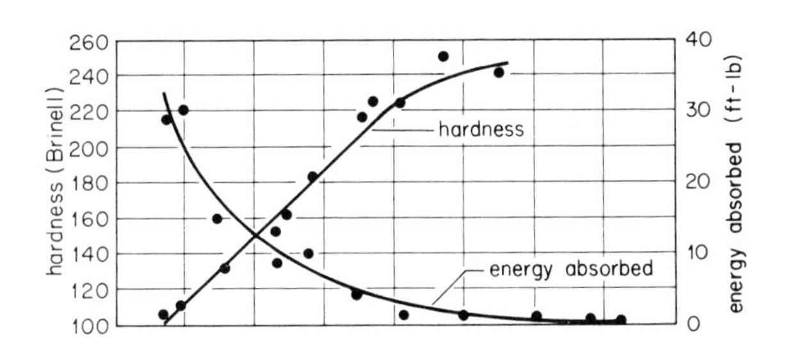

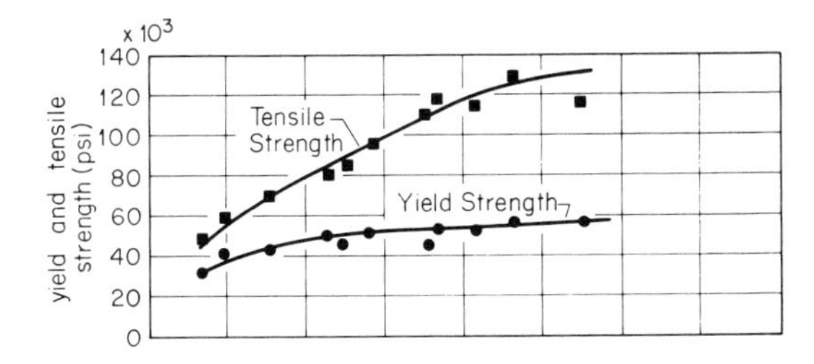

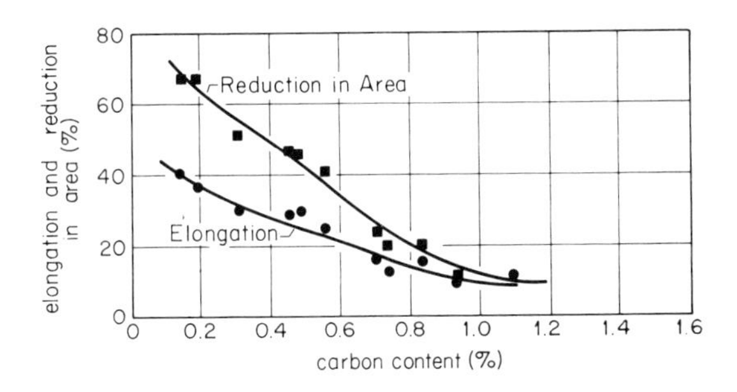

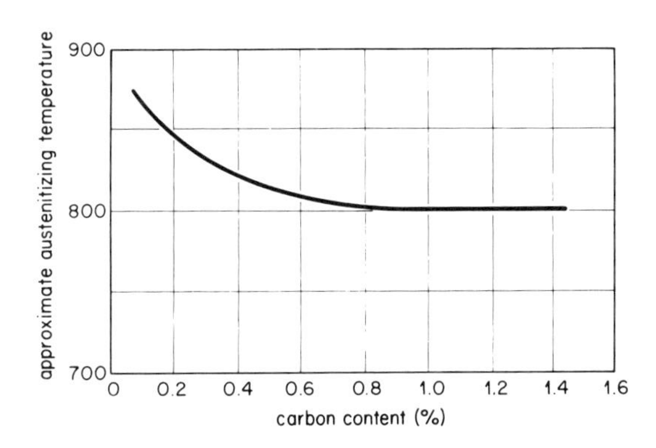

Figure 1-10 Mechanical properties of annealed plain carbon steels. *(Adapted from J. H. Nead,* Trans. AIME, *vol. 53, p. 218, 1916; H. Meyer and W. Wesseling,* Stahl und Eisen, *vol. 45, p. 1169, 1925; F. C. Langenberg,* Chem. Met. Engr., *vol. 25, p. 910, 1921.)*

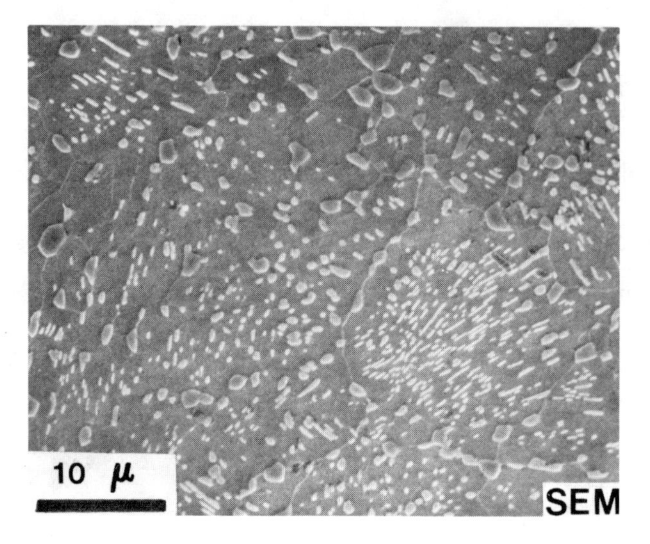

Figure 1-11 Microstructure of spheroidized 1080 steel, obtained by heating the steel in the annealed condition to 690°C for 96 hours.

of pearlite and either primary ferrite or primary carbide will alter their shape with time at a given temperature. Experimentally, it is observed that the carbides will spheroidize in the ferrite grains, and at the same time the ferrite grains will increase their size. The microstructure in Fig. 1-11 illustrates the behavior upon heating an annealed steel to 690°C for 96 hours. This spheroidization process is relatively slow, even at high temperatures.

In principle, the hardness and strength will be reduced upon spheroidization as the interparticle spacing is decreased, which reduces the dispersion hardening effect of the carbides; however the effect is found to be rather small.

1-3 THE ISOTHERMAL DECOMPOSITION OF AUSTENITE

It was seen in Fig. 1-1 that the products that form from austenite control the properties of steels; in this section we lay the groundwork for understanding how these products form from the austenite. To focus the discussion, attention will be centered around an eutectoid steel, 0.8% carbon. We will assume that the steel has been heated into the austenite region until equilibirium has been attained, say for 1 hour at 1000°C. Now we wish to observe how the austenite decomposes with time upon holding at some temperature below 723°C, say 650°C. This can be accomplished by cooling a steel specimen very rapidly to 650°C, allowing its temperature to attain the temperature desired very rapidly, and then observing the transformation of austenite. We will assume (and justify or clarify later) that the cooling rate is extremely rapid, so that the time necessary to go from 1000°C to the desired transformation temperature (650°C) is so short that no alteration occurs in the austenite until after the specimen has reached 650°C. The process is illustrated in Fig. 1-12.

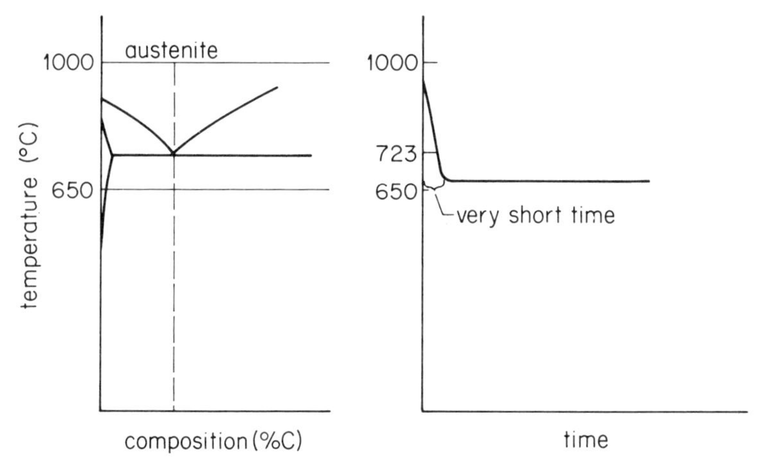

Figure 1-12 Schematic representation of the method to be used to study the isothermal decomposition of austenite. A 0.8% carbon steel is austenitized at 1000°C for about 1 hour, then cooled very rapidly to 650°C, the temperature at which the decomposition of austenite is to be studied. It is assumed that the time necessary to cool from 1000 to 650°C is so short that the austenite undergoes no change during the cooling.

In practice, the observation of the decomposition of austenite is not a simple matter, since we cannot "see" into the steel directly in order to follow the changes. However, let us assume that we are going to monitor some mechanical or physical property with time at 650°C, say electrical resistivity. What would be observed (Fig. 1-13) would be a constant resistivity for a period of time after the steel specimen has reached 650°C, then the resistivity would begin to change, and continue to change for a period of time, and then reach a value that would remain constant with further time. Thus, at time t_s, austenite has just begun to decompose, and at time t_f the decomposition process has apparently just ceased.

Formation of Pearlite

Now let us see if we can rationalize a possible decomposition process. We know that if the austenite is slowly cooled from 1000°C, pearlite forms on cooling below 723°C. So it would seem reasonable to believe that the decomposition product upon isothermal transformation at 650°C might be pearlite. This turns out to be the case at this temperature; however, we will see that this does not hold true at all transformation temperatures. Now, what is a reasonable mechanism by which the pearlite can form? From studies of solidification and of recrystallization in cold-worked metals, we know that structural changes occurred by nucleation and growth. That is, when a liquid metal freezes, the mechanism of the decomposition of the liquid is the formation of small regions (nuclei) of atoms which take on the correct geometry for the crystal structure of the solid, and the growth of these nuclei by the addition of atoms in the liquid to the solid at the interface. In the formation of pearlite, then, it appears that the process might be one of nucleation and growth.

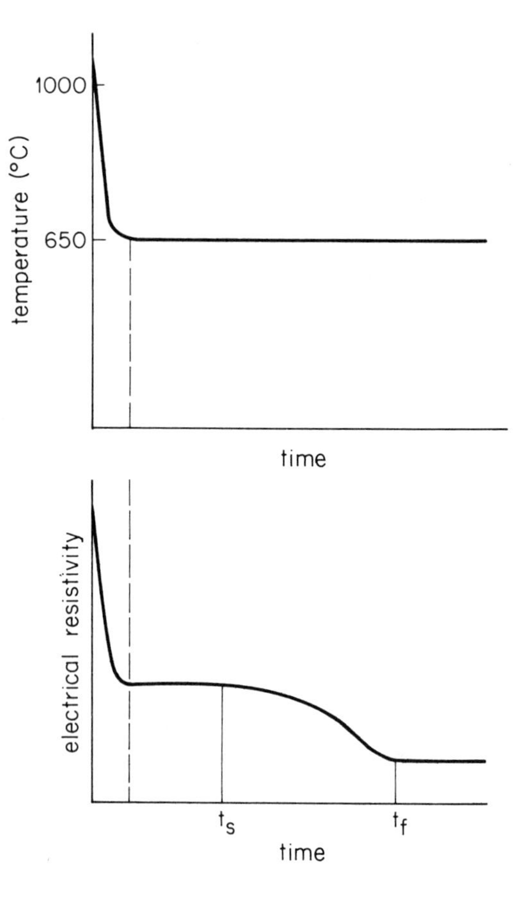

Figure 1-13 Schematic representation of the variation of the electrical resistivity of a steel specimen with time, upon quenching from 1000 to 650°C, and then holding at 650°C.

The pearlite will nucleate in the austenite grain boundaries, and in the grains. The grain boundary nucleation will occur by a process similar to that illustrated in Fig. 1-5. Whether a small volume of carbide nucleates first, or of ferrite, or of both simultaneously, is not clear. However, if ferrite nucleates first and begins to grow, the increasing carbon content in the ferrite-austenite interface favors the formation of carbide. Then as the carbide grows, the carbon content of the austenite at the carbide-austenite interface is very low, which favors the formation of ferrite. This repetitive growth description gives a rather general and qualitative picture of the formation of pearlite, but it does illustrate the general features.

Now, on the basis of the preceding description, we can visualize the microstructural changes that must be occurring during the formation of pearlite, even though it may be difficult to see these directly. These changes are depicted in Fig. 1-14, along with the variation with time at 650°C in the fraction of austenite decomposed.

The fraction of pearlite at any time depends on the rate of grain boundary nucleation (N_s), which is the number of nuclei formed per time per austenite grain-boundary area remaining, and on the rate of growth (G) of the nuclei, which is some

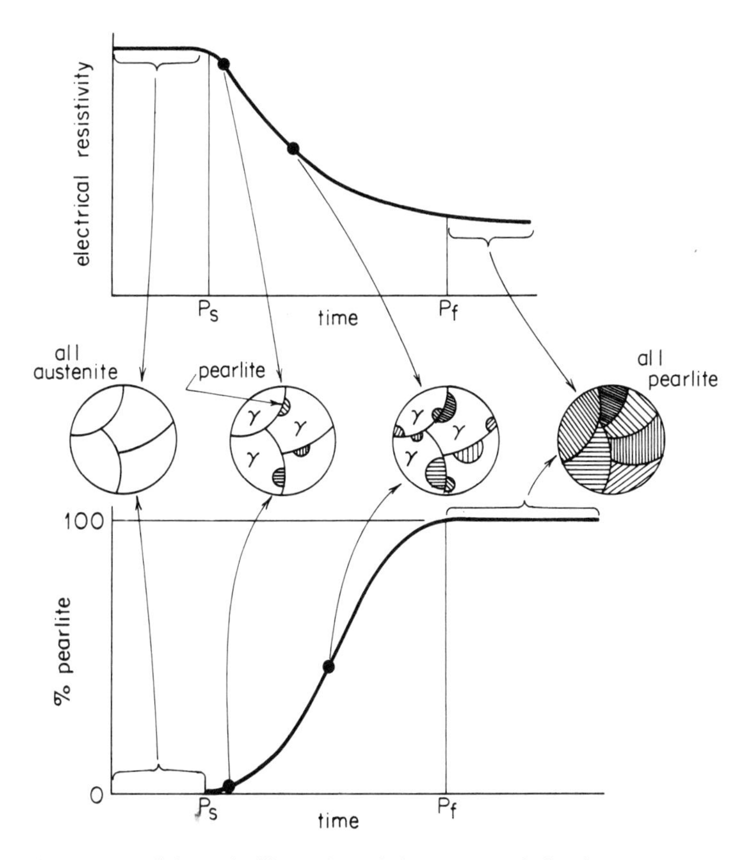

Figure 1-14 Schematic illustration of the process of the decomposition of austenite to pearlite at 650°C in a 0.8% carbon steel.

linear or volume measurement (e.g., cm/s). Theroretical treatments for the formation of pearlite lead to expressions that predict the general shape of the curve shown in Fig. 1-14 and that give the fraction of pearlite formed to be a function of the product $N_s G^3 t^4$.

In the above discussion it has been established that there is a time at 650°C at which pearlite starts to form (P_s), and a later time at which austenite has just finished forming pearlite (P_f). We will represent this time on a temperature-log time diagram (Fig. 1-15). Log time is used because, as will be seen, some of the times involved for the start and completion of the transformation are quite long. This type of diagram is an isothermal time-temperature-transformation diagram, or an *isothermal* TTT diagram.

Now we examine how the P_s and P_f times are dependent on temperature. We can predict qualitatively what to expect by the following consideration. In developing the TTT diagram above we assumed that upon cooling from 1000°C to the transformation temperature of 650°C, the cooling was so rapid that no change occurred in the austenite during the cooling time. If we assume that we can cool from 1000°C to absolute zero (or near this temperature) under the same conditions, we will have austenite

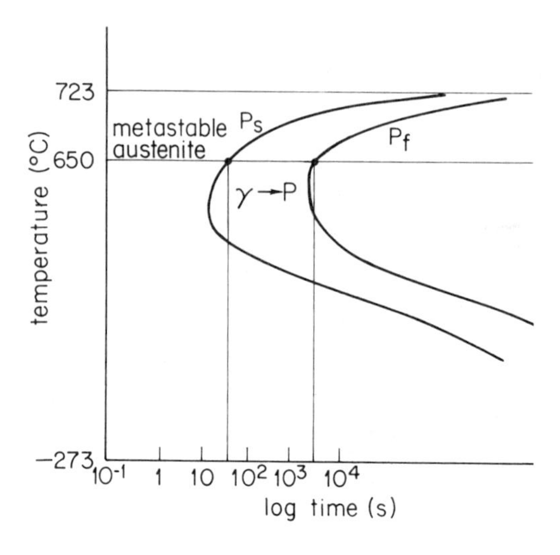

Figure 1-15 Schematic representation of an isothermal TTT diagram for the decomposition of austenite to pearlite in a 0.8% carbon steel.

present at 0 K, at which temperature we wish to follow the decomposition of austenite. At this low temperature, however, the atoms are immobile, diffusion is prevented, and the austenite cannot transform to pearlite. Thus, the rate of formation of pearlite is zero, and the P_s and P_f times are at infinity on the TTT diagram (Fig. 1-15).

Theories predicting the rate of growth and the rate of nucleation usually involve the energy difference between the austenite and pearlite (ΔE). This energy difference is zero at 723°C, where austenite, carbide, and ferrite are in equilibrium, and becomes greater the lower the temperature. However, ΔE has a finite value at absolute zero. In terms of the growth rate of pearlite, the theories lead to a dependency of the type

$$G = \text{function } [(\Delta E)e^{-A/T}]$$

where A is a constant and T is the absolute temperature. Note that as the temperature approaches absolute zero the term $e^{-A/T}$ approaches zero, whereas ΔE approaches a finite value. Thus G approaches zero, as was predicted in the preceding paragraph. Now as the transformation temperature approaches 723°C, $e^{-A/T}$ approaches a finite value but ΔE approaches zero. That is, the "driving force" ΔE for the transformation approaches zero, and thus G approaches zero. The same general argument is valid for the nucleation rate. This means, then, that the P_s and P_f are both very long at very low temperatures and at very high temperatures, but relatively short at intermediate temperatures. This is illustrated in Figure 1-15.

Formation of Bainite

The preceding discussion leaves the impression that austenite decomposes to pearlite at any temperature below 723°C. However, this turns out not to be the case, and we now examine what does form instead of pearlite and, to some extent, why. Turning to the latter problem first, it is to be noted that there are many solid-state tranformations in which a phase decomposes, but the initial products are not the equilibrium products; the equilibrium phases will form from the intermediate one at a later stage. One requirement for a phase, or phases, to be the product of the austenite is that there be a decrease in energy, even though it may not be of the magnitude of that associated with the formation of pearlite. However, it is the rate of formation of a given product which determines the sequence of the decomposition of a phase, and it is more difficult to rationalize (i.e., to theoretically predict) this. For that reason, in the case of the decomposition of austenite, we will present the experimental data without further detailed analysis concerning why the particular phases and distributions form.

In the case of the decomposition of austenite in a 0.8% carbon steel being considered here, it is observed that if the transformation temperature is below around 600°C, ferrite and carbide are still the decomposition products of austenite, *but* the distribution of the phases is not alternate plates. That is, pearlite does not form. Instead, the product consists approximately of plates of ferrite between which or inside of which short carbide rods form. This morphology of ferrite and carbide is called *bainite.* Around 600°C both pearlite and bainite form. Figure 1-16*a* shows the type of TTT diagram obtained for an 0.8% carbon steel in the temperature range in which bainite and pearlite form. The dashed lines indicate that the particular product would form if the austenite did not first decompose by the formation of the other product. As in the case of pearlite, the bainite should have a start and finish curve which approaches infinite time at both high and low temperatures. However, experimentally, in this plain carbon steel the curves for both bainite and pearlite

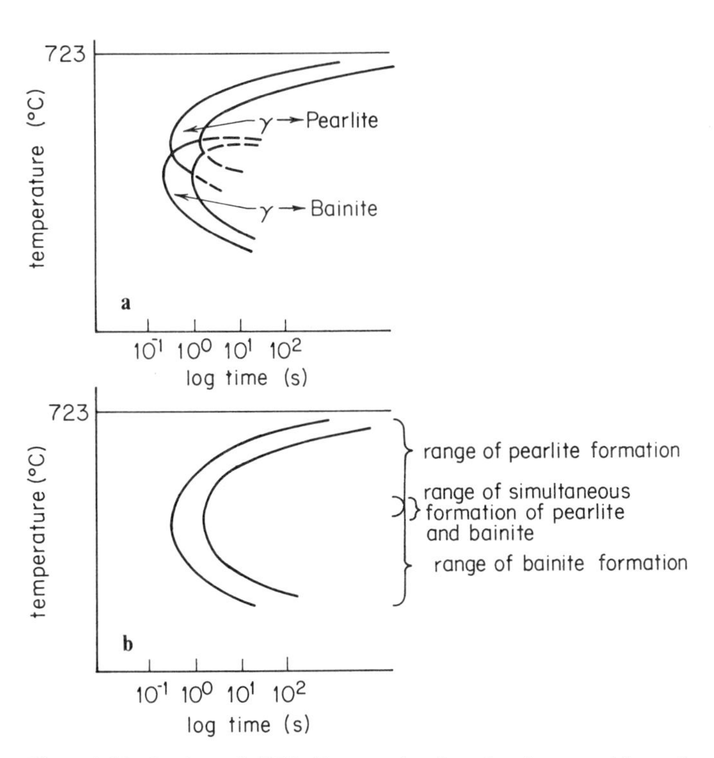

Figure 1-16 Isothermal TTT diagram showing the decomposition of austenite to pearlite and bainite.

minimize in the same temperature and time range, and thus frequently only one smooth curve is shown, as in Fig. 1-16*b*.

Like pearlite, bainite grows by a nucleation and growth process. However, the bainite plates are not randomly oriented but instead are approximately parallel to the {111} planes of the austenite crystal in which they form. The formation is illustrated schematically in Fig. 1-17.

Mechanical Properties of Pearlite and Bainite

Now the effect on the mechanical properties of the microstructure formed in the 0.8% carbon steel can be illustrated. Consider the austenite held at a given temperature until transformation is just complete (i.e., the finish line is crossed), then cooled to room temperature (e.g., air cooled) and the hardness measured. It is found that the value of the hardness depends on the transformation temperature as shown in Fig. 1-18. Note that, generally, as the transformation temperature decreases the hardness increases. This increase is associated with a decrease in the carbide lamellae spacing in the pearlite as the transformation temperature decreases, and with a decrease in the average carbide separation in the bainite as the temperature decreases. Thus the hardness increase is due to a dispersion hardening of the ferrite by the carbides.

Formation of Martensite

The discussions of the formation of pearlite and bainite leave the impression that austenite can be retained at sufficiently low temperature by rapid quenching from the austenite region. However, this is not necessarily the case, and we now examine what the correct picture is. To illustrate the decomposition of austenite upon rapid cooling to sufficiently low temperature, consider the curves shown in Fig. 1-19. If we begin at a low temperature (e.g., −50°C) with a cylinder of a 0.8% carbon steel in the pearlitic condition, and heat it slowly, the specimen will expand almost linearly along line *ab* up to 723°C. At this temperature, the pearlite will decompose to austenite. Since the steel is tranforming from a structure of 88% ferrite and 12% carbide to the more closely packed fcc structure of austenite, the steel specimen contracts (line *bc*). Further heating causes the austenite to expand almost linearly along line *cd*. Now, if the steel specimen is cooled very rapidly (e.g., 300°C/s), the austenite contracts along line *dce*, but it is observed that no transformation occurs upon passing 723°C. Instead, the austenite remains until some low temperature is reached (M_s, point *e*), at which temperature the specimen begins to expand, and it continues to expand until some lower temperature (M_f, point *f*) is reached. Upon further rapid cooling the steel contracts linearly along line *fg*. It is seen that the specimen has a greater length at the final low temperature (initial temperature T_0) than it had when it was pearlite. This indicates that upon rapid cooling the austenite has not reverted to pearlite, but to some other structure. X-ray analysis of the quenched steel would reveal the presence of only a body-centered tetragonal (bct) phase, which is called *martensite*.

The dependency of the lattice parameter of martensite on the carbon content is

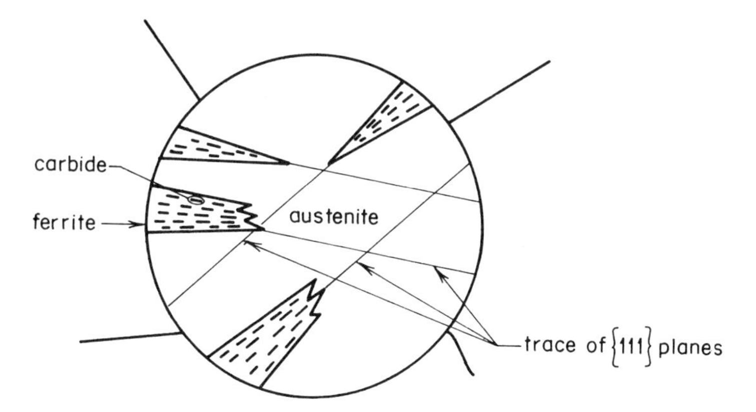

Figure 1-17 Schematic illustration of the formation of bainite in austenite in a 0.8% carbon steel. The magnification would be quite high here (e.g., 10,000X).

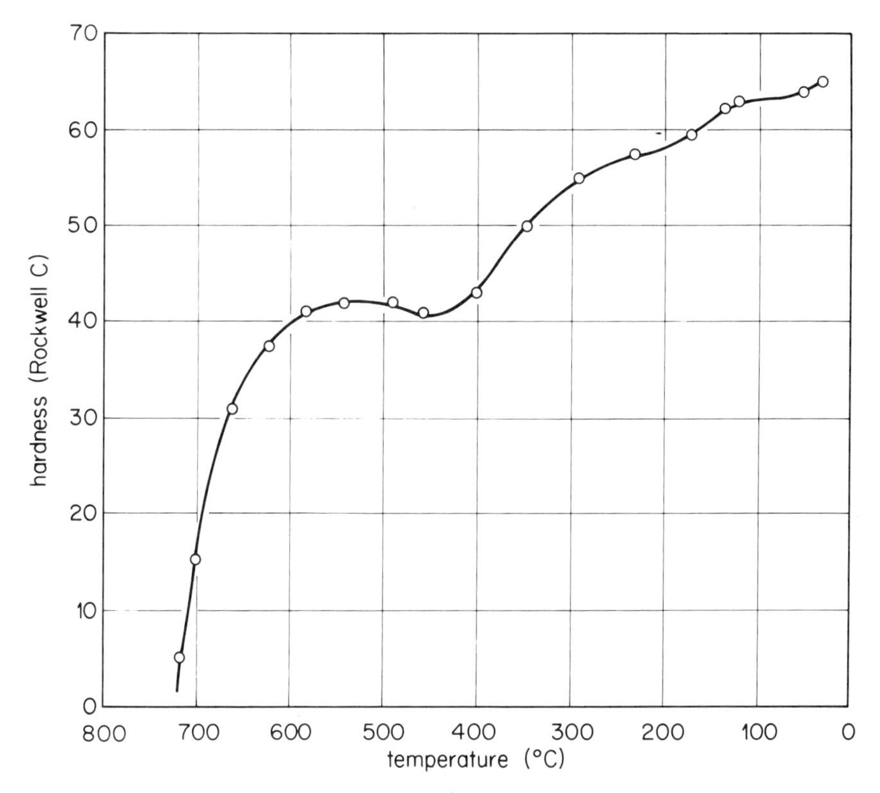

Figure 1-18 Variation of the hardness (at 25°C) with the temperature at which the steel (0.89% C, plain carbon) was isothermally transformed. The heat treatment involved isothermally transforming a sample at a given temperature until the finish line was reached, then cooling to 25°C, after which the hardness was measured. *(Adapted from E. S. Davenport,* Trans. ASM, *vol. 27, p. 837, 1939.)*

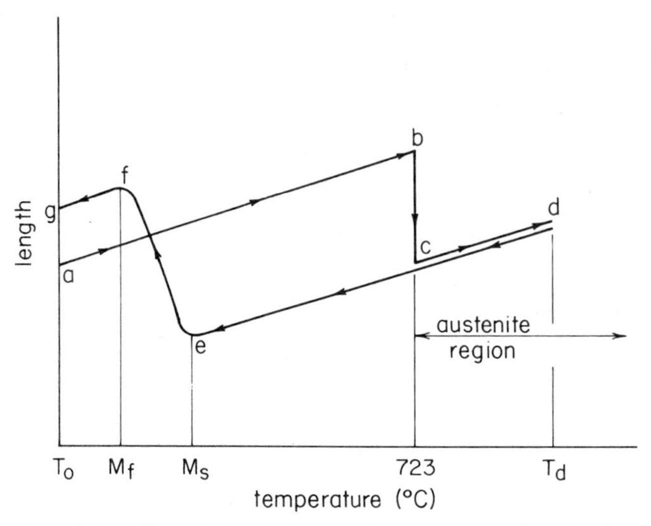

Figure 1-19 The variation in length with temperature for a 0.8% carbon steel cylinder. Along line *abcd*, the specimen is heated slowly (e.g., 2°C/min); line *defg* was obtained upon quenching the cylinder in water from the high temperature T_d.

given in Fig. 1-20. Thus, as carbon is removed from the tetragonal phase, the lattice expands slightly along the *a*-axis but contracts greatly along the *c*-axis. Indeed, extrapolation to zero carbon content gives an identical value for both axes, so the structure would be bcc; the *a*-axis value is also identical to that of ferrite. It appears, then, that martensite, formed upon such rapid quenching, is a supersaturated (with carbon) ferrite, expanded along one of the original axes of the cubic cell to become tetragonal.

Perhaps the formation of the tetragonal phase from the austenite should not be unexpected. At equilibrium, the fcc austenite decomposes to the bcc ferrite and a small amount of carbide. Thus, it might be expected that the formation of the bct

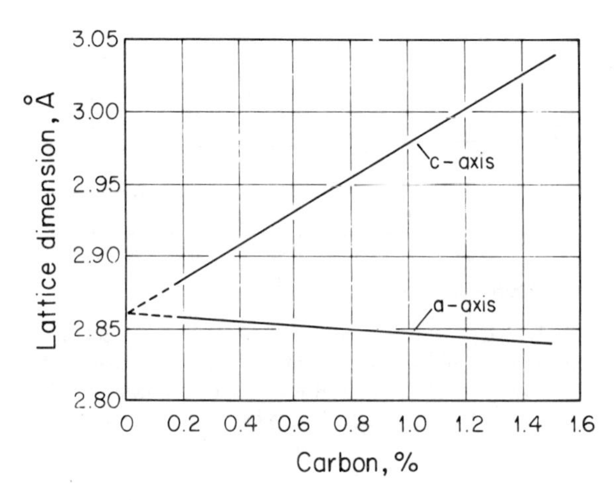

Figure 1-20 The variation of the lattice parameters of the body-centered tetragonal martensite with carbon content. *(Adapted from G. V. Kurdjumow,* J.I.S.I., *vol. 195, p. 26, 1960.)*

phase, supersaturated with carbon, would precede the formation of the ferrite and carbide.

The characteristics of the formation of the martensite phase differ radically from those of pearlite and bainite. In the formation of pearlite and bainite, carbon must move over many atom distances, away from the regions converting to the ferrite and toward those becoming carbide. This does not happen in the martensite formation; instead, each carbon atom maintains nearly the same location, relative to its neighboring iron atoms, as it had in the austenite. Thus, the mechanism of martensite formation must maintain this carbon location. In order to avoid the tendency for carbide precipitation and ferrite formation, it might be expected that the martensite transformation must occur rapidly, since the diffusion of carbon is relatively rapid. This indeed is the case. A transformation that could occur rapidly would be one involving cooperative movement of regions or layers of atoms (similar to mechanical twinning) and hence would be related crystallographically to the austenite lattice. Indeed, this is the mechanism of martensite formation from austenite. In an austenite grain, a small group of atoms shears (and readjusts the lattice dimension slightly) to the tetragonal structure; this shearing action continues across the grain, until it impinges upon a boundary.

The general features of the martensite transformation, as applied to a 0.8% carbon steel, are outlined below. It is emphasized that these comments are not all rigorously valid for all martensite transformations; however, they are reasonably valid for the type of steels to be discussed in this chapter.

1. Recalling from crystallography that the fcc lattice can be described equally well as a bct lattice gives immediately a way of producing a bct structure from the fcc austenite. This is illustrated in Fig. 1-21. However, in order to produce the actual dimensions of the martensite, some readjustment of the atoms occurs. The lattice changes involved are also illustrated in Fig. 1-21. The interstitial locations that can accommodate the largest sphere are indicated in Fig. 1-22. It is seen that in the bct cell derived from this fcc cell these locations all lie along the $\{100\}$ directions (Fig. 1-21). Thus, in a given volume of a single crystal of austenite which transforms to the bct martensite, *each* carbon atom is trapped in one of the sites shown in Fig. 1-22. If these carbon atoms were not present, the iron atoms would contract along the $\{001\}$ direction to become bcc. However, the location of the carbon atoms causes the cell to be elongated along this one direction, giving the bct structures.
2. The mechanism of the nucleation of martensite is still in doubt. But in a localized region, perhaps due to normal atom movement fluctuations (thermal vibrations) and/or stress (e.g., quenching stresses), a group of the fcc atoms shears into a group that has a bct structure (Fig. 1-23). This shearing, once begun, occurs very rapidly (approximately with the speed of sound in the steel), and proceeds from the point of origin until the advancing front encounters an obstacle, such as a grain boundary. In a simplified view the process is as shown in Fig. 1-24.
3. Because of the crystallographic relation described under comment (1), the martensite plates do not form with random orientation in the austenite. It is found

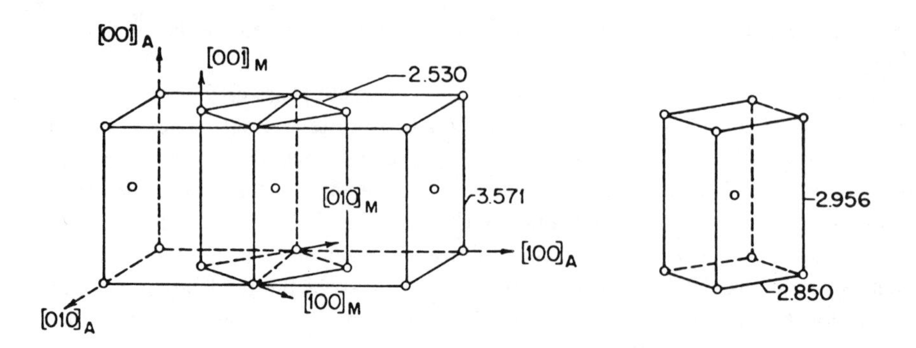

face-centered cubic austenite, showing body-centered tetragonal cell. A stands for austenite, M for martensite.

body-centered tetragonal martensite

lattice parameters of austenite: $a_0 = 3.571 \times 10^{-8}$ cm

lattice parameters of bct cell in fcc austenite: $a = 2.530 \times 10^{-8}$ cm, $c = 3.571 \times 10^{-8}$ cm

lattice parameters of bct martensite: $a = 2.850 \times 10^{-8}$ cm, $c = 2.950 \times 10^{-8}$ cm

Figure 1-21 Illustration of the relationship between the fcc austenite cell and the martensite bct cell. The lattice parameters are based on a 0.8% C alloy and 25°C.

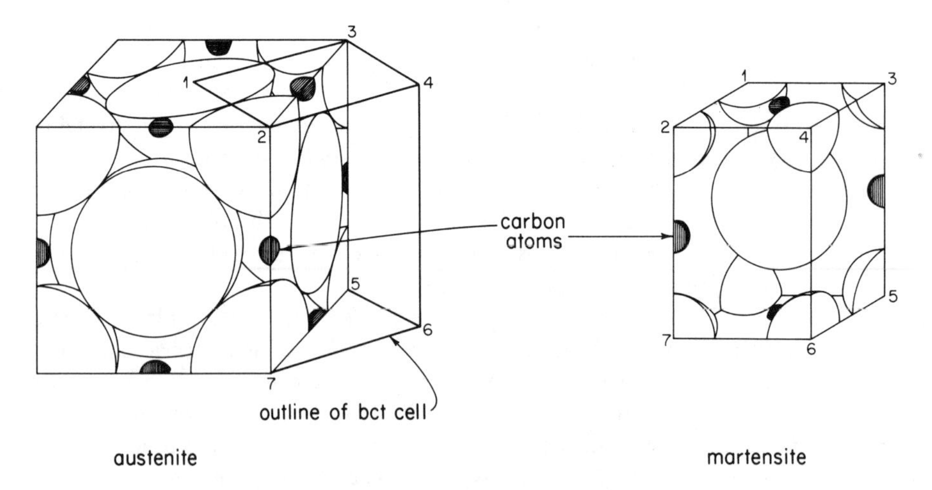

Figure 1-22 Illustration of the relation between the position of carbon atoms in the fcc austenite and the positions in the bct martensite inherited from this.

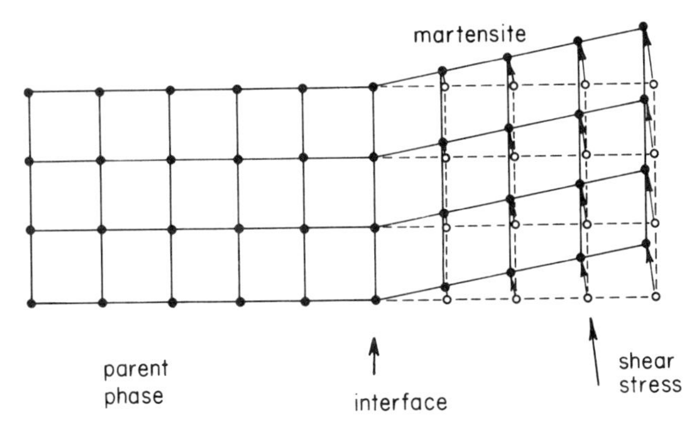

Figure 1-23 Schematic representation of shear transformation on an atomic scale.

that the "habit plane" of the martensite, or the plane along which the plate lies, is approximately parallel to the {111} planes of the fcc austenite. Thus, inside an austenite grain, the plates can form with four distinct orientations, and the microstructure appears as shown in Fig. 1-25. A martensite plate nucleates and grows across an austenite grain, then another plate grows, and then another. Each new martensite plate may grow parallel to any of the four {111} planes of the austenite. If the transformation is suspended, by some means, prior to complete conversion of the austenite to martensite, the type of microstructure seen is illustrated by the example in Fig. 1-26. If the transformation proceeds until only martensite is present, the microstructure is less revealing, since the clear austenite with which to contrast the martensite is no longer present. Instead, due to the localized strains, the microstructure etches in a rather confusing pattern, although the directional pattern is reflected in the "brush strokes" features of the micro-

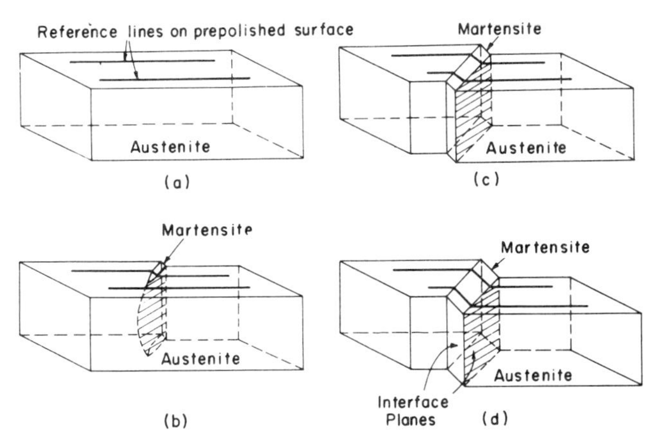

Figure 1-24 Shape change of reference lines in a crystal with martensite formation. *(From M. Cohen,* Trans. AIME, *vol. 224, p. 638, 1962.)*

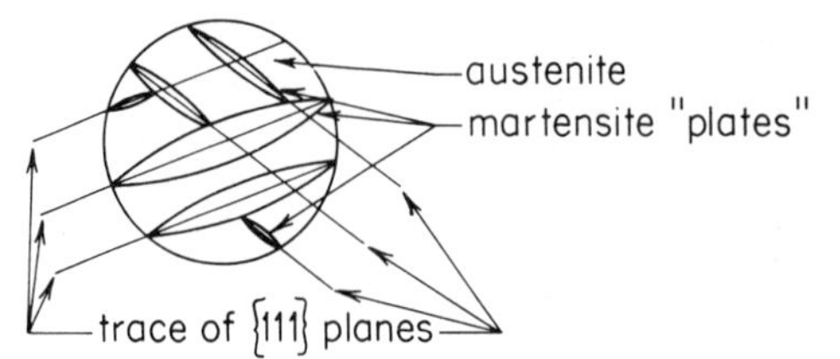

Figure 1-25 Schematic representation of the microstructure of martensite plates in an austenite grain.

structure (Fig. 1-27). Even though this microstructure appears to contain more than one phase, it is actually composed of a single phase, although each prior austenite grain is now composed of many martensite crystals.

4. The formation of martensite, in the 0.8% carbon steel being considered here, is athermal, meaning that the transformation does not proceed with time, but only with a decrease in temperature. This is illustrated in Fig. 1-28, which shows the amount of martensite obtained in the austenite upon cooling continuously to a low temperature. Upon cooling, martensite begins to form when a certain temperature, called the martensite-start temperature (M_s) (point *e* in Fig. 1-19), is reached. Upon further cooling, the amount of martensite increases rather linearly, but eventually approaches complete conversion (100%) asymptotically. The temperature at which complete conversion occurs, or at which some high degree (e.g., 95%) of conversion occurs, is called the martensite-finish (M_f) temperature (point *f* in Fig. 1-19). If the cooling is interrupted at a temperature between the M_s and M_f, and held at that temperature, the amount of martensite remains constant. More martensite will form only on continued cooling. This behavior is what is meant by an *athermal* transformation. Perhaps this characteristic should

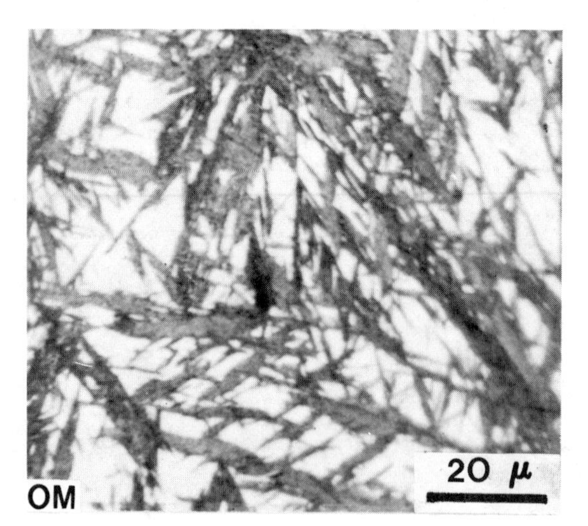

Figure 1-26 Micrograph of martensite and retained austenite (white background.)

Figure 1-27 Micrograph of martensite.

not be unexpected. Martensite forms with an increase in volume (Figs. 1-19 and 1-21). Thus inside an austenite grain both the austenite and the martensite become strained. A certain "driving force," or energy difference, must be attained to overcome the induced strain energy associated with the martensite formation, and this can be achieved by more undercooling. That is, to increase the amount of martensite, the austenite must be undercooled more to increase the energy difference between the martensite and the austenite.

5. The martensite formation is not reversible. That is, below 723°C reheating the martensite will not cause reversion to austenite. Instead, the martensite will

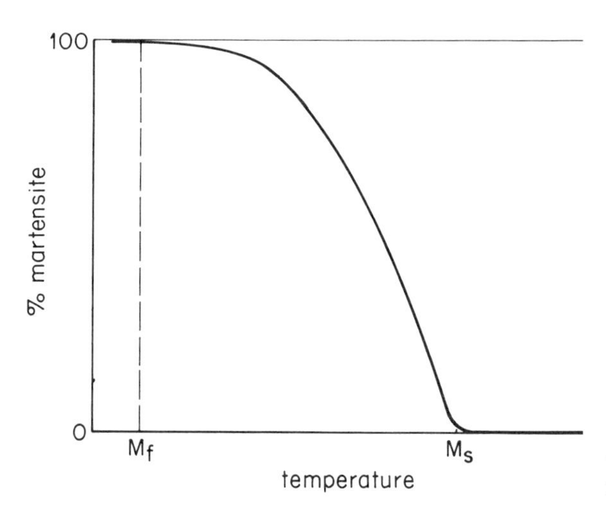

Figure 1-28 The amount of martensite in austenite versus temperature.

decompose to the equilibrium phases of ferrite and carbide. This will be discussed later.

6. The martensite transformation is depicted on the isothermal TTT diagram as shown in Fig. 1-29. This seems inconsistent with the C-shaped curve obtained for the formation of pearlite and of bainite. That is, the same arguments put forward to predict qualitatively the shape of the pearlite and bainite curves, leading to a maximum rate of formation at an intermediate temperature, and hence a C-shaped TTT curve, should be valid here. The answer may lie in the speed at which the martensite forms. That is, the martensite forms so rapidly that the steel specimen cannot be quenched from 1000°C to, say, −100°C without martensite forming upon cooling. The hypothesized situation is illustrated in Fig. 1-30. In some iron-based alloys of high alloy content the martensite transformation is sufficiently slow to be monitored, and it does exhibit a measurable C-shaped TTT curve.
7. The exact structural features of martensite depend upon the carbon content. For low-carbon (e.g., 0.2% C or less) martensites, the crystal structure is bcc. The martensite units are in the shape of laths, grouped into packets. The fine structure has a high dislocation density. This type of martensite is referred to as lath martensite. For higher carbon content, the martensite is usually bct. The individual crystals form as lenticular plates, and the fine structure consists of very fine twins. This type of martensite is called plate martensite. In some steels, both types of martensite can be present.

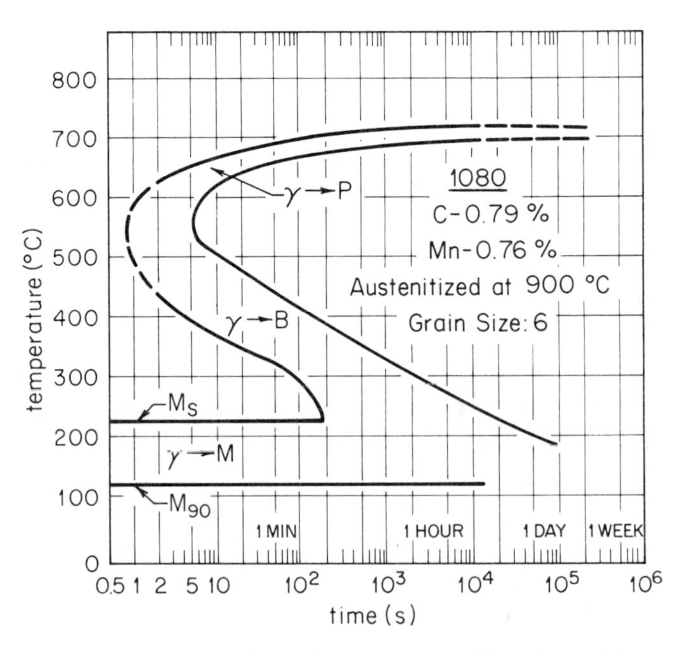

Figure 1-29 Isothermal TTT diagram for a 0.8% carbon, plain carbon steel. *(Adapted from "Atlas of Isothermal Transformation Diagrams," United States Steel, Pittsburgh, 1951.)*

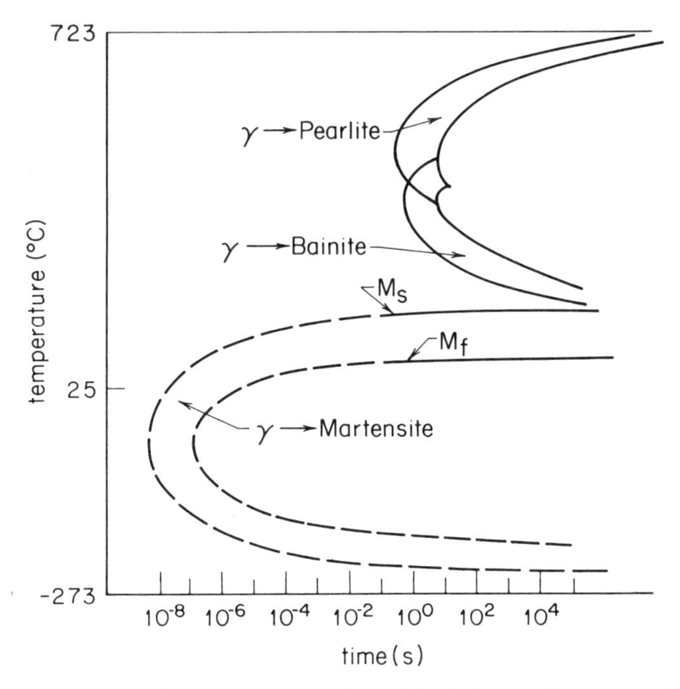

Figure 1-30 Schematic representation of TTT diagram showing possible "C" curves for martensite formation occurring at such short times that only the solid-line portion can be experimentally verified.

8. The most important feature of martensite is that it is extremely strong and hard, being the hardest form of steel. Upon straining the martensite plastically, when a dislocation encounters a carbon atom the strain energy is reduced if the carbon atom can remain in the dislocation core; hence, there is a strong binding energy between the dislocations and the carbon atoms, and it is difficult to force the dislocation to leave the carbon atom. Carbon also moves relatively rapidly, and certainly some of the carbon atoms in the martensite move to dislocation cores after the quench but before mechanical properties can be measured. Again, because of the large carbon atom-dislocation binding energy, this contributes to the hardening. Further, the martensite plates are rather fine themselves, and their boundaries inhibit dislocation motion. Thus, there are two general contributors to martensite strength. One is the impediment of dislocation motion by the fine structure, consisting of martensite plate interfaces and a high dislocation density induced by the transformation itself and very fine mechanical twinning. The other factor is the high binding force between dislocations and carbon atoms.

   The hardness of martensite increases with increasing carbon content of the martensite, although essentially becoming constant beyond about 0.6% carbon. Further, the dependency of the hardness on carbon content is virtually unaffected by the presence of other alloying elements (e.g., Cr, Ni, Mn, etc.). The dependency

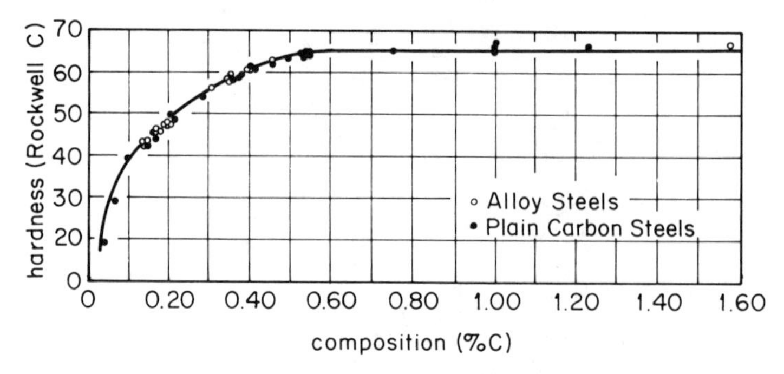

Figure 1-31 Hardness of martensite as a function of the carbon content of the martensite. *(Adapted from J. L. Burns, T. L. Moore, and R. S. Archer,* Trans. ASM, *vol. 26, p. 1, 1938; W. P. Sykes and Z. Jeffries,* Trans. ASST, *vol. 12, p. 871, 1927.)*

is shown in Fig. 1-31. Thus, for practical purposes, the hardness of martensite is a function only of the carbon content of the martensite. The maximum hardness a steel can develop can be predicted by the carbon content and the curve in Fig. 1-31.

1-4 DECOMPOSITION OF AUSTENITE UPON CONTINUOUS COOLING

It was seen in Fig. 1-18 that a given hardness can be obtained, for a given steel, by knowing the time necessary at the proper temperature to just complete the transformation of austenite. The temperature must be picked in accordance with a curve such as that shown in Fig. 1-18, and the time to give complete transformation is obtained from the isothermal TTT diagram, Fig. 1-29. In addition, if greater hardness is desired, the steel can be quenched to obtain martensite. Thus, it appears that a desired hardness can be obtained by isothermally transforming at the proper temperature. However, as will be seen, it is usually very difficult to transform steels isothermally because the size and shape of the part is sufficiently large to prevent isothermal transformation. Instead, most steels are heat treated by cooling continuously from the austenite, and it is the transformation characteristic of austenite under such a condition that is the subject of this section.

Cooling Curves

First, we want to establish clearly the temperature-time behavior when a steel in the austenite condition at high temperature is placed in a colder medium (e.g., water at room temperature). In general, the time-temperature curve (the *cooling curve*) is divided into three sections, corresponding to three relatively distinct mechanisms of heat transfer. When the hot steel first encounters the liquid, vapor bubbles are nucleated, and grow, at the hot interface. Initially, the nucleation rate and growth rate are so high that the surface of the steel is covered with a continuous vapor film. This film has a much lower thermal conductivity than the liquid, and heat transfer is relatively slow. However, as the steel cools, the rate of nucleation and growth of the vapor bubbles becomes smaller, and, when a region of the vapor breaks from the surface under the buoyant force and rises in the liquid, fresh liquid is swept into the region it occupied. This colder, fresh liquid, of course, cools the spot, the liquid heats to the boiling point, and vapor begins to form. Again, upon reaching a certain size, this bubble breaks free, and the process is repeated. Because in this stage the colder liquid is constantly brought into contact with the hot surface, the steel cools the fastest. This stage is sometimes referred to as the nucleate boiling region.

When the temperature of the steel reaches a point where convection of the liquid is sufficient to keep the liquid from boiling, heat transfer is controlled by conduction and convection in the liquid, and the rate of cooling again decreases.

The three stages are shown in Fig. 1-32*a* for a typical case. Heat transfer by

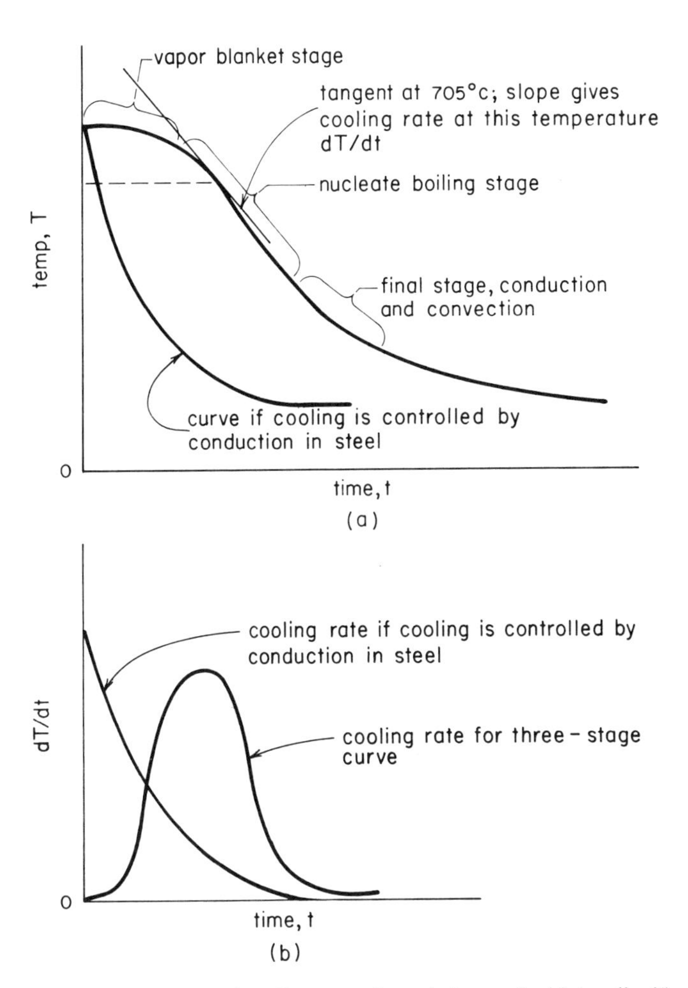

Figure 1-32 A typical cooling curve for a steel quenched into a liquid such as oil or water.

radiation is neglected in the previous discussion, because in general it is not too significant for the cases being discussed here.

It is very important to realize that the fastest cooling, for a given steel and for a given size and geometry of a part, is obtained when the surface of the steel is maintained at the temperature of the medium (e.g., 25°C). In such a case, the heat transfer is controlled entirely by conduction *in the steel*; *this is the fastest cooling rate attainable*. Such a case is illustrated in Fig. 1-32*a*.

A quantitative measure of the cooling rate can be given by the slope of the tangent to the cooling curve (i.e., the derivative dT/dt). It is seen in Fig. 1-32*b* that the cooling rate itself is not constant but depends upon the temperature at which the slope is taken. When a comparison of the cooling-rate capabilities of various quenching media is made, it is common to compare the cooling rate (i.e., the slope) at some specified temperature such as 705°C.

The range of times required to cool to 25°C which are encountered in the heat treatment of steels is sufficiently large that it is common to utilize temperature-log time plots. A comparison of the temperature-time and the temperature-log time plot is shown in Fig. 1-33.

Continuous Cooling TTT Diagrams

We will now develop the *continuous cooling* time-temperature-transformation diagram (CCT diagram) for the decomposition of austenite. In this case, it is desired to determine the sequence of events of the transformation of austenite when a steel is cooled continuously from the austenite region in a rather smooth fashion, such as would be obtained by typical quenches of water, oil, or air.

The procedure for determining the CCT diagram can be visualized as follows. A sample of the steel is austenitized for the desired time at the desired temperature (e.g., 1 hour at 1000°C). It is then placed in a quenching medium (e.g., oil, water, air) to give a cooling curve similar to the one in Fig. 1-32. When the steel has reached a certain temperature (T_1, see Fig. 1-34), it is then cooled very rapidly (e.g., water quenched) to room temperature. It is assumed that any austenite at temperature T_1 will transform only to martensite upon the fast quench from T_1, so that any martensite in the microstructure will be interpreted to have been austenite at T_1. This procedure is repeated for the same cooling curve for a number of different temperatures (such as T_2 and T_3 shown in Fig. 1-34), and the sequence of events of the decomposition of austenite is mapped out along this cooling curve. The process is then repeated using different cooling curves, until, over the range of cooling curves of interest, the transformation is determined. In practice, the method described here is not always convenient to use, and other methods are employed; however, the description gives the essential features of the method for developing CCT diagrams.

A typical CCT diagram is shown in Fig. 1-35. This is for a 0.38% carbon, plain carbon steel (1038) which has been austenitized at 870°C for about 1 hour. Look first at curve 1. Upon cooling the steel along this curve, the decomposition of austenite begins at point *a* with the beginning of the formation of primary ferrite. Primary ferrite continues to form until a temperature corresponding to point *b* is reached,

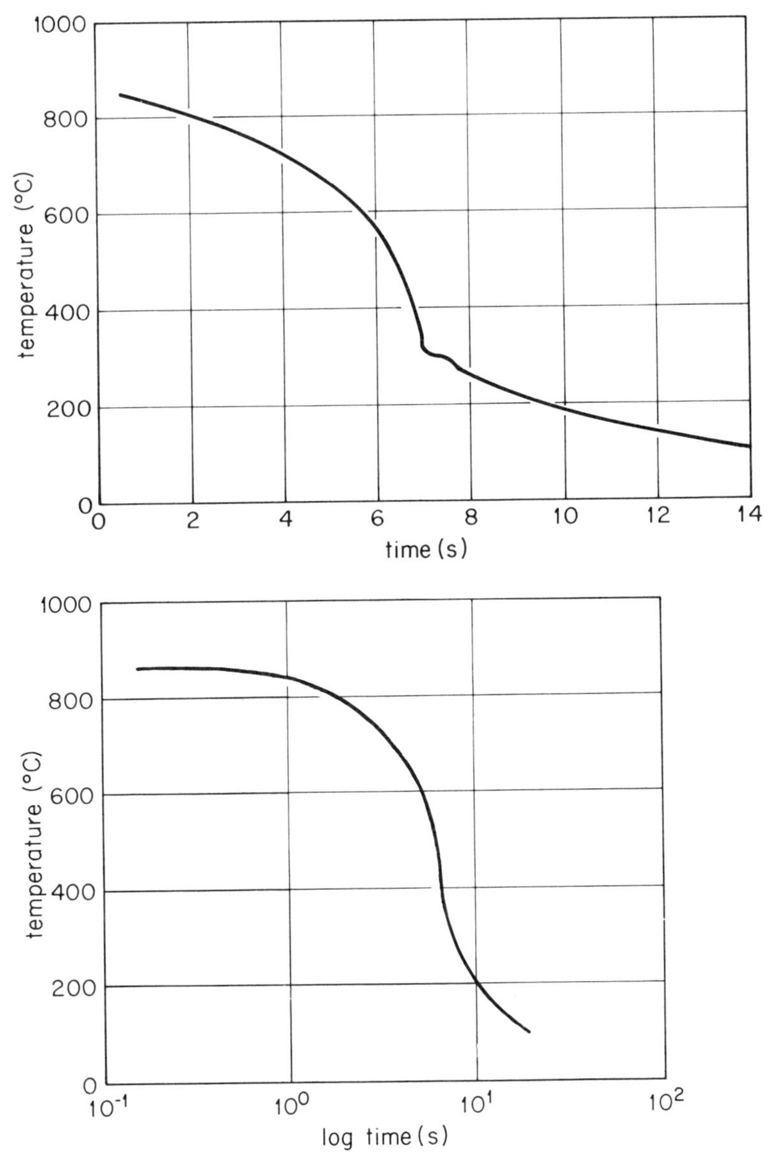

Figure 1-33 Comparison of the linear temperature-time curve and the semilog temperature-time curve.

where the transformation is essentially interrupted by the beginning of the formation of pearlite. Thus, the austenite decomposes to pearlite between points *b* and *c*. At point *c*, no more austenite remains, and the microstructure at point *c* (and at room temperature) is primary ferrite and pearlite. (The unusual appearance of the cooling curve between points *a* and *c*, during the transformation, is the effect of the heat of transformation on the cooling curve).

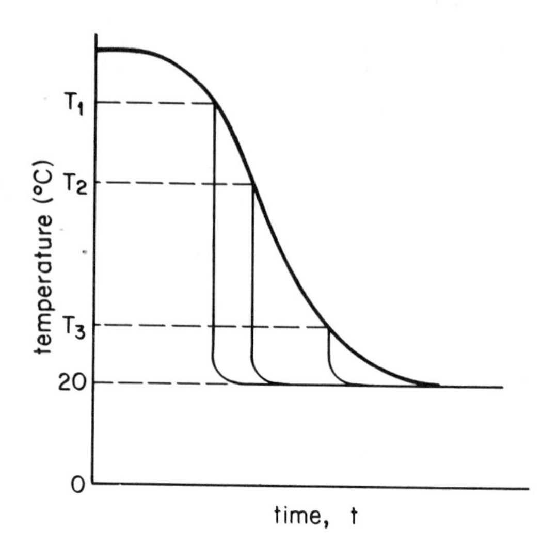

Figure 1-34 Schematic illustration of the temperature-time relationship followed to map out the transformation of austenite during cooling along the cooling curve shown.

If the steel is cooled along curve 2, a more complex microstructure is developed. Primary ferrite begins to form at point *a*, and pearlite forms between points *b* and *c*. However, at point *c*, some austenite remains, and some of it decomposes to bainite between points *c* and *d*. At point *d* there still remains some austenite, and between points *d* and *e* the remaining austenite decomposes to martensite. Thus, the room-temper-

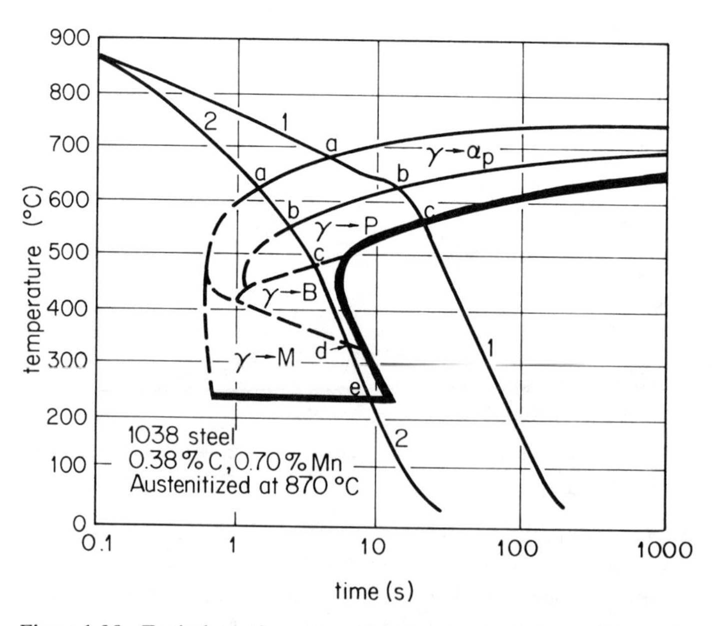

Figure 1-35 Typical continuous cooling time-temperature-transformation (CCT) diagram for the decomposition of austenite. *(Adapted from C. F. Zurlippe and J. D. Grozier,* Met. Progr., *vol. 92, no. 6, p. 86, 1967.)*

ature microstructure consists of primary ferrite, pearlite, bainite, and martensite—a complex microstructure indeed.

It is important to remember that the decomposition of austenite is complete if the heavy lines in Fig. 1-35 are crossed. Upon reaching these lines at any point, no more austenite remains. Also, for the temperature and times illustrated in Fig. 1-35, the subsequent cooling to room temperature after crossing the heavy lines has a negligible effect on the microstructure.

Typical microstructures developed are illustrated in Fig. 1-36. These were obtained for a steel similar to that for which the CCT diagram is shown in Fig. 1-35. Annealing the steel involves very slow cooling (e.g., 10 hours to 25°C) and corresponds to cooling similar to curve 1 in Fig. 1-35. The CCT diagram shows that the microstructure should be primary ferrite and pearlite, and Fig. 1-36*a* shows the actual microstructure. Cooling in air (normalizing) follows a curve similar to curve 1 in Fig. 1-35, which shows that the microstructure should still be primary ferrite and pearlite. This is substantiated in Fig. 1-36*b*, but notice that the amount of primary ferrite has decreased; also, although not seen in Fig. 1-36*b*, the pearlite has a finer spacing.

If the steel is quenched in oil, curve 2 is followed, which predicts primary ferrite, pearlite, bainite, and martensite. The microstructure is shown in Fig. 1-36*c*. Regions of pearlite, bainite, and martensite are indicated. However, the primary ferrite etches so similar to martensite that it is difficult to detect at this magnification.

Figure 1-37 summarizes the transformation behavior for the 1038 steel. The number at the end of each curve in Fig. 1-37*b* is the hardness at 25°C after completion of the cooling. It is seen that in the range in which only primary ferrite and pearlite form, the DPH hardness increases from 139 to 220 as the cooling rate increases. This is because of increased amounts of pearlite (see Fig. 1-36) and finer pearlite spacing. Upon cooling such that bainite and martensite form, the hardness increases further because of these harder constituents. Maximum hardness is obtained if the cooling curve misses the CCT diagram and the austenite forms all martensite.

The same factors that affect the isothermal TTT diagram affect the CCT diagram and in the same manner. Increasing austenite grain size displaces the diagram to longer times. Most alloying additions move the diagram to longer times. Increasing carbon and alloy content lowers the martensite start and finish temperatures. Thus, for a given CCT diagram, it is necessary to specify the austenite grain size as well as the carbon and alloy content.

The Jominy Test

Consider that a certain hardness is desired at the center of a 10-cm-diameter shaft. Assume that the choice of steel has been made, and it only remains now to determine the heat treatment (i.e., quench) to obtain the hardness. If a curve similar to Fig. 1-18 were available for the given steel, the temperature to *isothermally* transform the center of the shaft to give the hardness could be determined. However, experimentally, it may be difficult to transform the center isothermally; upon quenching the shaft to the chosen temperature, the center may cool so slowly that the applicability of the

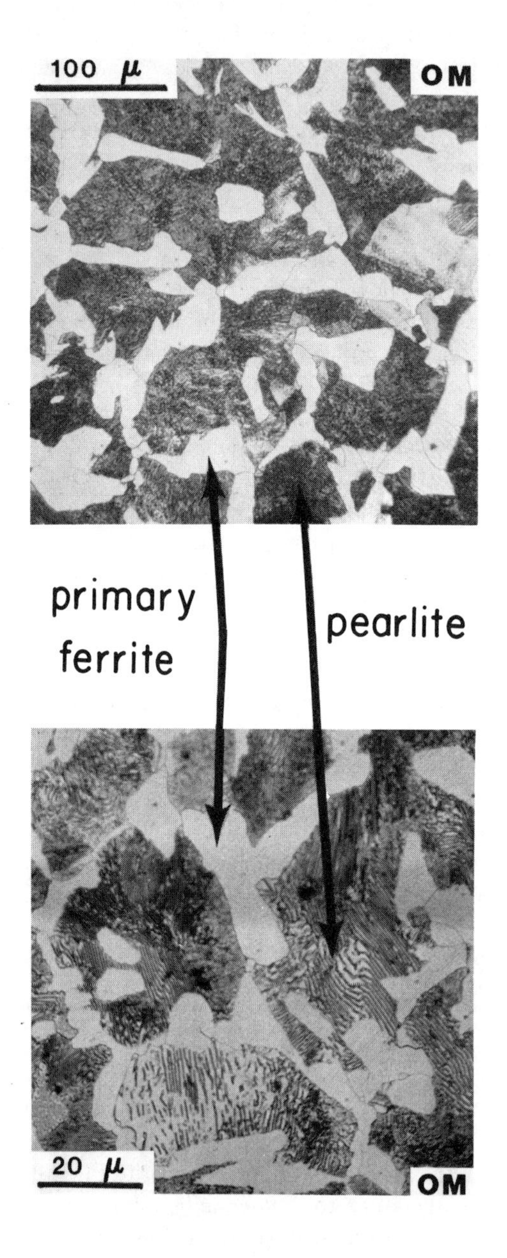

Figure 1-36 Illustration of microstructures formed upon cooling from 800°C a 0.38% carbon, plain carbon steel.

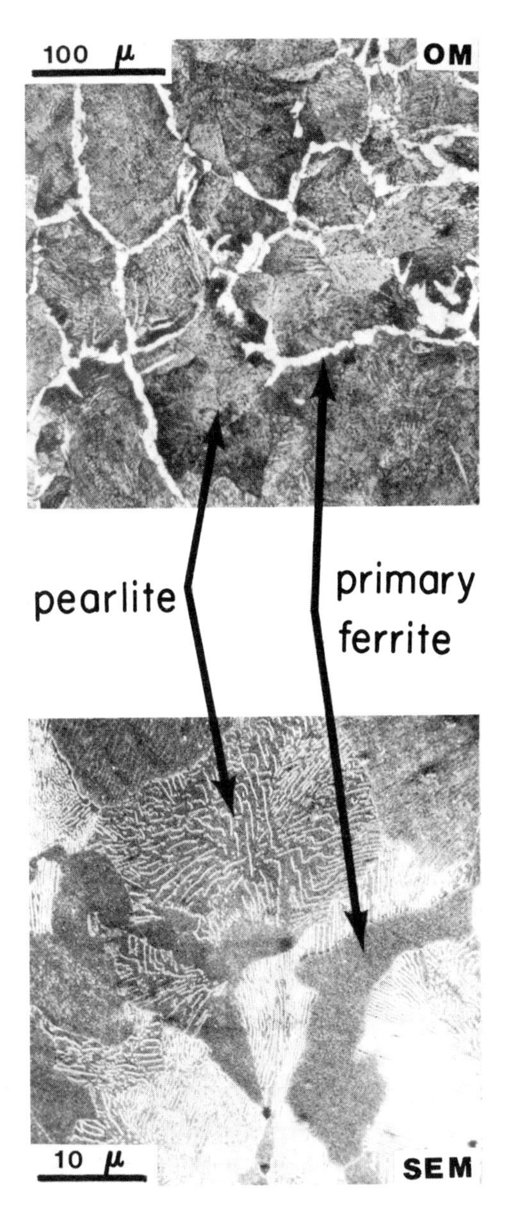

Figure 1-36 *(continued)*

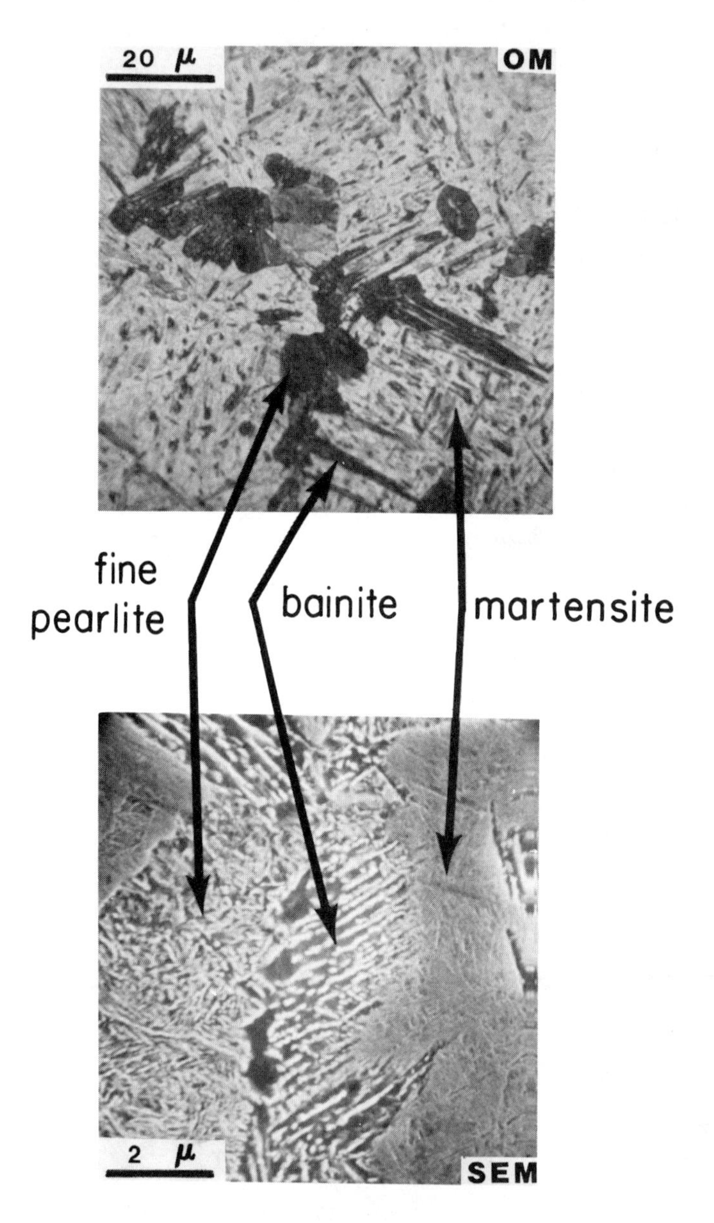

Figure 1-36 *(continued)*

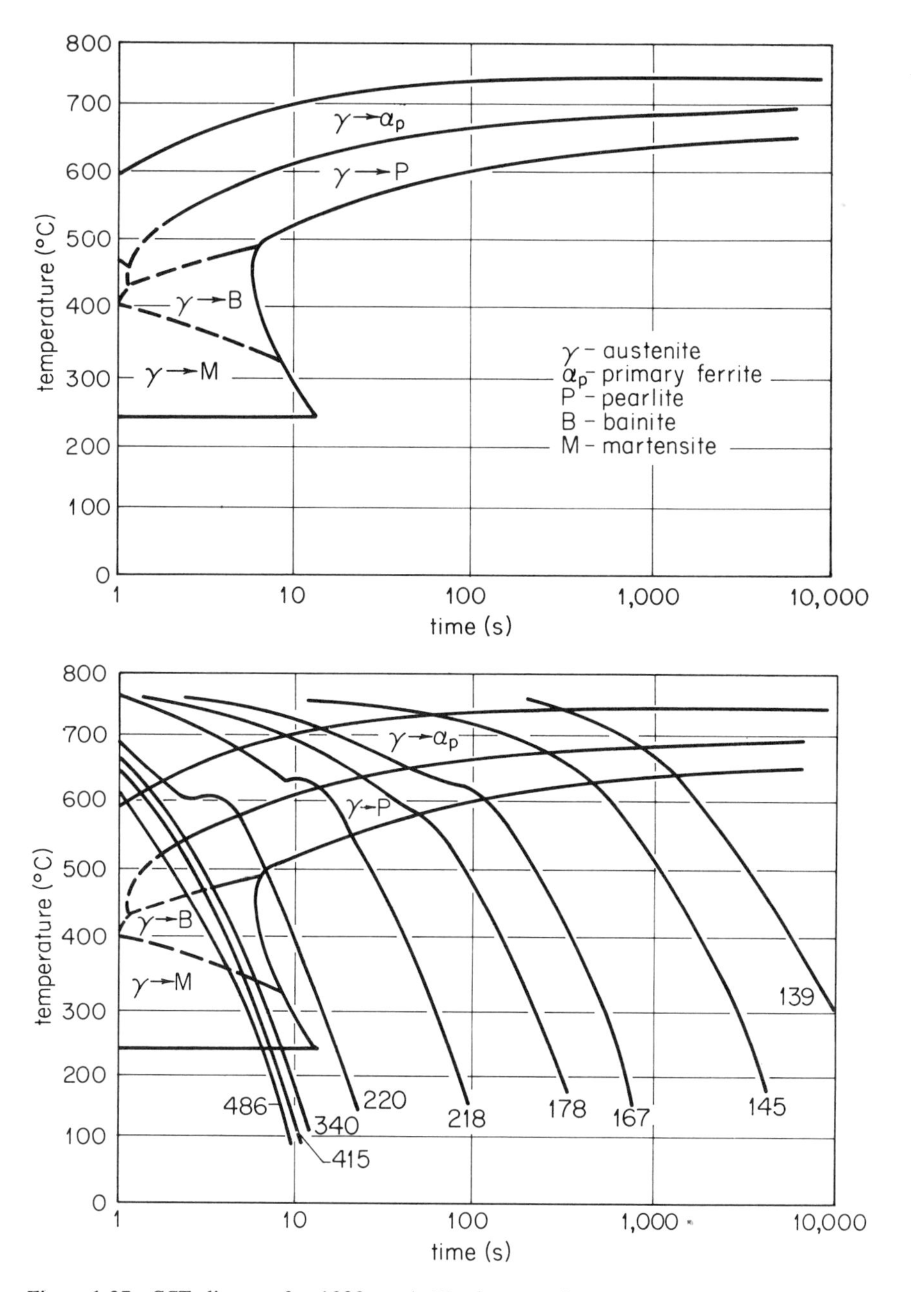

Figure 1-37 CCT diagram for 1038 steel. The bottom diagram is the same as the top diagram, except cooling curves are shown. The number at the end of each curve is the DPH hardness at 20°C after completion of cooling. *(Adapted from C. F. Zurlippe and J. D. Grozier,* Met. Progr., *vol. 92, no. 6, p. 86, 1967.)*

isothermal TTT diagram is questionable. In such a case, attention is focused on the CCT diagram, and the problem becomes one of choosing the cooling curve to give the desired final hardness, such as illustrated in Fig. 1-37. In fact, if it is not important to know the microstructure that gives the desired hardness, then the complicated procedures for determining the CCT diagram can be avoided. All that is required is to cool a sample of the steel of interest at different known rates to room temperature and then measure the hardness associated with each known cooling rate. A curve of hardness versus some measure of cooling rate (e.g., slope of tangent to curve at 705°C) can then be used to determine the cooling rate required at the center of the shaft to give the desired hardness. This type of curve is obtained by carrying out a *Jominy test*.

The Jominy test is carried out as follows. A steel sample, 2.54 cm in diameter and 10 cm long, with a lip on one end, is given the desired austenitizing treatment. The bar is then suspended in a special quenching apparatus (Fig. 1-38) and water is sprayed on one end. This end quench maintains the end of the bar at the temperature of the water (i.e., room temperature). Heat transfer to the air is negligible relative to that extracted by the water; thus heat transfer is unidirectional, and any cross section perpendicular to the bar axis has a constant temperature profile. That is, the temperature-time response at a given location from the quenched end can be monitored at the center or at the surface and the same result is obtained. Of particular importance is that the thermal diffusivity of steels is very similar, so that the cooling curves obtained are independent of the steel. This fact, coupled with the test-procedure specifications (Fig. 1-38), means that essentially the same set of cooling curves along the bar is obtained no matter which steel is tested or where it is tested. At a given distance from the quenched end the cooling rate is fixed, independent of the steel (see Fig. 1-39).

Upon completion of the end quench, a flat is ground on the surface of the bar, and the hardness is obtained as a function of distance from the quenched end. The curve obtained, the *Jominy curve*, is illustrated in Fig. 1-39. Note that the bottom abscissa is the distance from the quenched end, but the cooling rate, shown at the top, could just as well have been used.

The actual hardness-distance curve depends upon the steel, since the hardness is determined by the location of the CCT diagram. This relationship is illustrated in Fig. 1-40. Thus, those factors which must be specified for a TTT diagram (e.g., austenite grain size, chemical composition) must also be specified for a Jominy curve.

1-5 HARDENABILITY

The ability of a steel to reach its maximum hardness (i.e., to form martensite) is referred to as *hardenability*. The Jominy curve provides a visual measure of hardenability (although for detailed heat-treating procedures hardenability can be placed on a more quantitative basis), and this will be illustrated by consideration of two specific steels. Suppose that the center of a 10-cm-diameter shaft must have a hardness of 50 Rockwell C (50 R_c). Two steels are to be considered for the part: a plain carbon, eutectoid steel (i.e., 0.8% carbon), type 1080, and a high alloy steel (i.e., 0.99%

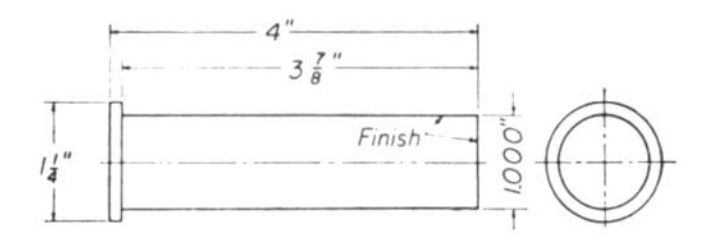

The test specimen is normalized at about 150 °F above the Ac_3 point, then machined to remove any decarburization and to obtain the correct dimensions. Following this, the specimen is heated to about 75 °F above the Ac_3 point in a closed container that has a layer of cast-iron chips in the bottom. The specimen is held at this temperature 30 min and is cooled immediately thereafter on the hardenability fixture. The time spent in transferring to the fixture should not be more than 5 sec. This fixture is constructed so that the test specimen is held 1/2 in. above the water opening, in order that a column of water may be directed against the bottom of the piece. The water opening is 1/2 in. in diam and adjustment is made so that before the specimen is placed over it, a column of water 2½ in. high comes from the opening. The water temperature is kept at 75 F plus or minus 5 °F and a condition of still air is maintained around the specimen during cooling. The piece is permitted to remain on the fixture until cold, or at least for 10 min, and is then quenched in cold water.

After cooling, two parallel flats, 180 deg apart and 0.015 in. deep, are ground along the entire length of the bar, and hardness measurements are taken at intervals of 1/16 in. for the first inch, then usually at intervals of 1/8 in. for the second inch, and 1/4 in. for the remainder of the bar. These hardnesses may be plotted to give a standard hardenability curve.

Jominy Test Specifications

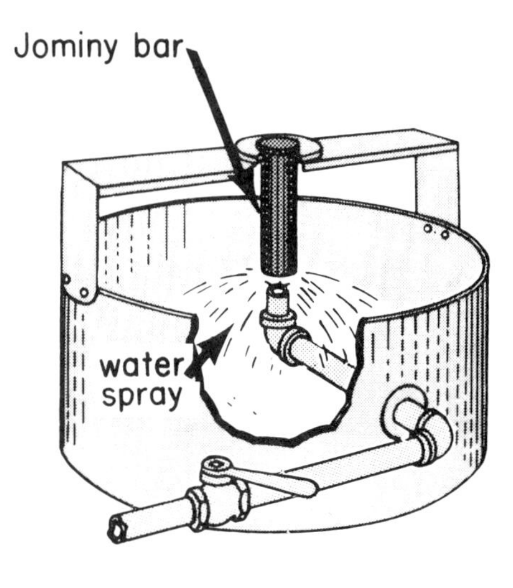

Schematic Illustration of Jominy Test

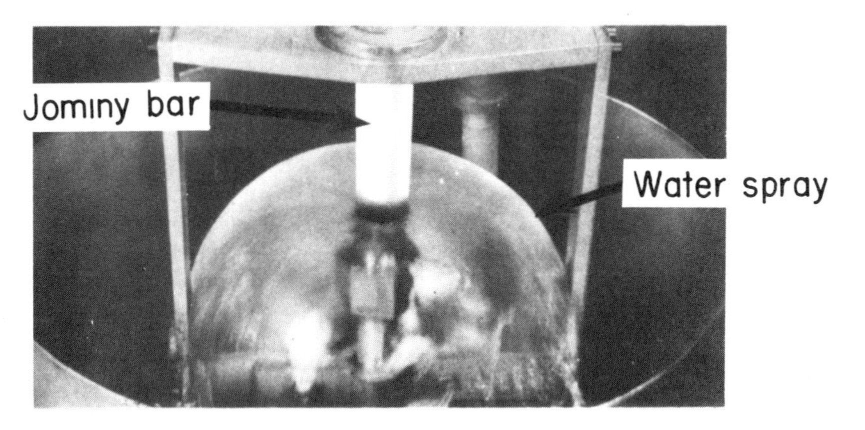

Photograph of Jominy Test

Figure 1-38 The Jominy test. *(Adapted from H. E. McGannon (ed.), "The Making, Shaping and Treating of Steel," 9th ed., United States Steel, Pittsburgh, 1971; "Metals Handbook," 1948 Edition, American Society for Metals, Metals Park, Ohio, 1948; "U.S.S. Carilloy Steels," United States Steel, Pittsburgh.)*

chromium, 0.2% molybdenum, containing 0.4% carbon), type 4340. For the given austenitizing conditions, the CCT diagrams are shown in Fig. 1-41 and the corresponding Jominy curves in Fig. 1-42. Although the plain carbon steel has a higher hardness *if* the microstructure is all martensite (see Fig. 1-31), the alloy steel has the higher hardenability (but not necessarily the higher hardness at every cooling rate) since it

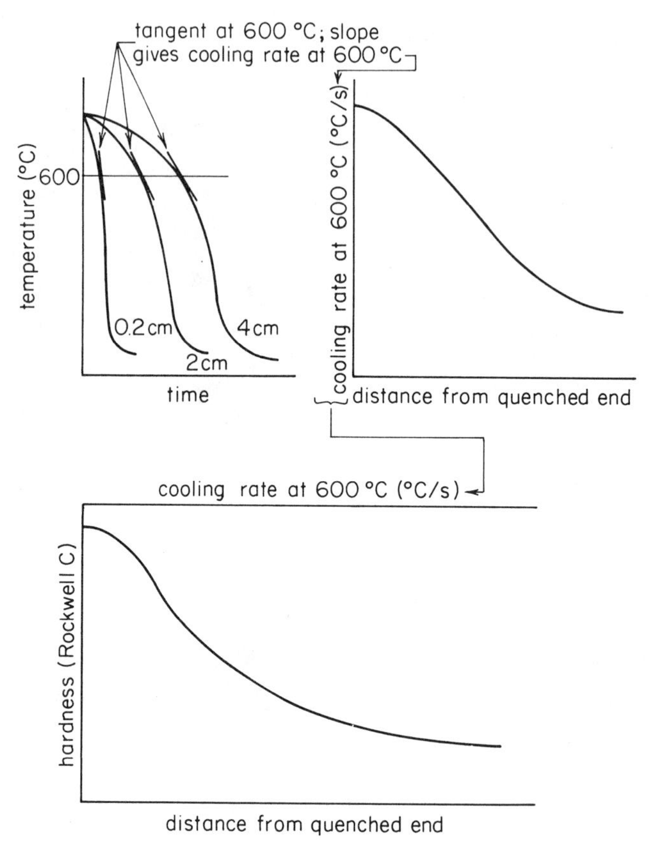

Figure 1-39 Illustration of relationship between cooling rate and position along Jominy bar.

requires a lower minimum cooling rate (*critical cooling rate*) to achieve all martensite. Indeed, the greater ability of alloy steels to harden is associated with their greater hardenability, and not with the effect the alloy additions have on the hardness of martensite, as this is independent of the alloy content and controlled by the carbon content (Fig. 1-31).

The comparison of the hardenability of the 1080 steel and the 4340 steel, as reflected in the Jominy curves in Fig. 1-42, illustrates an important point with regard to the addition of alloying elements to steels. In terms of steels to be heat treated for high hardness or strength, the alloying elements usually should not be thought of as contributing to the strength directly but through the contribution they make to hardenability, or the ability to harden. In terms of martensite, this is particularly clear. Figure 1-31 has shown that the hardness of martensite is not dependent on the alloy content, but only on the carbon content. Thus, if a hardness of 65 R_c is required, the martensite in a 0.6% carbon steel will have this hardness. The problem arises in the low hardenability of this steel; it may be difficult to obtain this hardness in thick sections. For such cases, a more expensive alloy steel of this carbon content will be required in order to allow the formation of martensite.

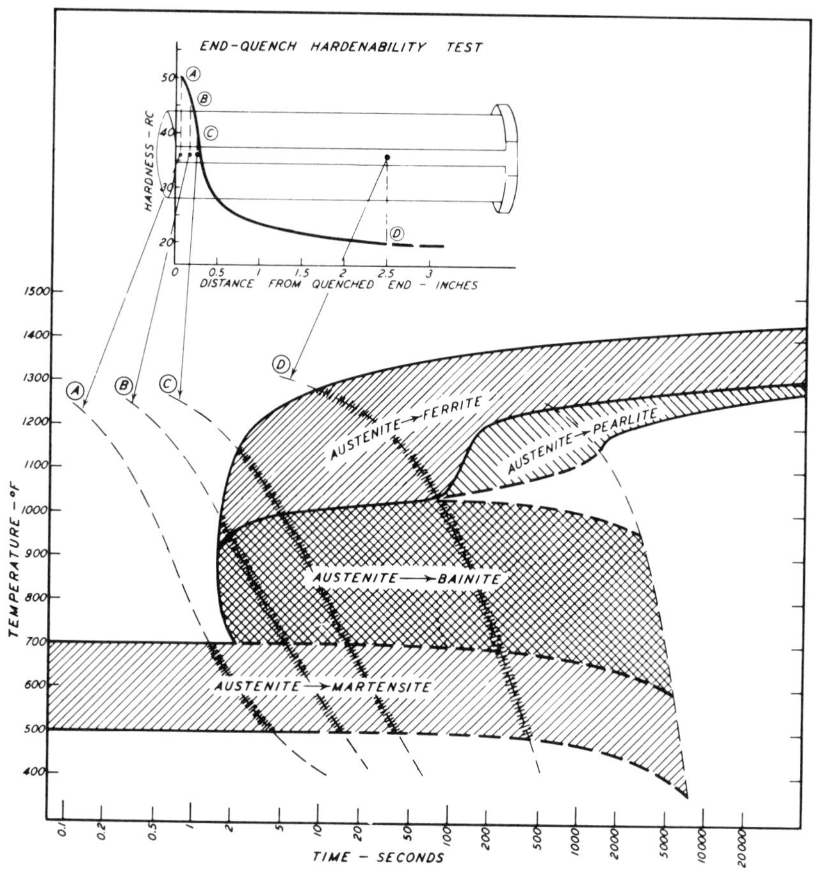

Figure 1-40 Illustration of the relationship between the Jominy curve and the CCT diagram. (8630 steel). *(Adapted from "U.S.S. Carilloy Steels," United States Steel, Pittsburgh.)*

1-6 TEMPERING

In many applications, it is necessary to use a rather ductile material (e.g., greater than 10% elongation at fracture in a tensile test), and hence the brittle structures (e.g., martensitic) are not usable. In principle, it is possible to obtain a curve of ductility for various isothermal transformation temperatures similar to that for hardness shown in Fig. 1-18. Thus, keeping in mind both the hardness and the ductility requirements, for a given steel these two curves will yield the required isothermal transformation temperature and time. However, the preceding sections emphasized the difficulty in isothermal transformation, which lead to an analysis of transformation upon continuous cooling. It appears that what is needed is accompanying ductility values for the hardness values shown at the end of the superimposed cooling curves in Fig. 1-37. From such information, the desired cooling curve can be chosen. However, a more convenient method and one that is more frequently used involves the reheating of the

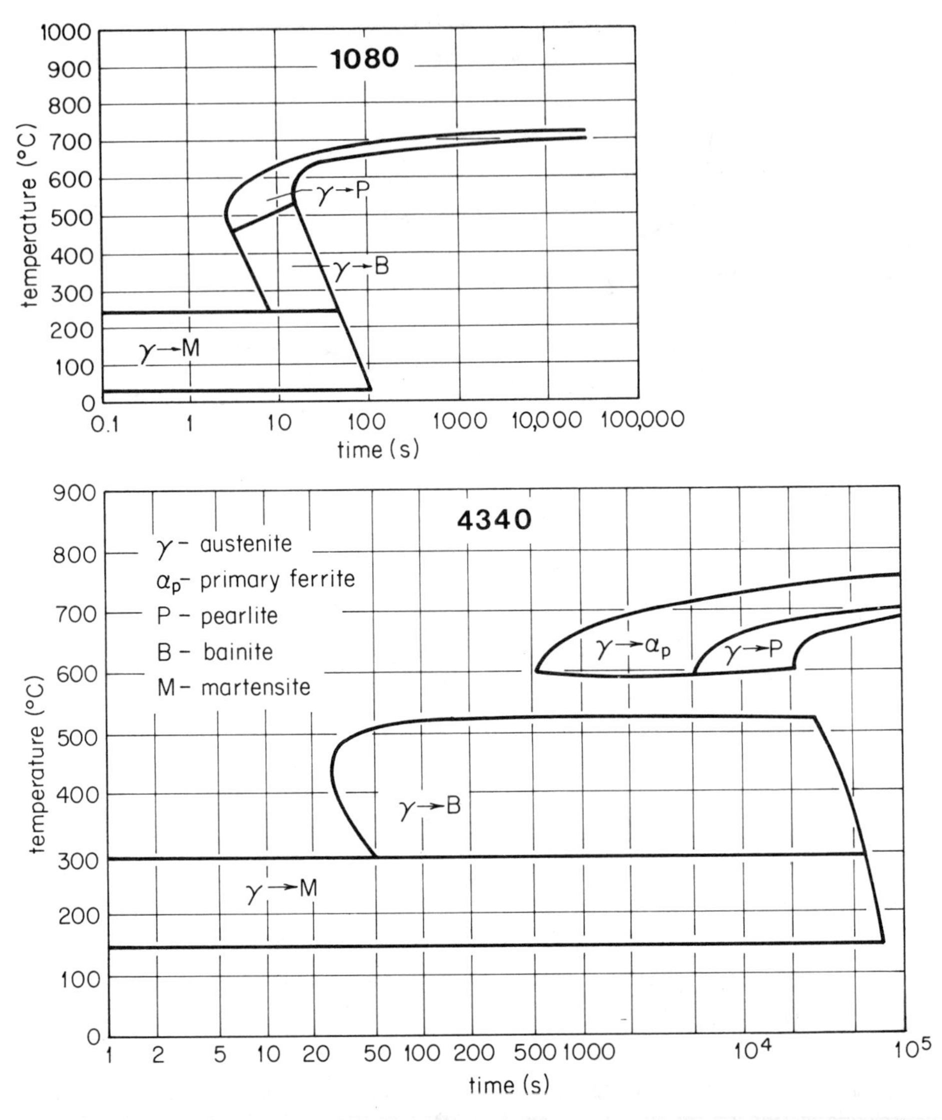

Figure 1-41 CCT diagrams for 1080 and 4340 steels. *(1080: adapted from "De Ferri Metallographia," vol. II, W. B. Saunders Co., Philadelphia, 1966; 4340: adapted from R. A. Grange and J. M. Kiefer,* Trans. ASM, *vol. 29, p. 85, 1941, and E. P. Klier and T. H. Yeh,* Trans. ASM, *vol. 53, p. 75, 1961.)*

quenched structure below the transformation temperature (e.g., below 723°C, Fig. 1-3); this treatment in the area of heating of steel is referred to as *tempering**.

* This term is another of those which have different meanings. In this book, as applied to the heat treatment of steels, it will be used as defined above. The word *drawing* is sometimes used to mean tempering in this sense.

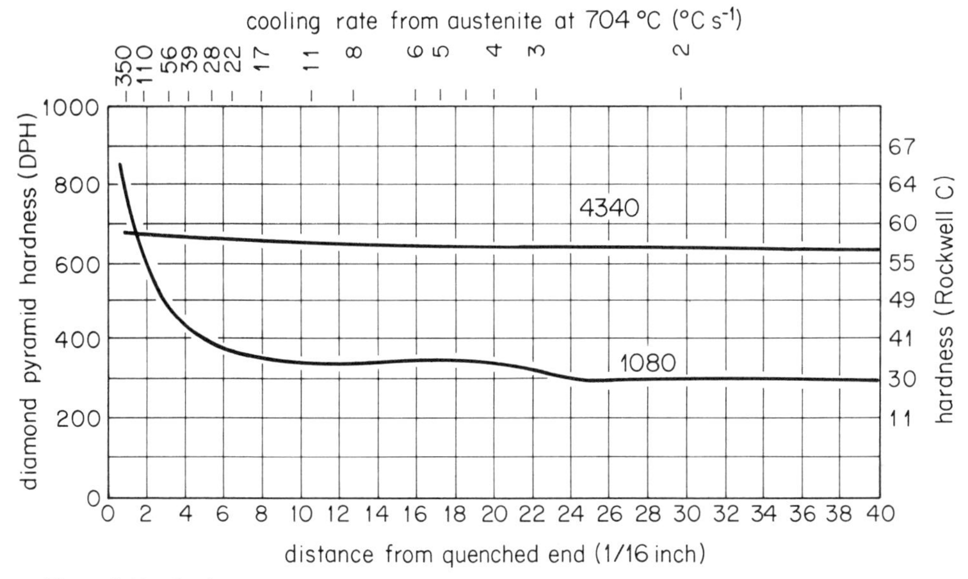

Figure 1-42 Jominy curves for 1080 and 4340 steels.

To illustrate tempering, we will only consider the heat treatment below about 700°C of steels that have been previously quenched to produce martensite. It is emphasized that we will be looking at the isothermal decomposition of the martensite, not the austenite, in this case. The equilibrium phase diagram shows that eventually ferrite and carbides will be present. It appears, then, that the reheating of the martensite permits sufficient carbon mobility to allow carbides to precipitate, leaving behind essentially carbon-free ferrite. Because of the increased ductility of the ferrite, relative to that of martensite, an improvement in the brittleness of the martensitic steel is expected.

Figure 1-43 illustrates the type of data obtained. The 4340 steel is a high-hardenability steel (see Fig. 1-42), and the samples were sufficiently small that quenching in oil allowed the formation of martensite. The properties shown at about 25°C are those of the as-quenched, martensitic structure, for tempering for 1 hour. It is seen that tempering reduces the strength and hardness, yet generally improves the ductility. Of more importance is the improvement in the impact energy. In general, for safe use such steels are required to have 15ft-lb energy absorption at the lowest temperature of applications. The data shown in Fig. 1-43 are the values at 25°C and show that at 25°C this steel is considered to be brittle until tempered above about 100°C. (From 100 to around 400°C, the steel borders on being brittle.)

Because of the highly strained nature of martensite, a multitude of sites exist for the nucleation of the carbides, and hence the carbides form rather homogeneously and with a high density. Further, this carbide formation, due to the relatively high mobility of the small carbon atom, occurs rather rapidly even at low temperatures (e.g., 100°C), and hence the structure after tempering the martensite for 1 hour at 200°C is no longer martensite, but ferrite and carbides. The relatively high strength

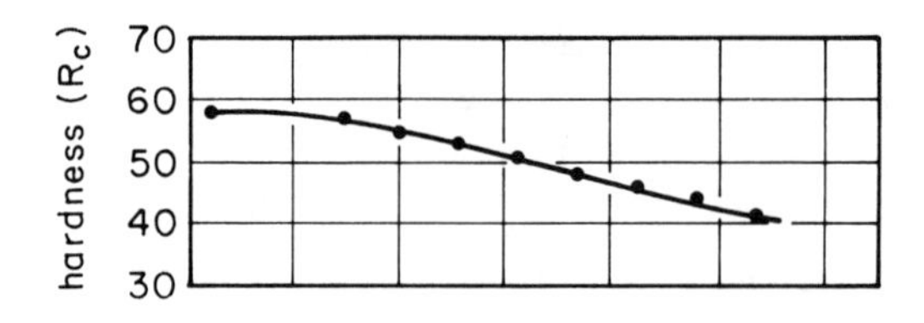

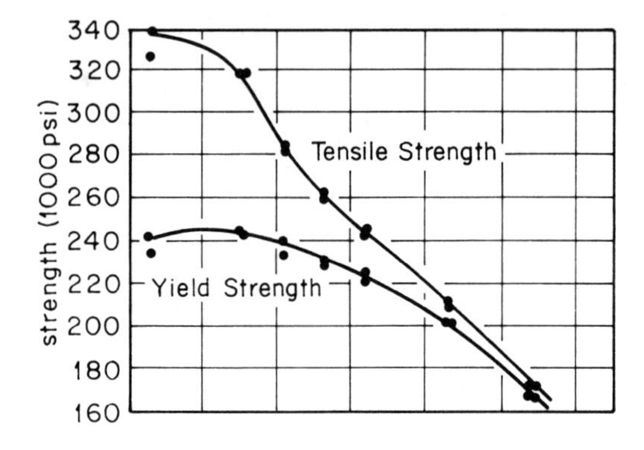

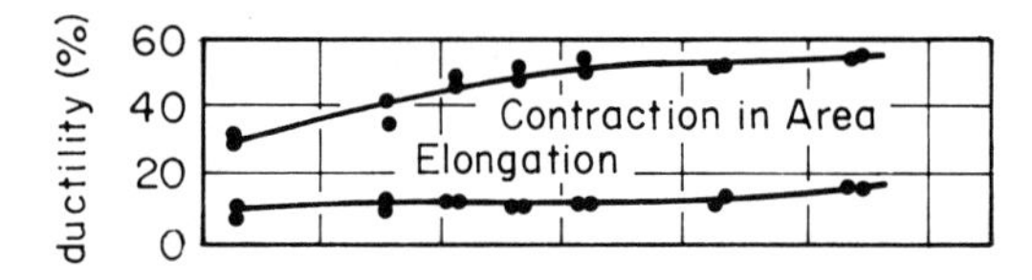

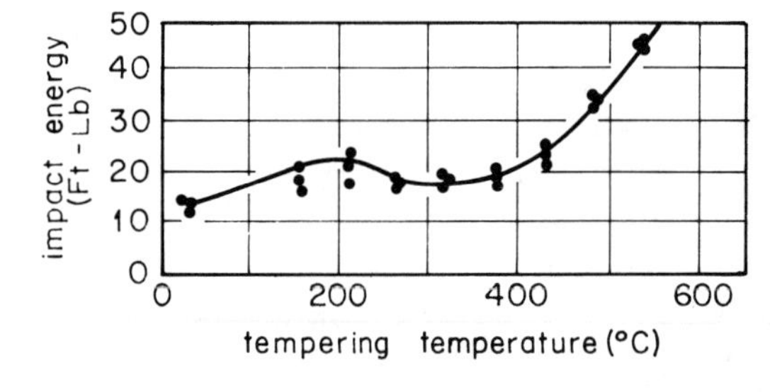

Figure 1-43 Effect of tempering on the mechanical properties of an original martensitic structure. The steel was 4340, oil quenched, and tempered 1 hour. *(Adapted from L. J. Klingler, W. J. Barnett, R. P. Frohmberg, and R. R. Troiano,* Trans. ASM, *vol. 46, p. 1557, 1954.)*

that still exists is due to a dispersion-hardening effect of the fine and numerous carbides. The continued loss in strength upon tempering at higher temperatures is due to the coarsening of these carbides, which causes an increase in the mean carbide spacing and reduces their effectiveness as a dispersion strengthener. The types of microstructural changes seen are illustrated in Fig. 1-44.

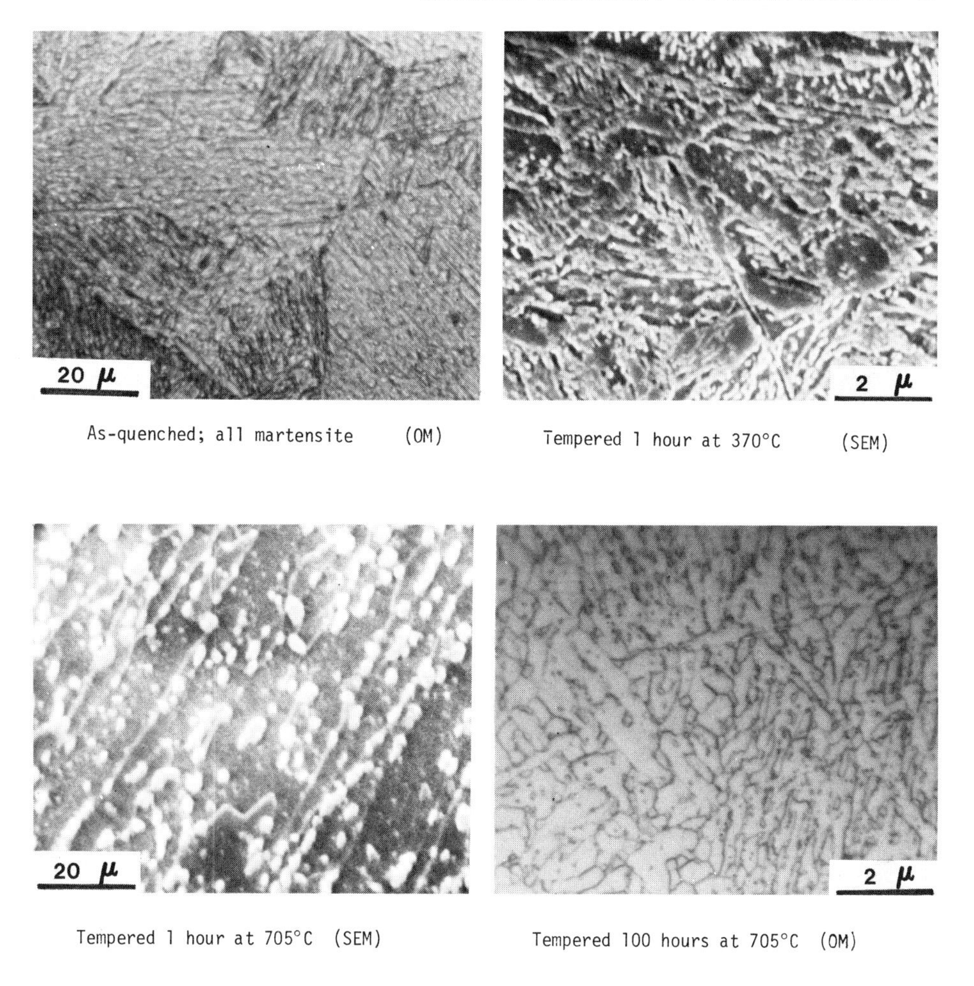

Figure 1-44 The microstructure of martensite and tempered martensite for a 1045 steel (0.47% C, 0.88% Mn). Note the different magnification used. The coarsening of the carbides for higher tempering temperature and longer times is to be noted.

CHAPTER

TWO

HARDENABILITY AND HEAT TREATMENTS

A problem in developing a steel part to meet property requirements is the choice of the steel *and* the necessary heat treatment. In this chapter we consider the questions: how is a steel chosen, and how is the steel part to be quenched and tempered, to give the desired properties? In Chap. 1 it was seen that from the Jominy curve the cooling rate necessary to produce a desired hardness in a given steel (i.e., a given chemistry and austenite grain size) can be obtained. If a relationship between hardness and the property, such as yield strength, can be found, then the problem is reduced to determining how to cool the part from the austenitizing temperature so that the position in the part requiring this property will have the desired cooling rate. This is a heat-transfer problem and for simple geometries (e.g., round bars) a satisfactory treatment is available.

Thus in this chapter the subjects of quenching, hardenability, and tempering are covered in detail, leading toward providing the necessary background and information requisite to specifying steels and heat treatments to give desired properties. Some specialized heat treatments are also described.

2-1 QUENCHING

The process of cooling a steel bar quenched from high temperature into a liquid at 20°C was described in Chap. 1. The temperature-time curve of a point inside or on the surface of the part can be divided into three regions. The high-temperature stage involves a relatively slow heat-transfer rate due to the hot part vaporizing some of the fluid to form a continuous gaseous blanket between the steel and the cooling fluid. As the steel cools, however, the nucleate boiling region is entered in which the vapor forms discrete bubbles, which are removed rapidly from the surface, allowing fresh fluid to contact the surface. During this stage, the cooling rate is at its highest. When the temperature is sufficiently low, boiling ceases; heat is then transferred by conduction and convection in the fluid, and the rate of heat transfer decreases. Figure 2-1 illustrates the temperature response. In such quenching processes the relative importance of the three stages depends upon a number of factors, such as the surface condition of the part and the flow pattern of the fluid past the part. In general, it is the second stage that is most important.

What is desired, then, is a quantitative measure of the ability of the quenching medium to extract heat, so that for a known geometry and size of a part, the cooling rate can be estimated at any position in the part. Studies of the heat extraction during quenching have shown that the rate of heat conduction away from the steel surface into the liquid can be approximated by a constant H' (film coefficient) multiplied by the difference between the temperature of the surface and that of the bulk fluid. This heat removal is equal, of course, to that being conducted from the hot metal to the metal surface; in one-dimension heat conduction (say, along the radius of a large

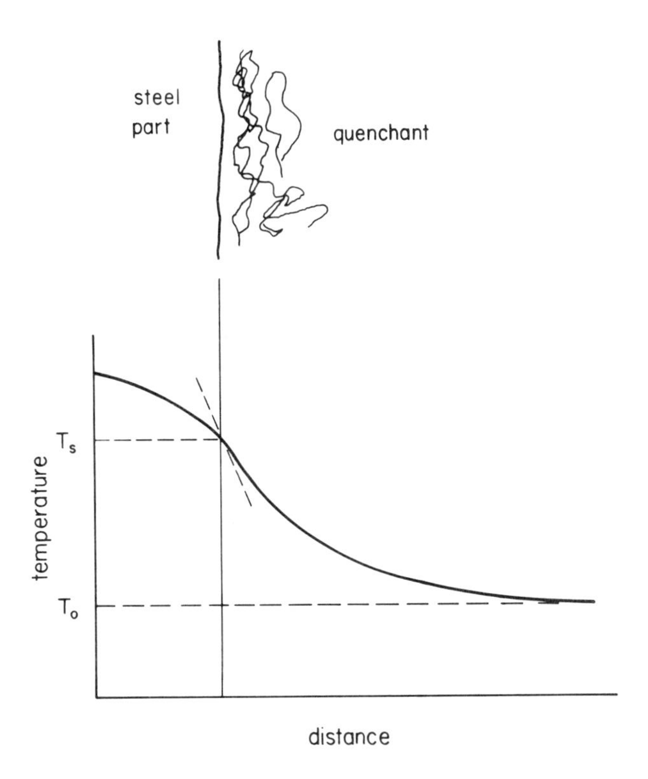

Figure 2-1 Temperature-distance curve from the center of a steel part into the quenchant. T_s is the temperature at the surface of the steel, and T_o is the temperature of the quenchant at a distance from the part such that the temperature is unaffected. The temperature gradient at the surface of the steel is shown.

round bar) this is given by the product of the thermal conductivity K of the steel and the temperature gradient at the surface. The relationship is then

$$-K\left(\frac{dT}{dx}\right)_{\text{surface}} = H'(T_s - T_0) \tag{2-1}$$

Figure 2-1 illustrates this relationship. It can be rewritten to give

$$-\left(\frac{dT}{dx}\right)_{\text{surface}} = h(T_s - T_0) \tag{2-2}$$

where

$$h = \frac{H'}{K} \tag{2-3}$$

This concept was developed by Grossmann, who referred to $H = h/2$ as the *severity of quench*.

In Eq. (2-2), the temperature T_s decreases with time. The relationship for the non-steady state heat conduction in the steel is

$$\frac{\partial T}{\partial t} = \alpha^2 \left(\frac{\partial^2 T}{\partial^2 x} + \frac{\partial^2 T}{\partial^2 y} + \frac{\partial^2 T}{\partial^2 z}\right) \tag{2-4}$$

in rectangular coordinates. Here α is the thermal diffusivity of the steel. Using boundary conditions, this equation must be solved for temperature as a function of time and substituted in Eq. (2-2). Then to determine the H value of a quenchant, a steel sample of simple geometry (e.g., cylinder or sphere) is cooled in the bath and the temperature-time curve at a known point (e.g., surface or center) of the sample is obtained. From this curve, the value of H can be determined at any temperature.

Determination of H by the method outlined above has shown that it is not constant. This is not surprising as a number of assumptions are inherent in the method. The thermal diffusivity and thermal conductivity of steels are temperature-dependent; both decrease with decreasing temperature for austenite, but for ferrite-carbide mixtures and for martensite both increase with temperature. In cooling a steel, not only are these properties temperature-dependent, but the phases present change as the austenite transforms. There is also a heat release associated with the transformation. The film coefficient H', which affects H, is dependent both on temperature and on the size of the steel part that is cooled to obtain data from which to calculate H.

This problem is illustrated in Figs. 2-2 and 2-3. Figure 2-2 shows calculated H values as a function of temperature (and for two different-size cylinders). Thus H is a function of temperature and of size of the steel being heat treated, not just a function of the quenchant. The influence of the type of steel is shown in Fig. 2-3, which shows cooling curves at the center of two different-size cylinders for three steels. Using the cooling rate at 400°C, the H values shown were calculated. For the 0.5-in.-diameter cylinder quenched into oil, the H value at 400°C varies from 2.3 to 1.6, depending on which steel is chosen for the calculations. For the 3-in.-diameter cylinder, the H value varies less, from 1.6 to 1.3.

Most reported H values using this method are based on cooling rates in the range of 540 to 700°C; the lower temperature includes the range where austenite decom-

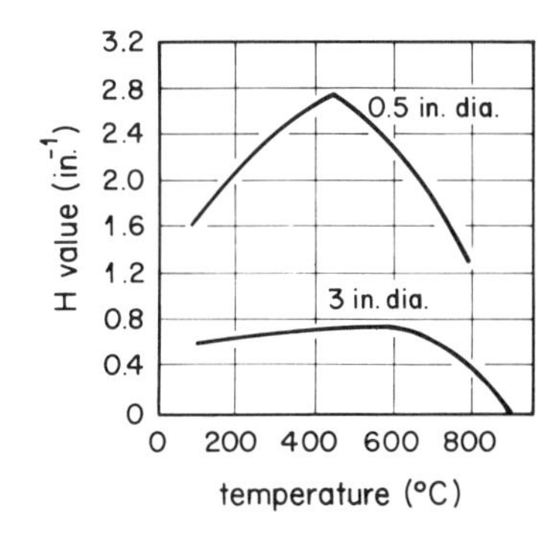

Figure 2-2 Variation of the calculated H value with temperature and size of cylinder. The cylinders were quenched in water from 845°C, and the temperature is that in the center of the bar. *(Adapted from D. J. Carney and A. D. Janulionis,* Trans. ASM, *vol. 43, p. 480, 1951.)*

position is most rapid for many steels. It is found that hardenability correlations with H values are more consistent using this temperature range. In actuality, most reported H values are derived from the hardness gradient along the radius of quenched cylinders. The H values from this method have similar uncertainties in them as discussed above. Table 2-1 lists typical H values for various quenching media.

In the following discussion of heat treating, calculations will be made assuming a constant value of H. However, the restriction illustrated in this section must be considered, and this point will be emphasized in Chap. 5 in examining heat treatments.

2-2 THE JOMINY TEST

The Jominy test was described in Chap. 1. It involves the cooling of a steel bar about 2.5 cm in diameter and 10 cm long by spraying water on one end. This end of the bar is thus maintained at the temperature of the water (e.g., 20°C) and most of the heat is extracted out this end, very little being lost to the air. A distribution of cooling curves is generated, with decreasing cooling rate as the distance from the quenched end increases. Once the bar has cooled to 20°C the hardness is measured along the axis, from which the Jominy curve is plotted. The Jominy curve depends on how the cooling curves cross the CCT diagram of the particular steel (see Fig. 1-40).

The reproducibility of the Jominy test is illustrated in Fig. 2-4. This figure shows the time-temperature data at 0.5 in. from the quenched end of a Jominy bar of the same steel, measured in 14 different tests. Note that at a given temperature the time is reproducible to about ±5 s around 540°C; the cooling rate at this temperature varies from about 11 to 17°C/s. Let us examine the significance of this uncertainty. In Fig. 2-5 are two Jominy curves. The reproducibility of the hardness measurements we will take as about ±1 R_c, or about ±18 DPH. It is seen that an uncertainty of ±3°C/s in reproducing the cooling rate of 140°C/s does not affect the hardness significantly. Indeed, for the high-hardenability alloy steel 4340, the resultant Jominy curve is not affected by slight variations in the Jominy test procedure. However, the hardness of the plain carbon steel 1045 is susceptible to uncertainties in the cooling rate from about 20°C/s or higher. For example, the hardness (from the curve) at 0.25 in. from the quenched end, at a cooling rate of 42°C/s, is 320 DPH. If the cooling rate had

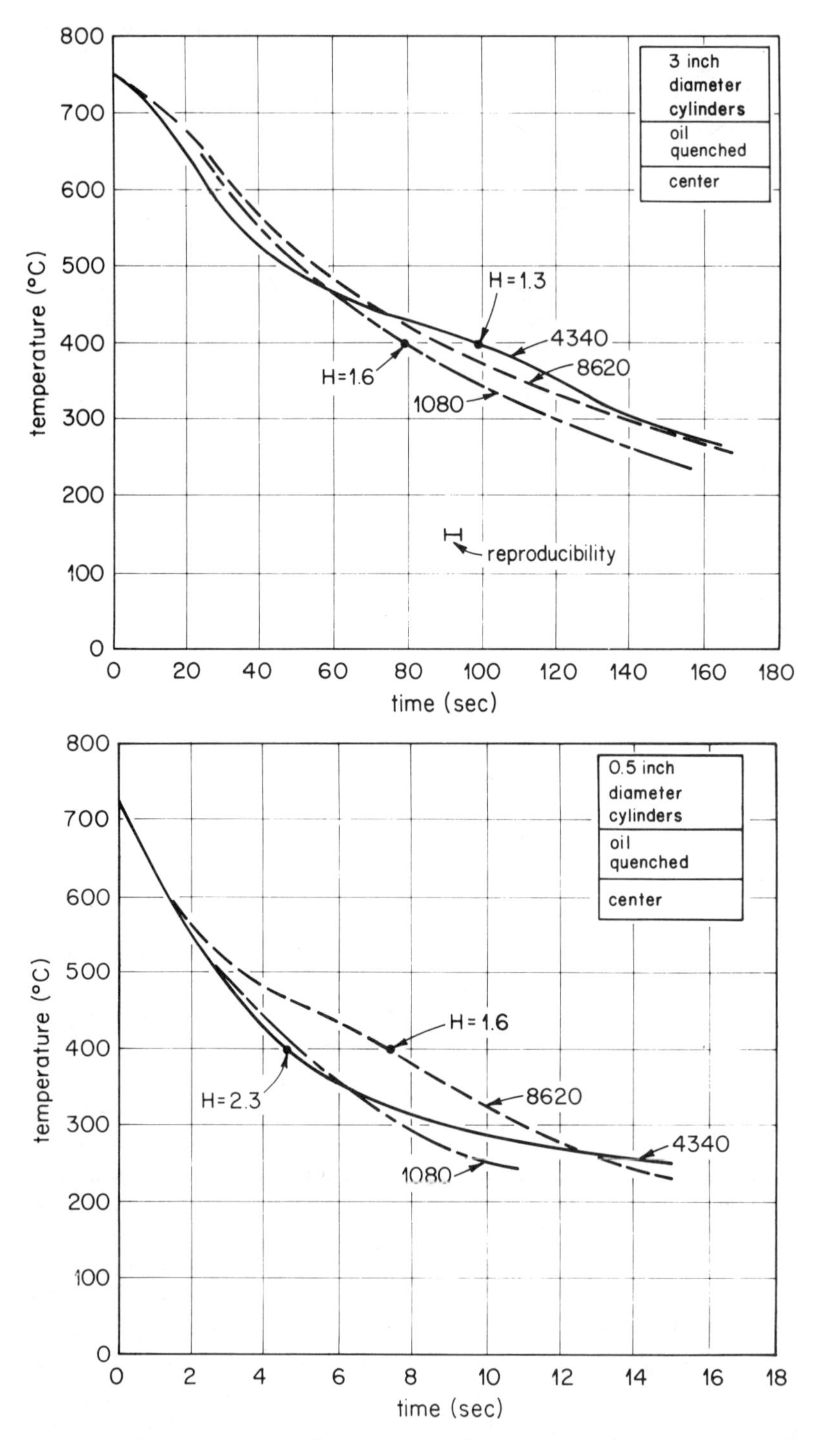

Figure 2-3 Cooling curves for the center of cylinders quenched into the same oil. The H values shown were calculated from the cooling times at 400°C. *(Adapted from D. J. Carney,* Trans. ASM, *vol. 46, p. 883, 1954.)*

Table 2-1 Approximate severity of quench values for various quenching conditions

H Value	Quenching conditions
0.20	Poor oil quench–no agitation
0.35	Good oil quench–moderate agitation
0.50	Very good oil quench–good agitation
0.70	Strong oil quench–violent agitation
1.00	Poor water quench–no agitation
1.50	Very good water quench–strong agitation
2.00	Brine quench–no agitation
5.00	Brine quench–violent agitation
∞	Ideal quench

Source: "U.S.S. Carilloy Steels," p. 42, United States Steel Corporation, Pittsburgh.

been about 38°C/s, the hardness would have been about 300 DPH, a difference in hardness of 20 DPH (or about 2 R_c units); this is about the level of reproducibility of the hardness readings themselves.

The conclusion from the above information is that the Jominy test has sufficient reproducibility so that the scatter produced in the hardness data of the Jominy curve is about the same as that intrinsically present in the hardness measurements.

It is necessary to examine how sensitive the cooling rates are along the Jominy

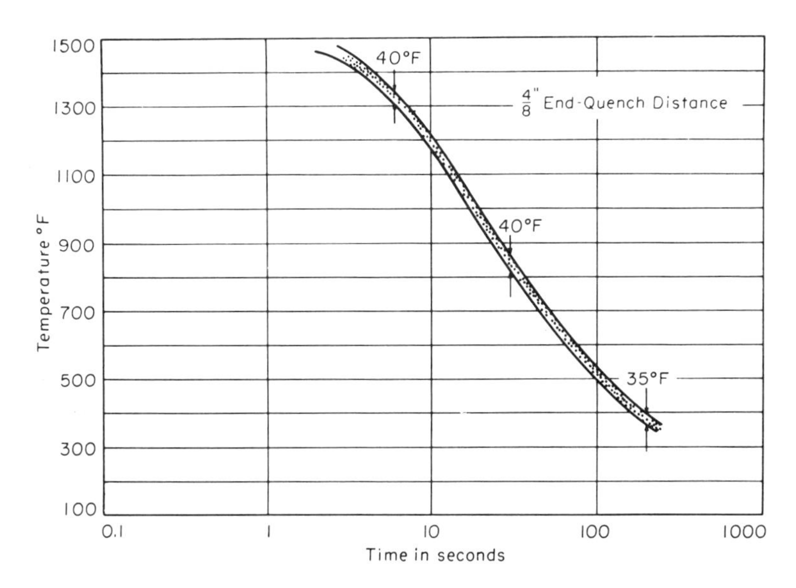

Figure 2-4 Temperature-time data obtained at 0.5 in. from the end of a Jominy bar. The points represent data from 14 different tests using the same steel. *(From J. Birtalan, R. G. Henley, and A. L. Christenson,* Trans. ASM, *vol. 46, p. 928, 1954.)*

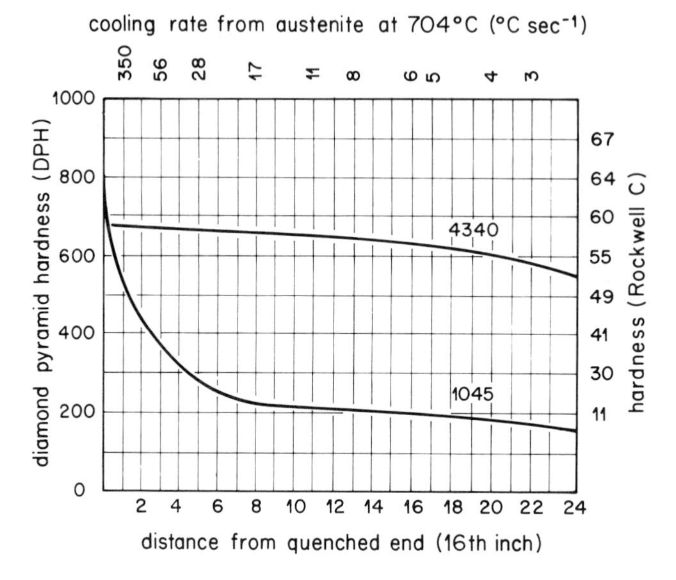

Figure 2-5 Jominy curves for a plain carbon steel 1045 (0.47% C, 0.75% Mn, 0.24% Si) and an alloy steel 4340 (0.39% C, 0.74% Mn, 0.31% Si, 1.73% Ni, 0.87% Cr, 0.24% Mo).

bar to the type of steel being cooled. In the Jominy test, the cooling rate depends on the thermal diffusivity of the steel and on any heat release as the austenite decomposes. Both of these factors are dependent on the type of steel. (This is expected from the data of Fig. 2-3.) Figure 2-6 compares the temperature-time curves for two steels, measured at the same point in an end-quench test. The curves differ, and the cooling rate at 540°C varies from about 1 to about 2°C/s.

The reproducibility of the Jominy test itself, and the dependence of the cooling rate on the steel, lead to a reproducibility in the cooling rate of about ±5°C/s at 700°C, and about ±1°C/s around 500°C. The significance of this in terms of choosing quenchants will be examined in Chap. 5.

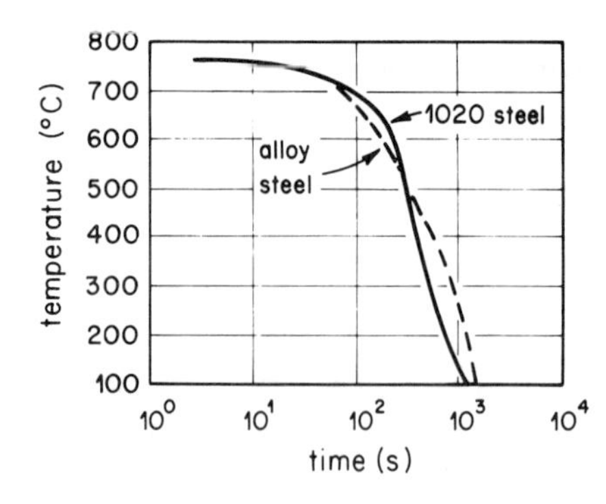

Figure 2-6 Cooling curves for a 1020 steel (0.20% C, 0.47% Mn) and an alloy steel (0.32% C, 1.62% Mn, 0.48% Cr, 0.34% Mo), both quenched from 900°C. The temperature was measured at the center of a 20-cm-long, 2.5-cm-diameter cylinder and quenched from both ends. *(Adapted from C. R. Wilks, E. Cook, and H. S. Avery,* Trans. ASM, *vol. 35, p. 1, 1945.)*

2-3 COOLING CORRELATIONS

In this section we examine correlations that exist between the cooling rate at positions along the Jominy bar and positions in various shapes, such as cylinders, when these shapes are quenched into a given medium. First we must get straight the methods used to make this correlation. One method would be to take a cylinder of a given steel and of a given size, place in the center a thermometer, austenitize the steel, and then quench it into a given medium. A temperature-time curve is obtained for this center position, from which a measure of the cooling rate can be obtained. (This cooling rate can be defined several ways; for example, the time necessary to cool from a specified high temperature to a specified low temperature can be used.) We then examine the temperature-time curves for various positions along the Jominy bar. Using the selected criteria for cooling rate, these data are used to obtain a relationship between cooling rate and distance along the Jominy bar. Comparison of this result with the cooling rate for the cylinder gives the position along the Jominy bar which cools at the same rate as the center of the cylinder of the given diameter.

The correlation can be calculated, making certain assumptions, or it can be based on measurements. The results do depend upon the criteria taken for cooling rate, and Fig. 2-7 illustrates the discrepancy. We will examine the influence of this uncertainty in developing heat treatments in Chap. 5.

In Fig. 2-8 are shown cooling-rate correlations for several geometries (round, square, and rectangular bars) and for several different H values.

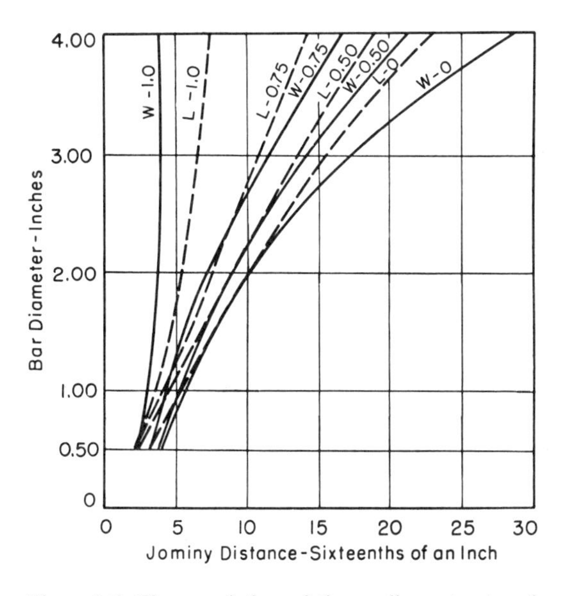

Figure 2-7 The correlation of the cooling rate at various fractions of the radius of round bars with position along the Jominy bar. The results were obtained by two different methods (W and L). *(Adapted from W. Wilson,* Trans. ASM, *vol. 44, p. 836, 1952.)*

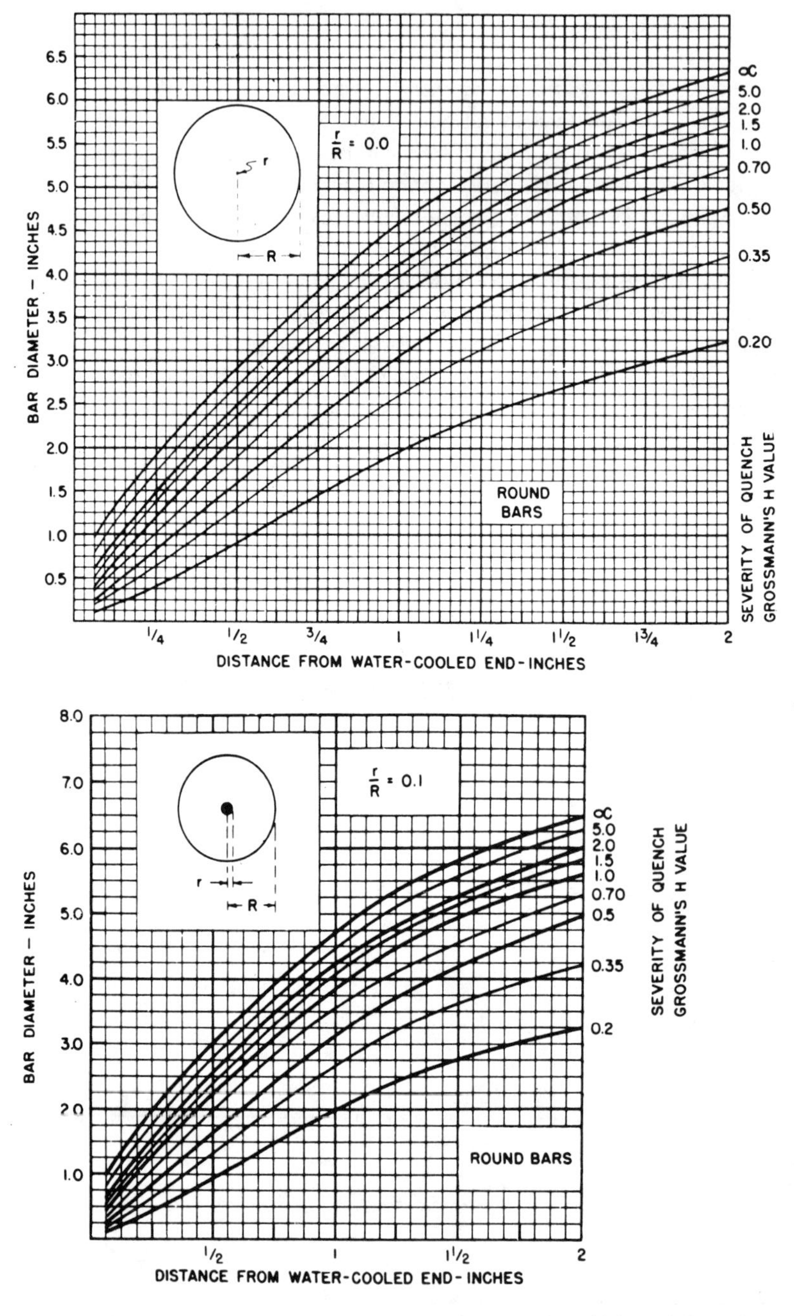

Figure 2-8 Cooling-rate correlations for several geometries. *(Adapted from J. L. Lamont,* Iron Age, *vol. 152, no. 16, p. 64, 1943.)*

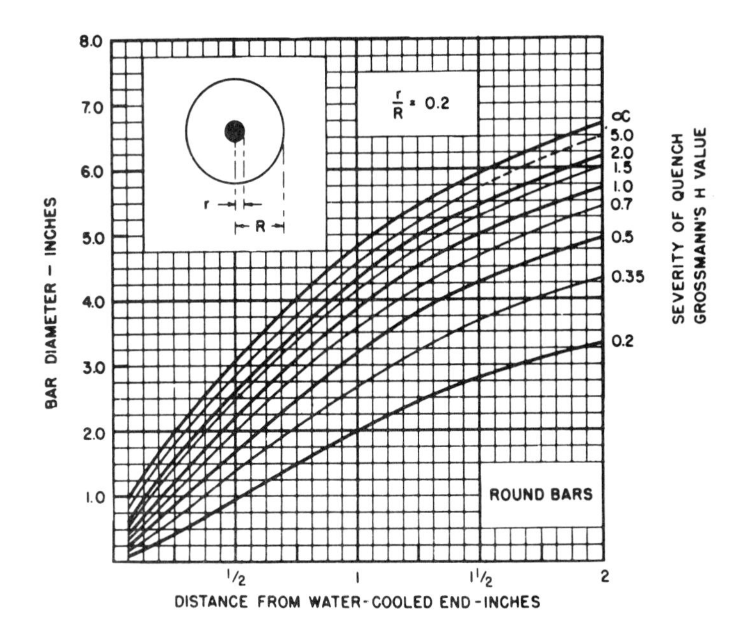

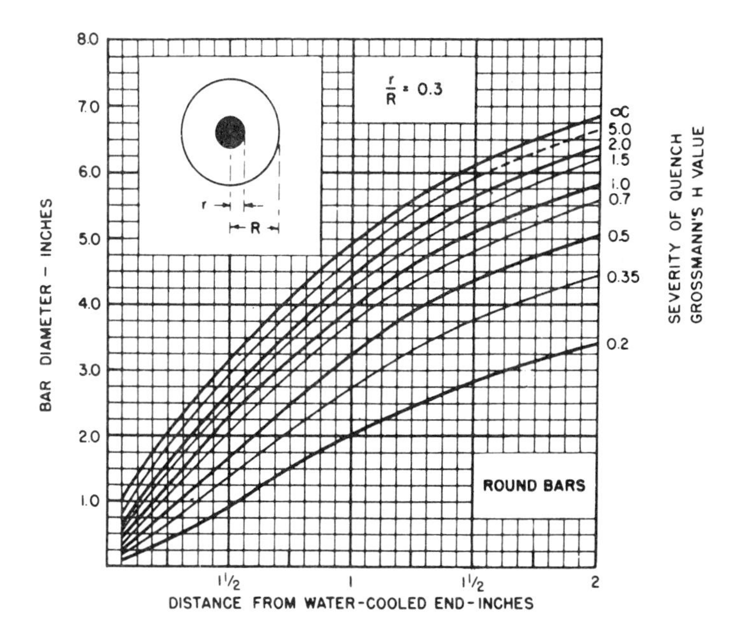

Figure 2-8 *(continued)*

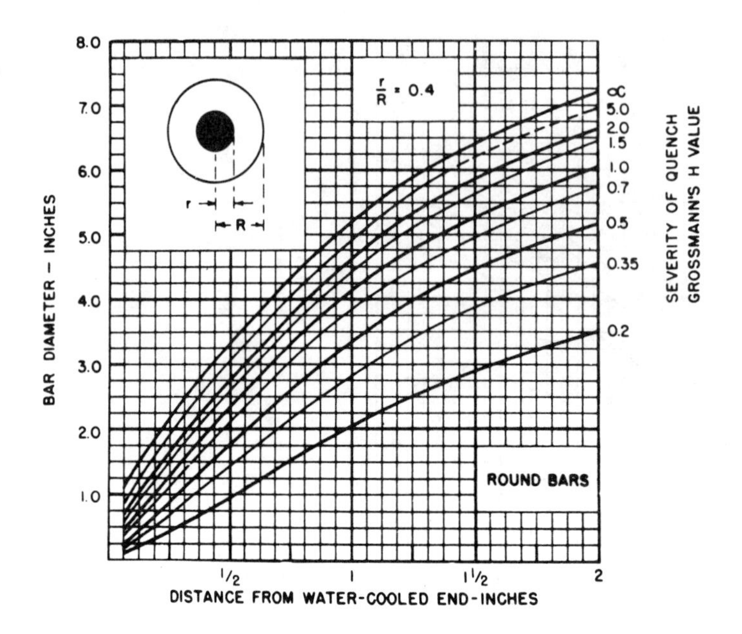

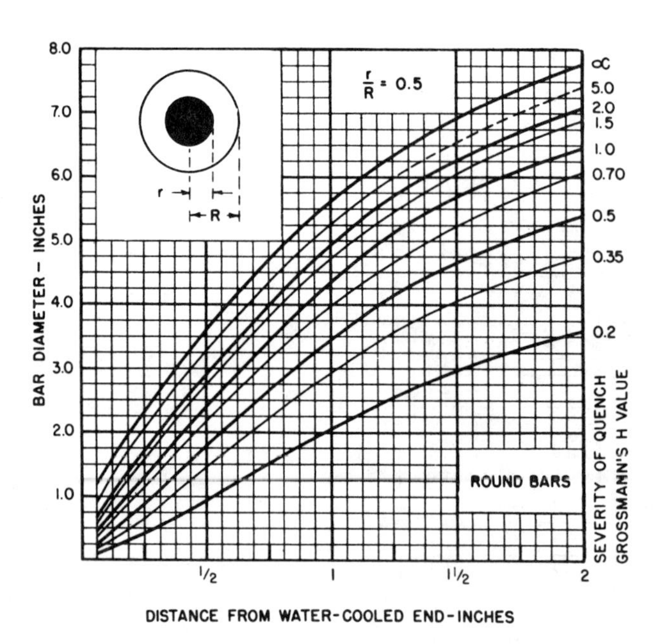

Figure 2-8 *(continued)*

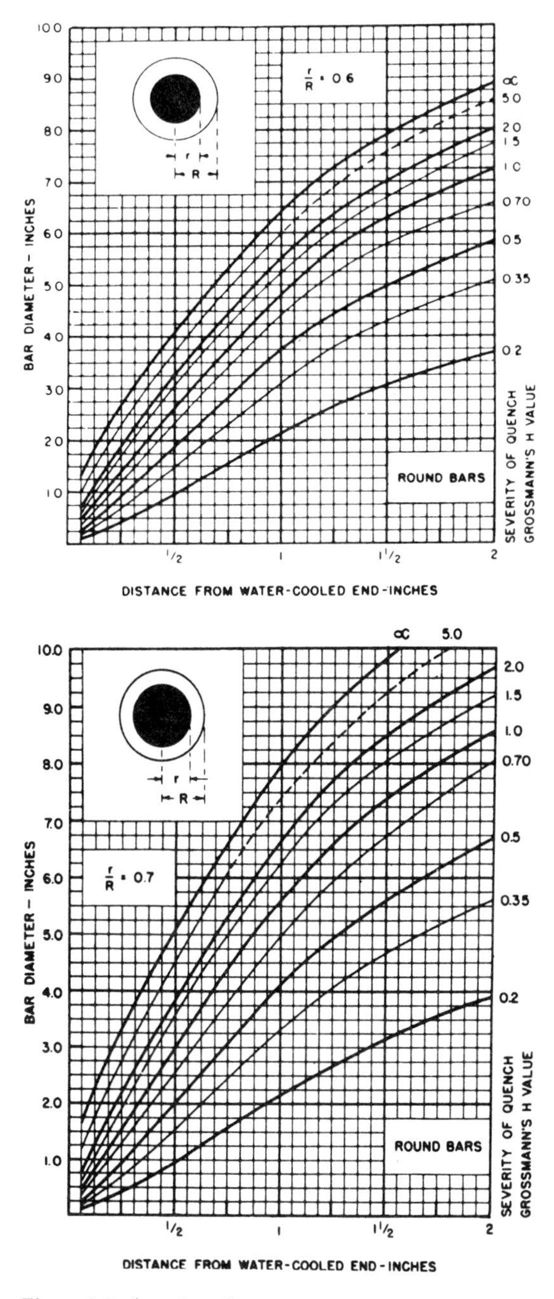

Figure 2-8 *(continued)*

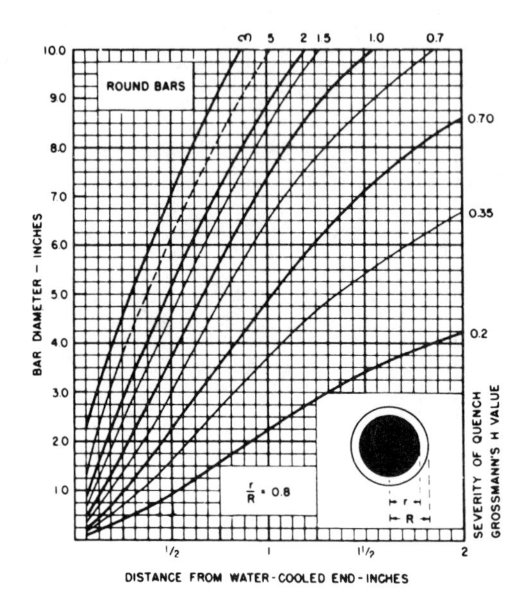

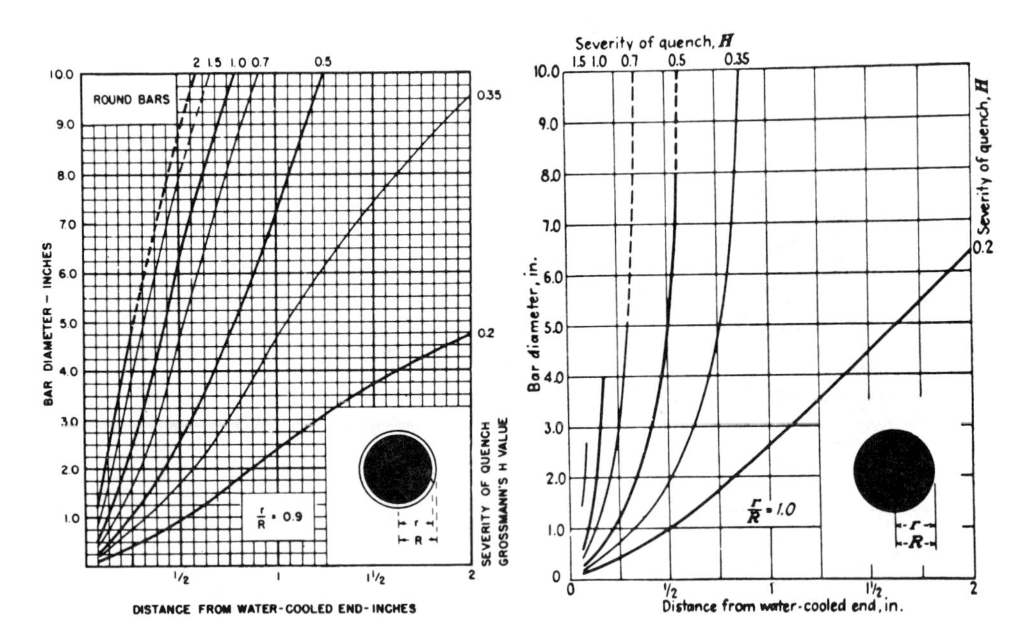

Figure 2-8 *(continued)*

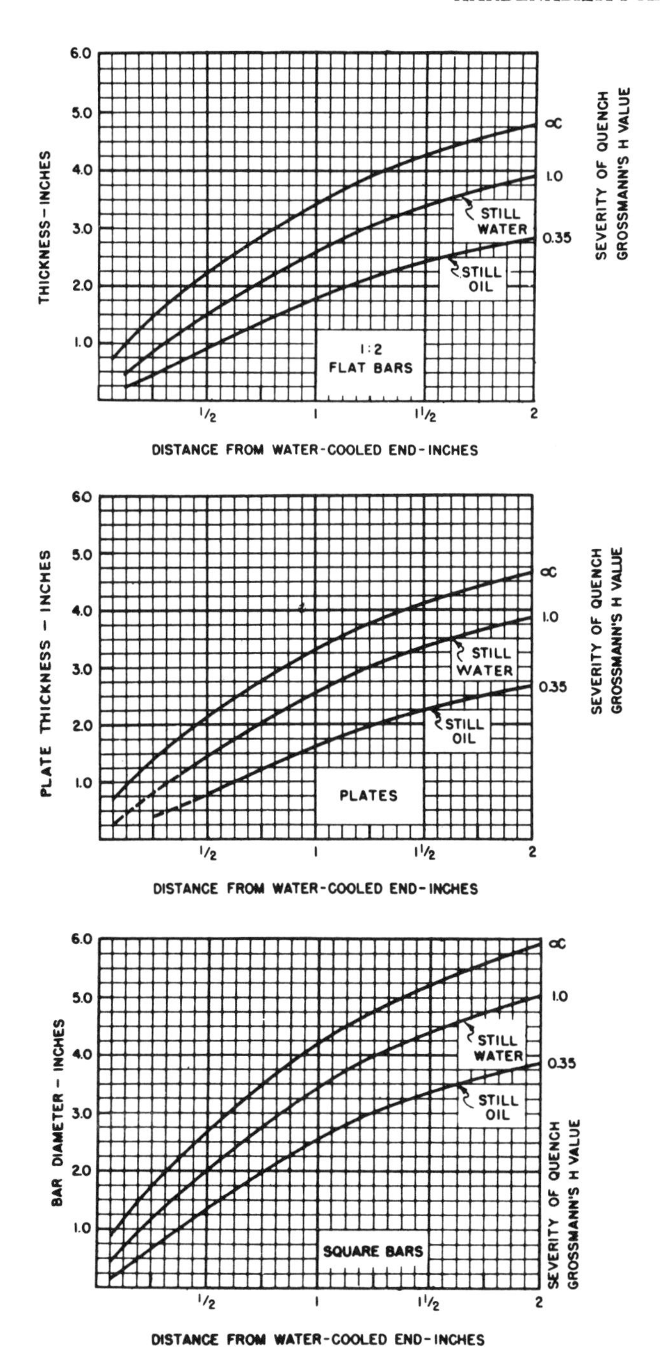

Figure 2-8 *(continued)*

2-4 HARDENABILITY

Hardenability is an intrinsic quality of the steel, depending only on chemical composition, austenite grain size, and chemical homogenity. However, the choice of the definition of hardenability is wide, and several different descriptions are used. For example, the Jominy curve is a measure of hardenability, and this curve is sufficient, along with the cooling-rate correlations, to allow estimates of the hardness developed in a part made of the steel for which the given Jominy curve is know. However, it is frequently desirable to use a quantitative definition of hardenability, and then develop a method of predicting the hardenability from chemical composition and austenite grain size. Methods of doing this are discussed in this section.

What is desired is to extract a definition of hardenability so that the influence, independently, on the hardenability of the amount of the elements present (e.g., carbon, chromium) and the austenite grain size can be evaluated. This will allow prediction of hardenability of a steel without the necessity of measuring it. A common definition is the diameter of a cylinder which will just give 50% martensite at the center when quenched into an ideal quench ($H = \infty$). This is referred to as the *ideal critical diameter*. The choice of 50% martensite is based on the success of predicting the hardenability. Another choice is the distance from the quenched end of a Jominy bar at which the inflection point occurs in the Jominy curve. There are others, but instead of describing several in detail, we will use one to examine the method of calculating hardenability.

We use as a measure of hardenability the ideal critical diameter D_I, the diameter of a right cylinder that has 50% martensite at the center when quenched into an ideal quench. This can be measured experimentally; for example, the amount of martensite can be obtained by metallographic observations of the microstructure. A systematic study of the influence on D_I of varying only one element is carried out. These measurements are repeated for several important alloying elements, and for various austenite grain sizes, and the results are examined to develop an *empirical* method of predicting the hardenability.

For steels of a range of carbon contents up to about 0.7% and alloying elements ranging up to about 3.5% for some elements, the following approach is usable (with limitations). The empirical correlation makes use of multiplying factors, and to illustrate its use, we will calculate the hardenability of a 4140 steel containing 0.40% C, 0.83% Mn, 0.31% Si, 0.20% Ni, 1.00% Cr, 0.19% Mo, 0.018% P, and 0.030% S and an austenite grain size of ASTM 7. (The meaning of this grain-size designation is discussed in Section 2-6.) Figure 2-9 shows the multiplying factors we will use to calculate D_I. For the base value we use the carbon content of 0.40 and the grain size of 7 to obtain $D_I = 0.63$. This value is then multiplied by the factors obtained for each element for the other figure. This gives $D_I = (0.63)(1.6)(1.2)(1.2)(2.8)(1.2) = 4.9$ in. where the factors are, in order, for Mn, Si, Ni, Cr, and Mo. Note that a 0.63-in.-diameter cylinder, of plain carbon steel containing 0.4% C and having an austenite grain size of 7, will have 50% martensite at the center when given an ideal quench. Adding just 0.2% Mo increases this to 0.76 in. (0.63×1.2), and adding all of the elements increases the ideal diameter to 4.9 inches. This increase is due to the in-

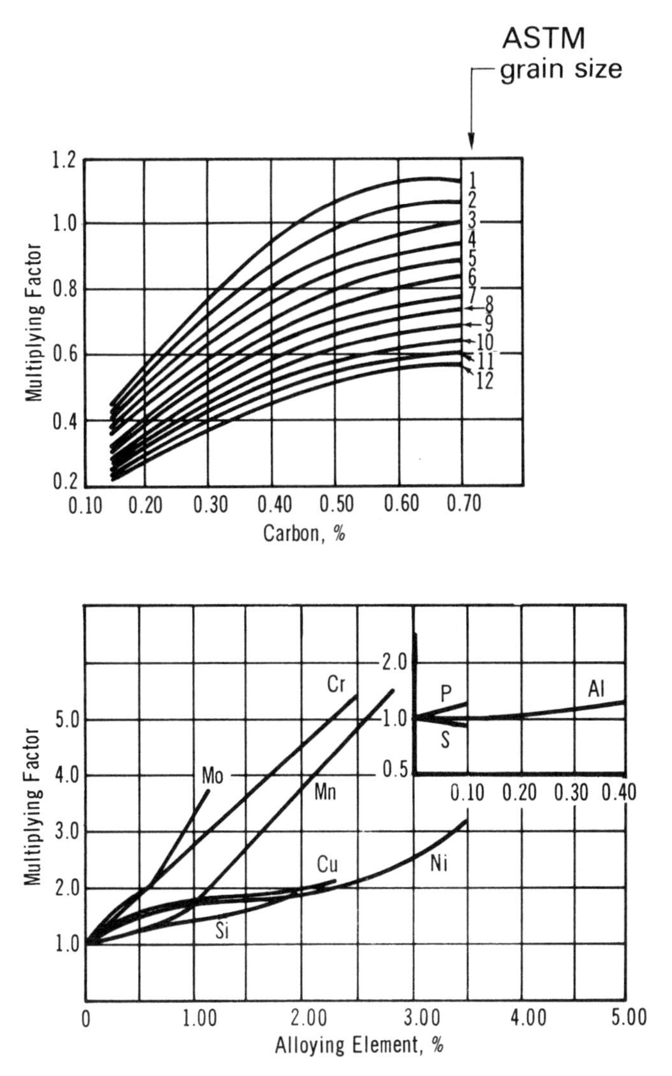

Figure 2-9 Multiplying factors for determination of the ideal critical diameter. *(From C. F. Jatczak,* Met. Progr. *vol. 100, no. 3, p. 60, 1971.)*

fluence of the alloying elements on the kinetics of the decomposition of the austenite during the cooling process. The alloying elements reduce the rate of formation of the pearlite and bainite, and they allow more of the harder martensite to form than in the plain carbon steel.

From experimental data, the curves in Fig. 2-10 were developed, which can be used to calculate a Jominy curve. It is assumed that the structure of the Jominy bar at 0.0625 in. from the quenched end is completely martensite, and therefore the hardness depends only on the carbon content of the steel. The value can be obtained

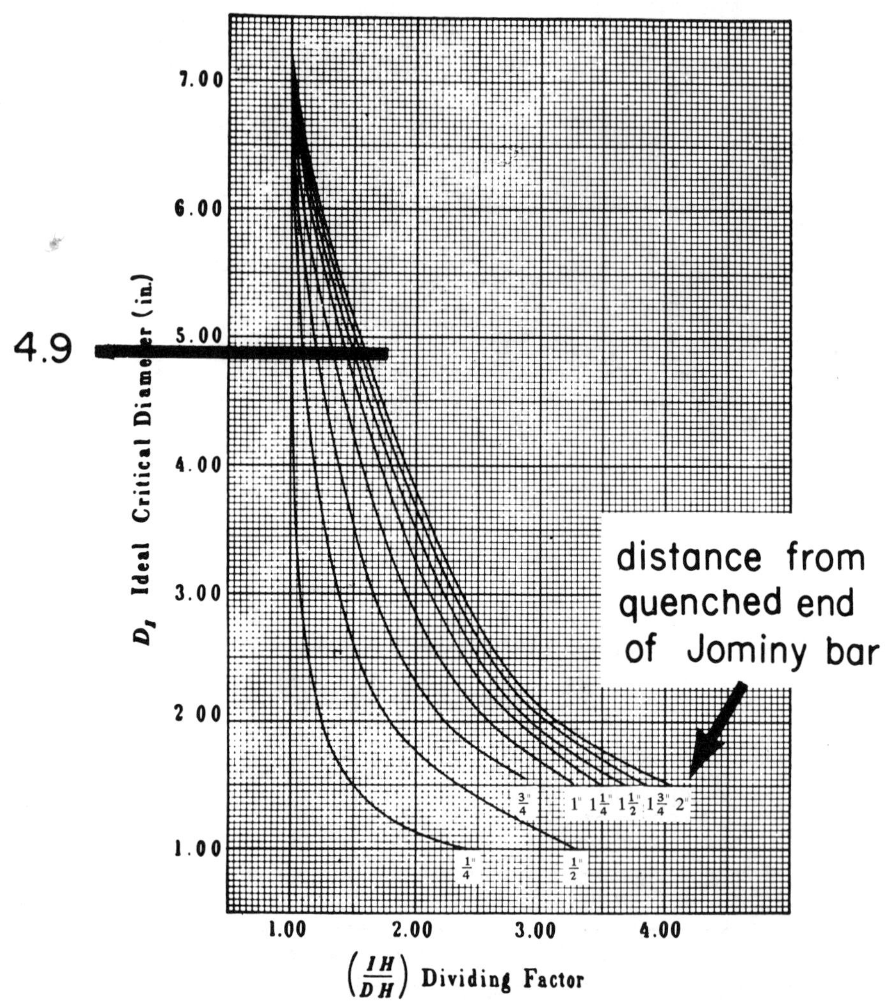

Figure 2-10 The ideal critical diameter as a function of the ratio of the hardness at 0.0625 in. from the quenched end of a Jominy bar to that at a given distance. *(Adapted from "Physical Metallurgy for Engineers," D. S. Clark and W. R. Varney, ©1952 by Litton Educational Publishers, Inc. Reprinted by permission of Van Nostrand Reinhold Company.)*

from Fig. 2-11. Knowing the ideal critical diameter of the steel, the curves in Fig. 2-10 can be used to obtain the ratio of the hardness at 0.0625 in. (*IH*) to that at any other distance (*DH*). Using this procedure for the 4140 steel having D_I = 4.9 in., the Jominy curve in Fig. 2-12 was obtained. The method is not valid for a wide range of steels, but it is a useful way to approximate the Jominy curve.

It is emphasized that the procedures outlined above are valid only for the range of steels from which they were derived. Other correlations have been developed which have the same limitation. Thus, the procedures described must be used with this

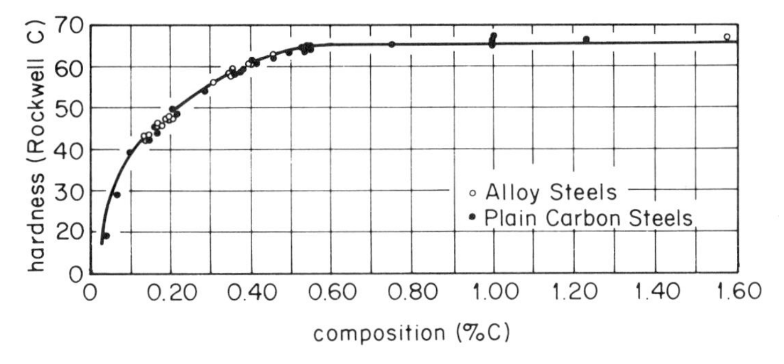

Figure 2-11 The hardness of martensite as a function of the carbon content of the martensite. *(Adapted from J. L. Burns, T. L. Moore, and R. S. Archer,* Trans. ASM, *vol. 26, p. 1, 1938; W. P. Sykes and Z. Jeffries,* Trans. ASST, *vol. 12, p. 871, 1927.)*

understanding. Attempts to improve hardenability calculations are undergoing active scrutiny, and the latest status is described in detail in a recent publication (*Hardenability Concepts with Applications to Steels;* see Appendix I, p. 250).

2-5 H-STEELS

The chemical specifications of steels have a range of composition for each element within which the actual value can fall and the steel still be classified as a given type (e.g., 4340). When a steel is bought, the producer guarantees the chemistry to meet the specifications. Table 2-2 gives values for the steel 4140. The hardenability of the specific steel (e.g., plate or bar) of this type depends upon the exact chemistry, and on the austenite grain size. To illustrate the wide variation obtainable, the Jominy curve has been calculated, using a grain size of ASTM 6, assuming that the steel con-

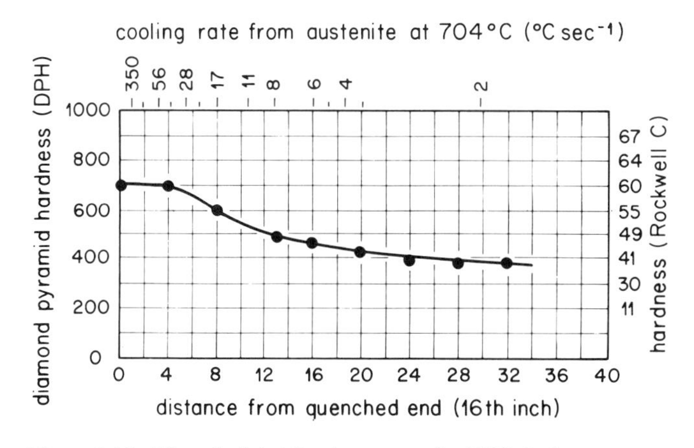

Figure 2-12 The calculated Jominy curve of a 4140 steel.

Table 2-2 The allowable chemical composition ranges of elements critical to hardenability for 4140 and 4140H steels

Element	Chemical composition, wt. %	
	4140H	4140
Carbon	0.37–0.45	0.38–0.43
Manganese	0.70–1.05	0.75–1.00
Silicon	0.20–0.35	0.20–0.35
Chromium	0.80–1.15	0.80–1.10
Molybdenum	0.15–0.25	0.15–0.25

tained the maximum allowable value for each element, and then assuming the minimum allowable value for each element. The results are shown in Fig. 2-13. The actual steel will have a hardenability somewhere within this approximate band depending on the austenite grain size.

During the Second World War, the shortage of elements critical to hardening led to a careful analysis of the influence of the variation of these elements on the hardenability. From these studies came the development of H-Band steels. When an H steel (e.g., 4140H) is used, it is guaranteed to have a Jominy curve between specified limits (using the normally encountered range of austenite grain sizes). For most

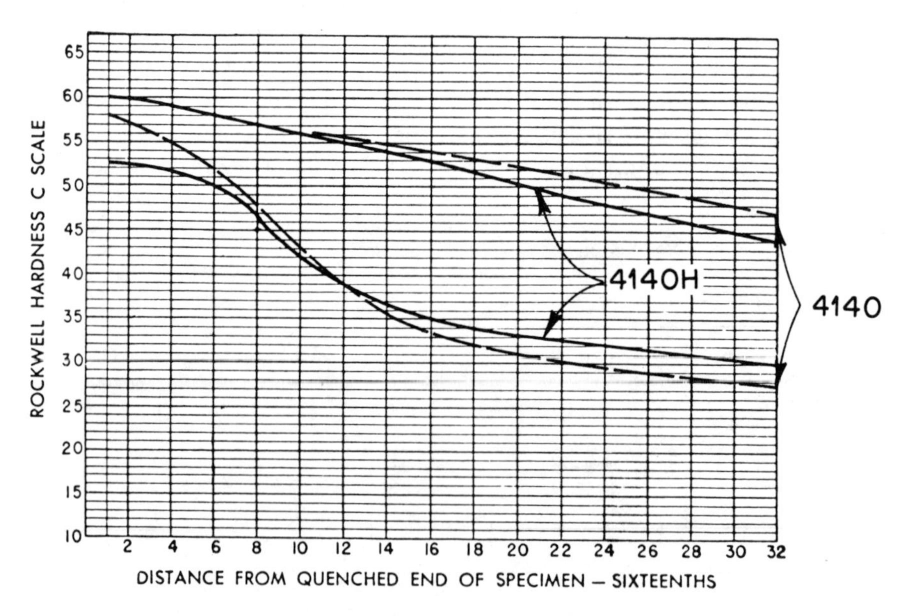

Figure 2-13 Calculated Jominy curves (dashed curve) for 4140 steel, using a grain size of ASTM 6, and the maximum and minimum allowable amount of each element (Table 2-2), compared to the H-Band for 4140H steel (solid curve).

applications, the exact chemistry of a steel is not as important as the hardenability, and thus the H-Band chemistry specifications will frequently be broader than the equivalent normal steel. However, the hardenability specifications are more restrictive. A comparison of the chemistry specified is shown for 4140 and 4140H steels in Table 2-2. Note that the 4140 steel has closer chemical specifications on carbon, manganese, and chromium. The Jominy curves calculated for the extremes of the analysis for 4140 are compared in Fig. 2-13 to the guaranteed H-band for 4140H steel. It is clearly seen that there is closer control of the hardenability for 4140H steels than there is for 4140 steels, even though the latter have more rigid chemical composition specifications.

The minimum of the H-band can be used to judge the minimum hardenability attainable in these steels, and it thus serves as a conservative base for hardenability calculations.

2-6 AUSTENITE GRAIN SIZE

Hardenability depends upon the austenite grain size, and consideration must be given to its control and prediction during austenitization. Any crystalline material will undergo grain growth because the energy is lowered by the reduction in the amount of grain boundary area. Growth occurs by the atoms in the disordered region of the boundary taking up the periodic arrangement of one crystal, and the boundary continues to move by this atomic mechanism. Movement of the boundary is toward its center of curvature, because an atom on the concave surface has, statistically, more bonding neighbors. Smaller grains eventually disappear, and the grain size increases. The rate of growth depends upon the atom mobility, which is a function of temperature involving the form $e^{-Q/RT}$, where Q is an activation energy for atom movement, R is the ideal gas constant, and T is the absolute temperature. Examination of this expression shows that the atom movement increases exponentially with temperature, so that the rate of growth increases very rapidly with increasing temperature. At a given temperature the rate of growth decreases with time as the amount of grain boundary area is reduced, and the material eventually becomes a single crystal.

This description of grain growth is valid if the boundary is unimpeded in its movement. However, the presence of a second phase usually retards the boundary movement, so that the grains grow quite slowly except at relatively high temperatures. Most steels have insoluble particles present. For example, Al_2O_3 or SiO_2 particles are usually present from the deoxidation process during the latter stages of refining. As the liquid steel is adjusted in chemistry by oxidation, elements such as aluminum or silicon are added to react preferentially with the dissolved oxygen in the liquid and to prevent the oxidation of carbon and other alloying elements (e.g., chromium) as these additions are made to obtain the final desired chemistry. The aluminum will form Al_2O_3 and the silicon SiO_2, and these will be present as particles sufficiently small that they do not rise to the liquid slag on top of the liquid steel but remain in suspension. When the steel is cast, these particles are then distributed in the steel at random on a fine scale. Upon austenitization of the steel, these particles then inhibit grain growth. The effect is illustrated in Fig. 2-14 for austenitization at 940°C. These data are based on about four different steels. Each steel contained about 0.2% C but had various additions (e.g., Al, Ti, Nb) which developed insoluble particles (Al_2O_3) or at least particles soluble only at very high temperatures or that dissolved slowly (e.g., NbC). It is seen that the grain size remained between 0.01 and 0.04 mm for up to about 10^6 s; beyond this time, the grain size increased drastically. The usual austenitizing time is around 1 hour, so that the range in grain size is about 0.015 to 0.035 mm. (This corresponds to a variation in the ASTM grain size of about 9 to 7; the meaning of this designation is discussed below.)

The rate of coarsening will depend upon the microstructure prior to heating to the austenite range, as this will affect the nucleation rate of the austenite grains. This effect is illustrated in Fig. 2-15. Although the grain size does depend upon the prior structure, it is important to note that no coarsening occurs below 1000°C for an austenitizing times of 8 hours. (These steels were actually austenitized in a carburizing atmosphere to make the austenite grain boundaries more discernable in the microstructure.) Thus austenitizing this steel below 1000°C for 1 or 2 hours will keep the grain size between ASTM 7 and 8.

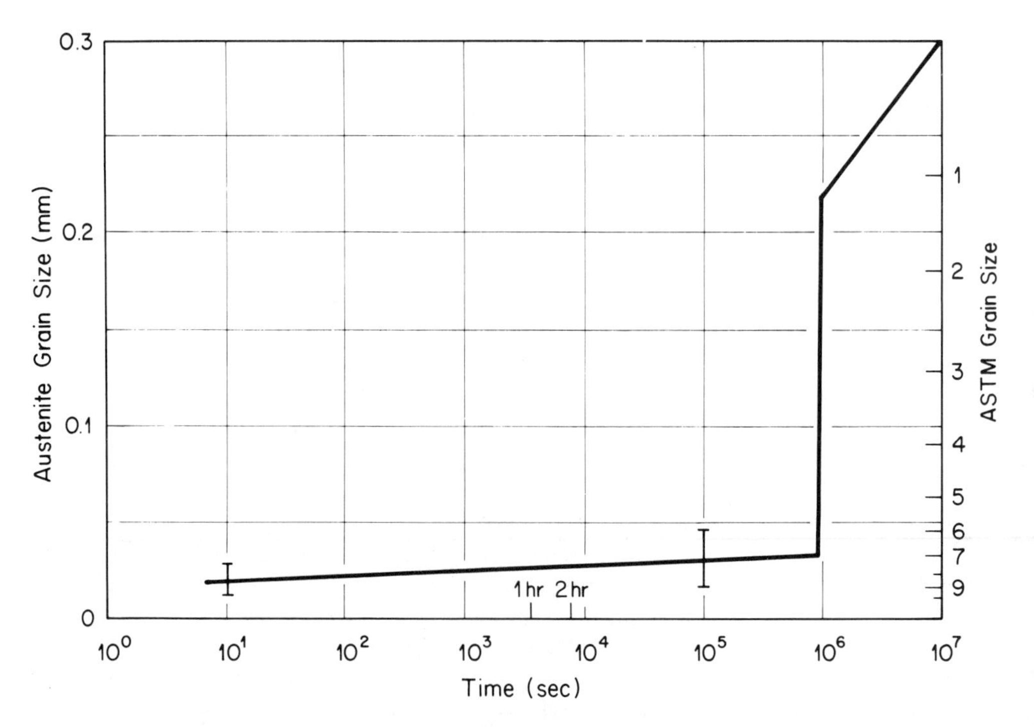

Figure 2-14 The effect of time at 950°C on the austenite grain size. This curve is based on a variety of steels containing about 0.2% C, 0.3% Si, and 0.7% Mn, with various minor additions (e.g., V, Ti, Al, Nb) for grain refinement. *(Adapted from N. E. Hannerz and F. De Kazinczy,* J.I.S.I., *vol. 208, p. 475, 1969.)*

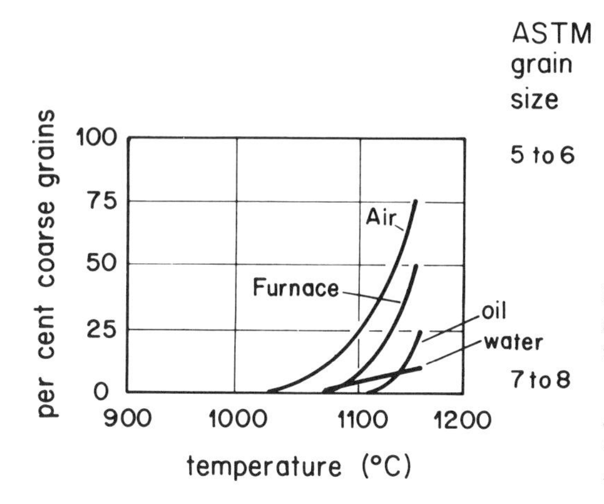

Figure 2-15 The effect of temperature and prior heat treatment on the austenite grain size of 4615 steel. The austenitizing time was 8 hours. *(Adapted from M. A. Grossmann in "Grain Size Symposium," American Society for Metals, Metals Park, Ohio, 1954.)*

Most steels have been manufactured so that they have insoluble particles present and hence are usually fine-grained steels, referring to the relative stability of the austenite grain size. For austenitization in the range 850 to 1050°C, the austenite grain size remains rather constant. Thus using an austenitization time of 1 or 2 hours ensures complete conversion of the prior microstructure to austenite and the chemical homogenization of the austenite, and yet the austenite grain size is insensitive to variation in the austenitizing time from one to several hours. This is the reason that on Jominy curves and TTT diagrams the austenitizing temperature is seldom listed. It is common to use 1 or 2 hours, and the austenite grain size is not very sensitive to the time chosen. It is important to remember that the exact austenite grain size established depends upon the steel; the value is difficult to predict and may have to be determined experimentally.

The most common designation of grain size is the ASTM grain size. It is defined by the relation

$$n = 2^{N-1}$$

where N is the *ASTM grain-size number* and n is the number of grains per square inch observed at a magnification of 100X. The relation of N to the actual size (diameter) of the grains is listed in Table 2-3. Note that as the size of the grain decreases, the ASTM number increases. The usual range of austenite grain sizes encountered is between (and including) ASTM numbers 1 and 9.

2-7 TEMPERING

In Chap. 1 tempering was reviewed briefly, where it was treated as a heat treatment designed to improve ductility by converting martensite to ferrite and carbide. Indeed, the basic goal of tempering is to make the steel less susceptible to brittle fracture. The effect of tempering on the mechanical properties is illustrated in Fig. 2-16 for a

Table 2-3 The relationship between the ASTM grain-size number and the average "diameter" of the grain

ASTM micro-grain size number	Calculated "diameter" of average grain mm	in.	ASTM micro-grain size number	Calculated "diameter" of average grain mm	in.
		$\times 10^{-3}$			$\times 10^{-3}$
00	0.508	20.0	7.5	0.027	1.05
0	0.359	14.1	–	0.025	0.984
0.5	0.302	11.9	8.0	0.0224	0.884
1.0	0.254	10.0	–	0.0200	0.787
–	0.250	9.84	8.5	0.0189	0.743
1.5	0.214	8.41	9.0	0.0159	0.625
–	0.200	7.87	–	0.0150	0.591
–	0.180	7.09	9.5	0.0134	0.526
2.0	0.180	7.07	10.0	0.0112	0.442
2.5	0.151	5.95	–	0.0100	0.394
–	0.150	5.91	10.5	0.00944	0.372
3.0	0.127	5.00	–	0.00900	0.354
–	0.120	4.72	–	0.00800	0.315
3.5	0.107	4.20	11.0	0.00794	0.313
–	0.090	3.54	–	0.00700	0.276
4.0	0.0898	3.54	11.5	0.00667	0.263
4.5	0.076	2.97	–	0.00600	0.236
–	0.070	2.76	12.0	0.00561	0.221
5.0	0.064	2.50	–	0.00500	0.197
–	0.060	2.36	12.5	0.00472	0.186
5.5	0.0534	2.10	–	0.00400	0.158
–	0.050	1.97	13.0	0.00397	0.156
6.0	0.045	1.77	13.5	0.00334	0.131
–	0.040	1.58	–	0.00300	0.118
6.5	0.038	1.49	14.0	0.00281	0.111
–	0.035	1.38	–	0.00250	0.098
7.0	0.032	1.25			
–	0.030	1.18			

Source: Adapted from "1966 Book of ASTM Standards," part 31, p. 228, American Society for Testing and Materials, Philadelphia, 1966.

4340 steel. Note that even though tempering at 300°C decreases the hardness considerably, the impact energy is still low. However, above 300°C the impact energy increases markedly.

When austenite is quenched sufficiently rapidly, and to a sufficiently low temperature, the martensite that forms retains the carbon in solid solution. For low-carbon content (0.2% C) the structure of the martensite is body-centered cubic, and the dislocation density is high, comparable to that of severely cold-worked metal. For higher carbon contents, the martensite is body-centered tetragonal, and the structure

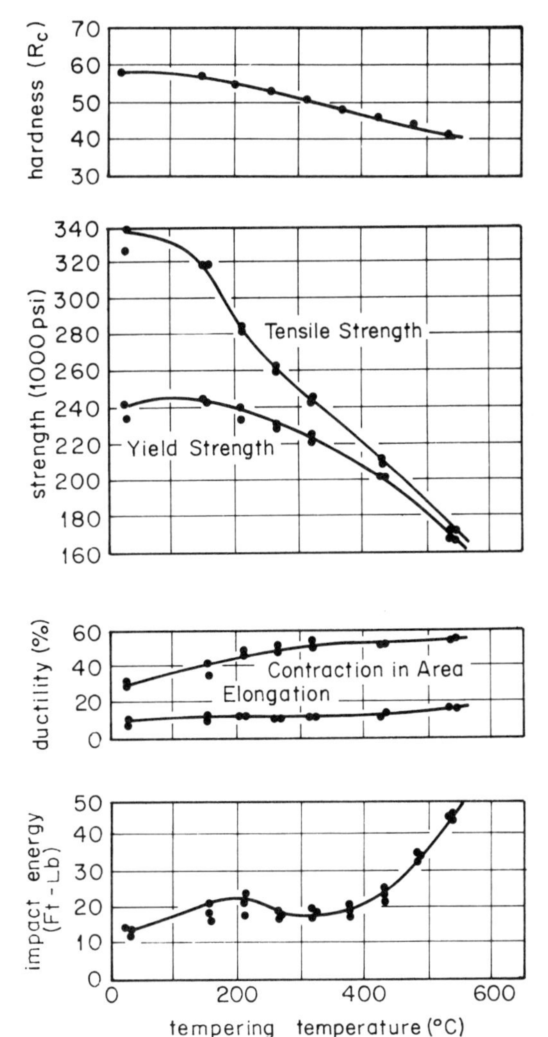

Figure 2-16 The effect of tempering martensite on the mechanical properties of a 4340 steel. All the properties were measured at 20°C. *(Adapted from L. J. Klingler, W. J. Barnett, R. P. Frohmberg, and A. R. Troiano,* Trans. ASM, *vol. 46, p. 1557, 1954.)*

consists of dislocations and fine twins. The appearance of the martensite in the optical microscope is illustrated by Fig. 2-17. The "brush-stroke" appearance is caused by the fact that the austenite, upon cooling, forms martensite suddenly in localized areas, in the form of lens-shaped regions whose mid-plane is approximately parallel to one of the {111} planes of the austenite. As the austenite continues to cool, more of the austenite is converted to martensite by the formation of more of these lens-shaped regions, until virtually all of the austenite is converted to martensite. Since there are four, nonparallel {111} planes in each austenite grain, the martensite plates delineate these planes on a surface. Due to elastic strain, an uneven surface is formed when a polished section is etched, which reveals this structure.

Due to the high atom mobility of carbon, even at relatively low temperatures (e.g., 20°C), it is difficult to form martensite with the carbon distributed at random in the interstitial sites of the iron lattice. Since most steels have the M_s temperature above 20°C some carbon atoms have sufficient time (although brief) to distribute themselves near lattice defects, such as dislocations. This effect is referred to as self-tempering or autotempering. Even if the martensite forms without this effect, it occurs as soon as tempering begins, even at 20°C. The degree of segregation is greatest for carbon contents below 0.2%; above this value, the dislocation density is lower, and the low-energy sites near the dislocations become saturated, with carbon left over to be distributed at random in the lattice. In this case, there is a tendency for these carbon atoms to "cluster" as a step preceding precipitation of carbides. In summary, the pre-precipitation stage involves carbon segregation to dislocations (autotempering), and clustering of carbon atoms into small groups.

Tempering of the martensite causes the precipitation of carbides and the concurrent formation of ferrite. Between 100 and 200°C a hexagonal close-packed carbide, ϵ carbide, precipitates for carbon contents greater than 0.2%. Its formation is kinetically favored, even though it is not the equilibrium carbide (e.g., Fe_3C). (In steels below 0.2% C, this carbide does not readily form.) From about 250 to 700°C, Fe_3C forms. The carbides are initially in the form of needles, nucleating along the interfaces between martensite plates. At higher temperatures, or for longer times, the needles gradually change to spheroids of the carbides in a low dislocation-density ferrite. Further tempering causes the smaller spheroids to dissolve and the larger ones to increase in size. Thus, in this latter stage there is a continued increase in carbide particle size and in the average spacing between carbides.

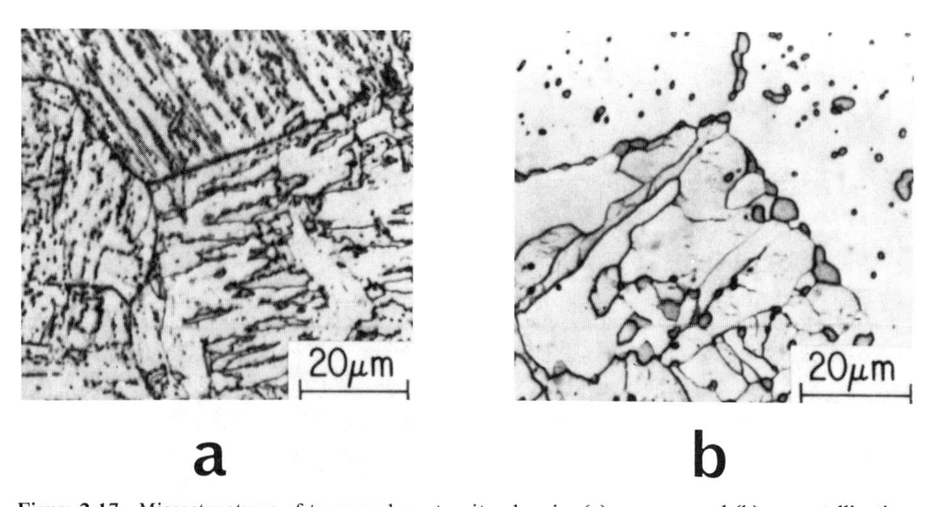

Figure 2-17 Microstructures of tempered martensite, showing (a) recovery and (b) recrystallization of the ferrite. Both microstructures are of a 0.18% C-Fe steel of martensite tempered (a) 10 min and (b) 96 hours at 600°C. In (a) the carbides are the small black dots, but in (b) the carbides are quite visible. *(Adapted from G. R. Speich,* Trans. AIME, *vol. 245, p. 2553, 1969.)*

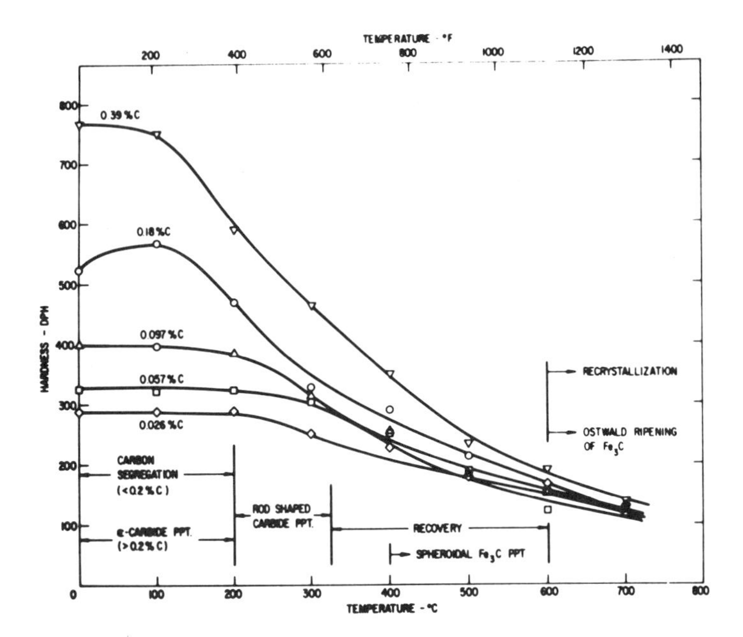

Figure 2-18 The hardness of tempered martensite for a 0.18% C-Fe steel. The tempering time was 1 hour. *(Adapted from G. R. Speich,* Trans. AIME, *vol. 245, p. 2553, 1969.)*

The martensite prior to tempering contains dislocations, and hence the tempering process can induce structural changes similar to those in a cold-worked structure. Recall that a cold-worked material will first undergo recovery, where dislocations are rearranged into alignment leading to a rather fine, low-angle boundary, substructure. Continued heating of the metal causes the formation and growth of essentially dislocation-free crystals, which is the recrystallization process. These crystals continue to nucleate and grow until the original cold-worked structure is consumed. Further heating involves only grain growth.

Recovery and recrystallization in martensite is complicated by the concurrent precipitation processes. However, the formation of low-angle subboundaries occurs, leading to the microstructure shown in Fig. 2-17; the structure is particularly clear because the boundaries are decorated with carbides. The formation of recrystallized grains is also shown in Fig. 2-17. In most steels, these processes are difficult to observe independently in the microstructure.

The influence of the tempering on the hardness of a plain carbon steel is illustrated in Fig. 2-18, where the microstructural changes discussed above are indicated. Figure 1-44 illustrates the microstructural changes during tempering for a plain carbon steel.

Alloying elements (e.g., nickel, molybdenum) generally lower the rate of carbide formation, so that the tempering curves remain higher than those of plain carbon steel, for the same carbon content. In alloy steels, elements are usually present (e.g., molybdenum) which have a strong affinity for carbon, so that the carbide contains a higher percentage of the alloying element than does the ferrite. Further, these carbides are usually of a different crystal structure than Fe_3C. In the early stage of tempering, the martensite still precipitates ϵ carbide and then (Fe, M)$_3$C. Here M represents elements other than iron, such as molybdenum, which are dissolved in the carbide. Because carbon diffuses more rapidly than the larger substitutional elements (e.g., molybdenum), the simpler carbides, although metastable, form first. The formation of the (Fe, M)$_3$C carbides causes the hardness to decrease with tempering temperature. A temperature range is finally attained in which, for the tempering times used, the alloy carbides begin to form, and, depending upon the exact chemistry of the steel, these carbides may be finer than the (Fe, M)$_3$C; this causes the hardness to increase, a phenomenon referred to as *secondary hardening*. Increasing the tempering temperature further allows these alloy carbides to coarsen, for the tempering time used, and so the hardness decreases again. Figure 2-19 illustrates this behavior. Secondary hardening is a particularly important process in tool steels, and it is discussed in more detail in Chap. 6.

Martensite is tempered to increase the ductility, but in some steels the precipitation of carbides causes embrittlement in certain temperature ranges (Fig. 2-20). This is sometimes referred to as *260°C (500°F) embrittlement*. It appears to be caused by a peculiar morphology of the (Fe, M)$_3$C carbides, which form as films along the grain boundaries. Alloying additions, especially silicon, prevent this embrittlement. Steels

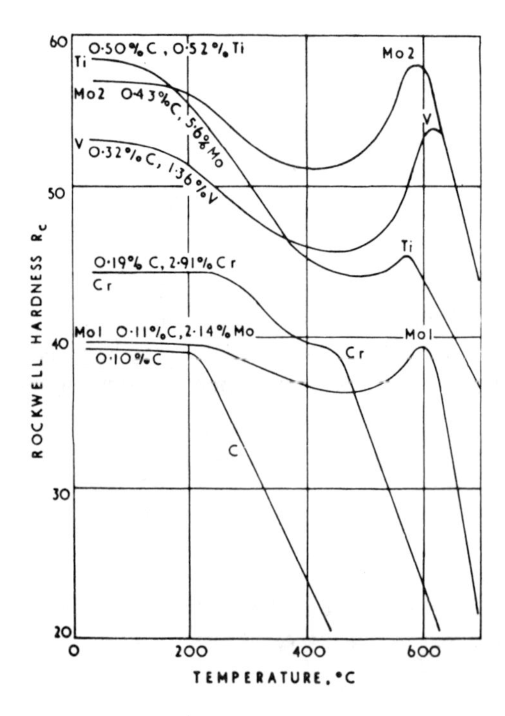

Figure 2-19 Tempering curves of steels containing chromium, molybdenum, vanadium, or titanium, showing secondary hardening. *(From K. Kuo,* J.I.S.I., *vol. 184, p. 258, 1956.)*

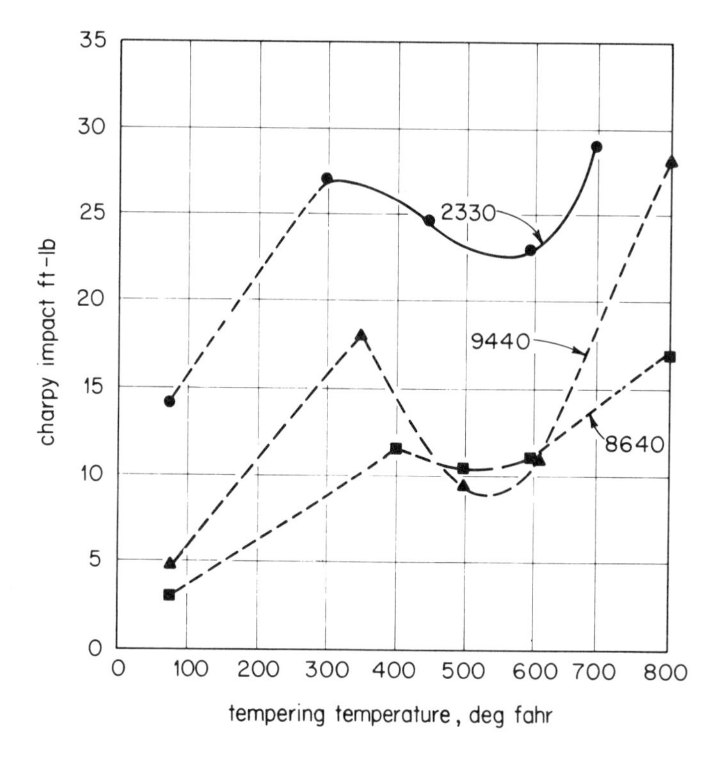

Figure 2-20 Impact curves for three commercial steels which show "260°C embrittlement." *(From R. L. Rickett and J. M. Hodge,* Proc. ASTM, *vol. 51, p. 931, 1951.)*

that show this behavior have been identified, but it is best in any case to avoid tempering in this range.

Another form of embrittlement encountered is referred to as *temper embrittlement*. Certain quenched and tempered steels will embrittle when slowly cooled through, or held in, the range 370 to 565°C. Fracture occurs along the former austenite grain boundaries, and it is believed to be due to segregation of elements such as antimony and phosphorus to the high-angle, prior austenite grain boundaries, where the cohesion of these boundaries is lowered. However, the phenomenon has not yet been clearly explained.

If the quenched steel has a sufficiently high carbon and alloy content, the M_f temperature will be below 20°C. This will result in retained austenited being present. Upon tempering the steel, this austenite decomposes, to bainite in the lower temperature range and to pearlite in the higher range. For the steels being discussed here, the presence of retained austenite is not common unless the carbon content exceeds about 0.4%. Retained austenite is especially important in tool steels and in carburizing, and its effect in these cases is discussed in subsequent chapters.

The preceding paragraphs have summarized the structural changes that occur upon tempering of martensite. The properties developed are controlled by the type of carbide and particularly the morphology. This in turn depends upon the specific steel and the tempering temperature and time. Unfortunately, the structural changes are not yet amenable to sufficiently general theoretical treatment to allow prediction of the mechanical properties from knowledge of the chemistry and the tempering time and temperature. Instead, properties must be measured and a correlation developed which is usable. We now turn our attention to these correlations.

Hollomon and Jaffe Correlation

Hollomon and Jaffe found that for a number of steels the dependence of tempered hardness on temperature and time can be combined into a single parameter. This approach is widely applicable, as illustrated in Fig. 2-21 for a plain carbon steel and for an alloy steel showing secondary hardening. It is interesting to compare the behavior of tempered martensite to that obtained upon reheating (below the austenite

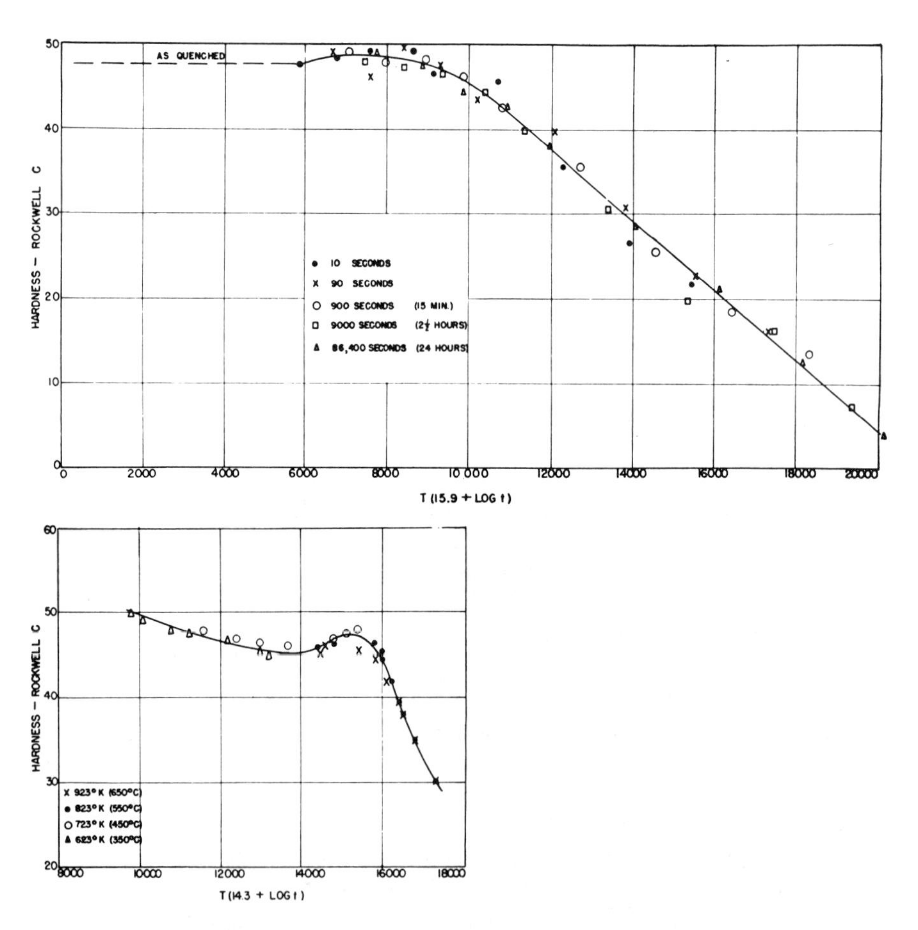

Figure 2-21 Tempering curves for a 0.31% C, plain carbon steel (top) and a 2% Mo, 0.35% C alloy steel (bottom) showing secondary hardening. T is in K, and t in hours. *(From J. H. Hollomon and L. D. Jaffe,* Trans. AIME, *vol. 162, p. 223, 1945.)*

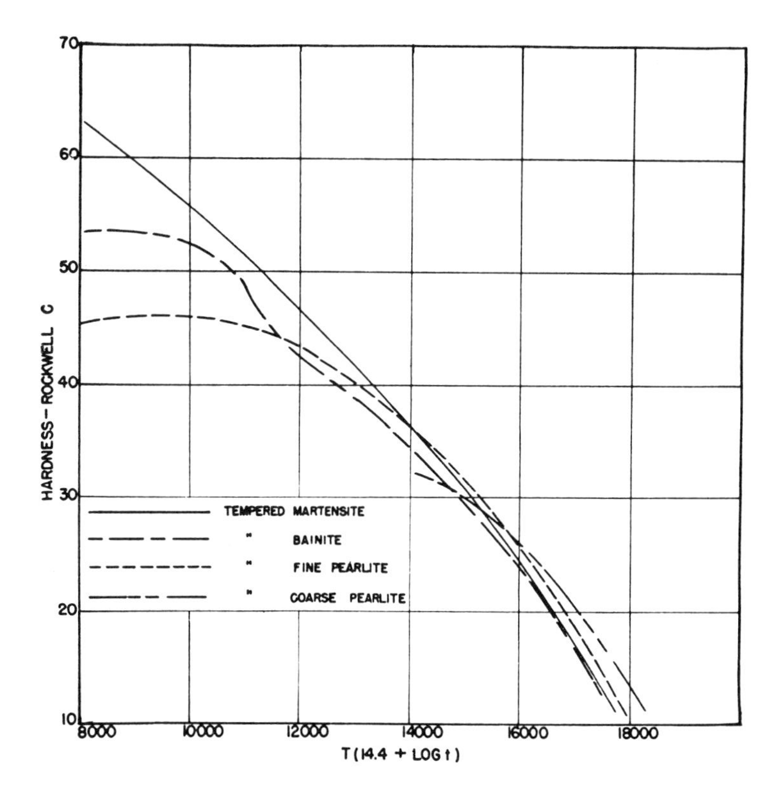

Figure 2-22 Hardness as a function of tempering parameter for a 0.94% C plain carbon steel, for various initial microstructures. T is in K, and t is seconds. *(From J. H. Hollomin and L. D. Jaffe,* Trans. AIME, *vol. 162, p. 223, 1945.)*

region) structures such as bainite and pearlite. In these cases, the structural change is only one of spheroidization of the carbides and their coarsening. In tempered martensite, especially at the higher temperatures, carbide coarsening is the main structural change. Thus it is to be expected that regardless of the initial microstructure the coarsening process would have similar time-temperature dependence. Such behavior is shown in Fig. 2-22, where the tempering curves for different beginning microstructures are approximately the same.

Jaffe and Gordon Correlation

Jaffe and Gordon, from examination of approximately 5000 samples, also developed a correlation to calculate tempered hardness. They present their data in such a way that it is convenient to determine the time and temperature for a desired hardness. The procedure can best be illustrated by example. The tempering temperature that will give a hardness H_a of 37.5 R_c, using a tempering time of 3.6 hours, is to be determined. The steel contains 0.32% C, 0.84% Mn, 0.99% Si, 1.52% Ni, 0.74% Cr, and 0.29% Mo and has an austenite grain size of ASTM 6 prior to quenching to form martensite. The necessary parameters are given in Fig. 2-23. The *compositional hard-*

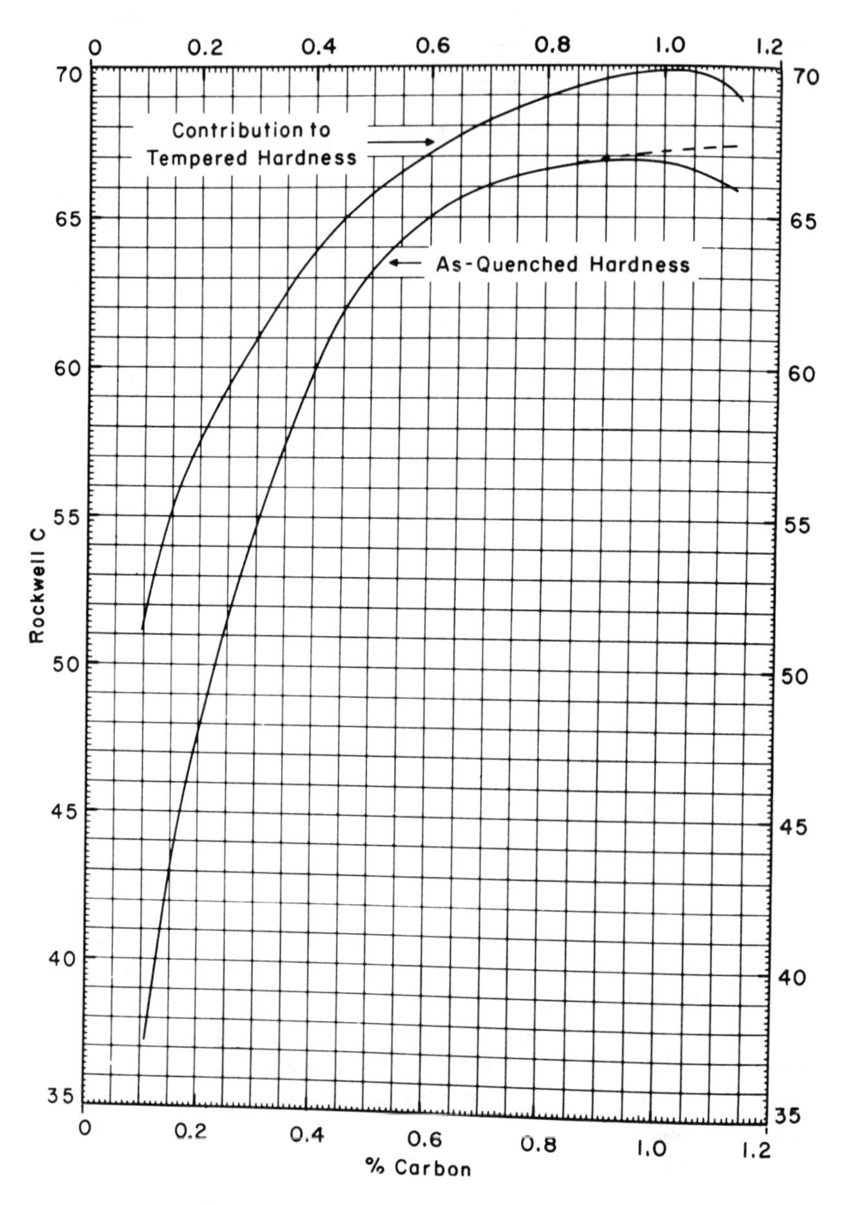

Figure 2-23 Parameters to estimate tempered hardness by the Jaffe and Gordon correlation. *(From L. D. Jaffe and E. Gordon,* Trans. ASM, *vol. 49, p. 359, 1957.)*

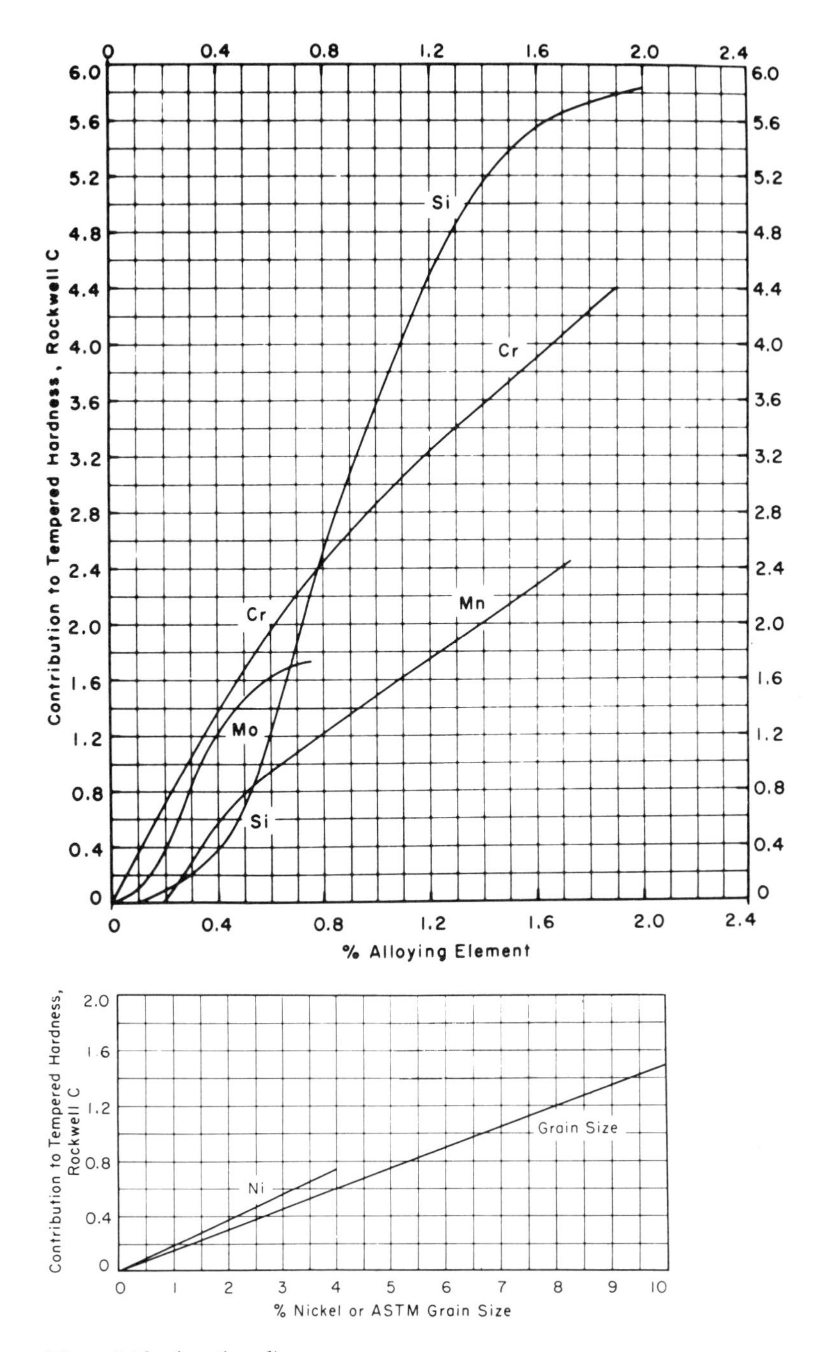

Figure 2-23 *(continued)*

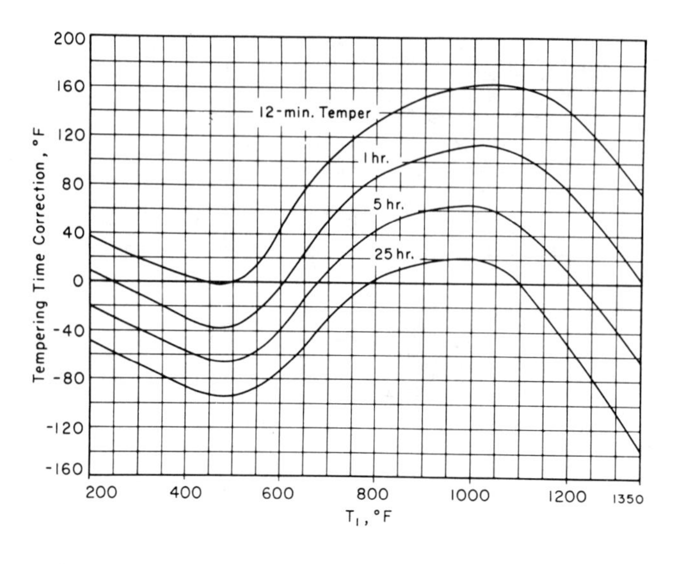

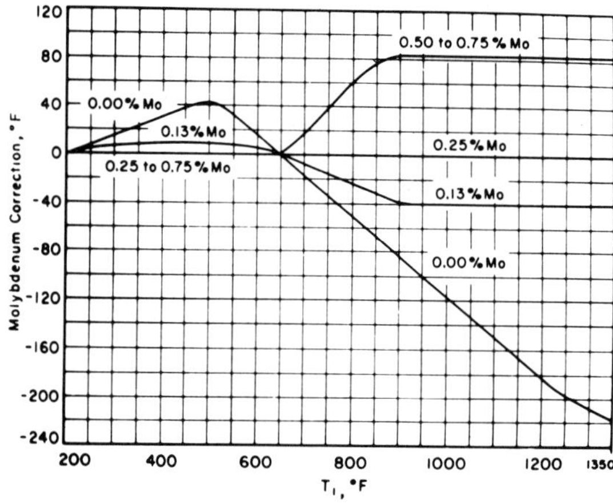

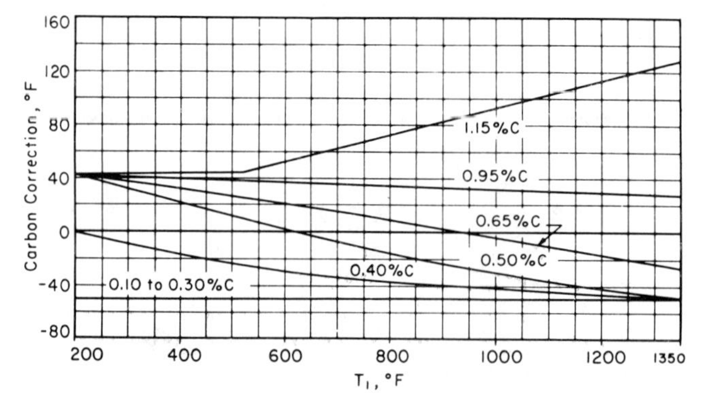

Figure 2-23 *(continued)*

Table 2-4 Values used to calculate the tempered hardness by the Jaffe and Gordon correlation

0.32% C	61.6	
0.84% Mn	1.3	
0.99% Si	3.5	
1.52% Ni	0.3	
0.74% Cr	2.3	
0.29% Mo	0.9	
ASTM grain size 6	0.9	
	70.7	Compositional hardness, H_c

$30(H_c - H_a) = 30(70.7 - 37.5)$ =	996°F
tempering-time correction	+ 75
0.29% Mo correction	+ 13
0.32% C correction	− 49
	1035°F

ness, H_c, is obtained by adding the contribution from the alloying elements. This is shown in Table 2-4. Then the equation $30(H_c - H_a) = 30(70.7 - 37.5) = 996°F$ gives the tempering temperature for a tempering time of 4 hours. This temperature must be corrected for the actual tempering time (3.6 hours) and for the nickel, molybdenum, and carbon content (Table 2-4). This gives 1035°F as the required temperature. If it is desired to determine the hardness for a given tempering temperature and time, the procedure is reversed and requires a trial-and-error approach. It is emphasized that this method is valid *only* for tempering martensite.

Crafts and Lamont Correlation

The previous two correlations are valid only for tempering martensite. The correlation developed by Crafts and Lamont is usable for other starting microstructures (e.g., bainite). Their procedure is illustrated by use of Fig. 2-24, which shows the tempered

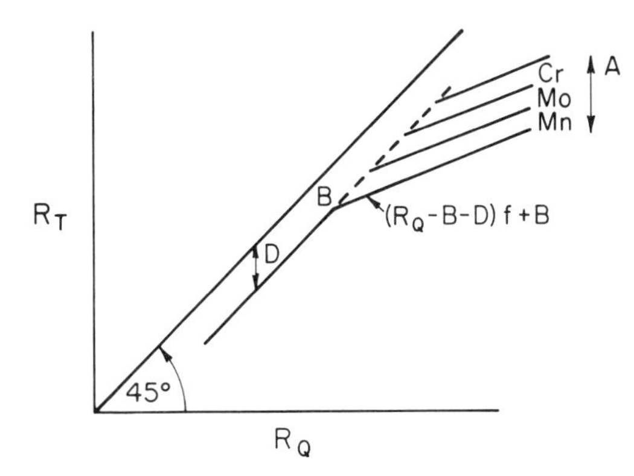

Figure 2-24 Schematic relationship between initial hardness R_Q and tempered hardness R_T and the parameters to calculate R_T. *(Adapted from W. Crafts and J. L. Lamont,* Trans. AIME, *vol. 172, p. 222, 1947.)*

hardness, R_T, plotted against the quenched hardness, R_Q. For low-carbon steels, the hardness will be little affected by tempering and the hardness will be $R_Q - D$, where D is the *temper decrement*. However, after a *critical hardness* B is reached the hardness drops more than the value D. This value depends on the carbon content and the tempering temperature. Then, after B,

$$R_T = (R_Q - B - D)f + B$$

where f is called the *disproportionate softening* which depends on the carbon content and tempering temperature. At a given temperature the alloying elements provide a greater hardness A, and

$$R_T = (R_Q - B - D)f + B + A$$

The required parameters can be obtained from the plots in Fig. 2-25. Then, knowing the initial hardness, the tempered hardness can be estimated for a given tempering temperature. However these correlations are valid only for a tempering time of 2 hours.

A Comparison of the Correlations

To compare the hardness values obtained by the three methods, we will determine the hardness of martensite tempered for 2 hours at 800°F (427°C) for a 4140 steel containing 0.40% C, 0.83% Mn, 0.31% Si, 0.20% Ni, 1.00% Cr, and 0.19% Mo and an austenite grain size of ASTM 7. According to the Jominy curve in Fig. 2-11, the martensite hardness will be 60 R_c. Figure 2-26 shows the Hollomon and Jaffe plot for a 4140 steel of similar composition, from which one obtains a tempered hardness of 42 R_c. The Jaffe and Gordon correlation yields a hardness of 42 R_c also. The Crafts and Lamont method gives a higher value, 46 R_c. Thus the three methods predict the tempered hardness to lie between 42 and 46 R_c. This uncertainty is within that of the methods themselves.

The importance of the correlations is that they provide methods of *estimating* the effect of tempering on the hardness, and this serves as a very useful guide in choosing heat-treating procedures.

2-8 SPECIAL HEAT TREATMENTS

There are a number of heat treatments that can be employed to obtain the desired properties. The choice is strongly controlled by the economics of the situation; not only the most economical steel, but the most economical heat treatment, must be considered. For example, for machining steels a structure of primary ferrite and pearlite may be desired. This can be obtained by cooling very slowly (e.g., furnace cooling) from the austenite region, or, for some steels, cooling in air. The air cooling will be more economical than furnace cooling. Or it may be less expensive to oil

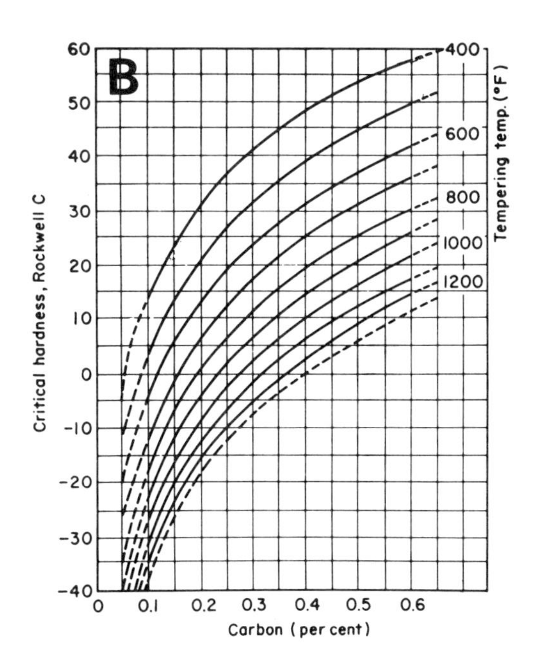

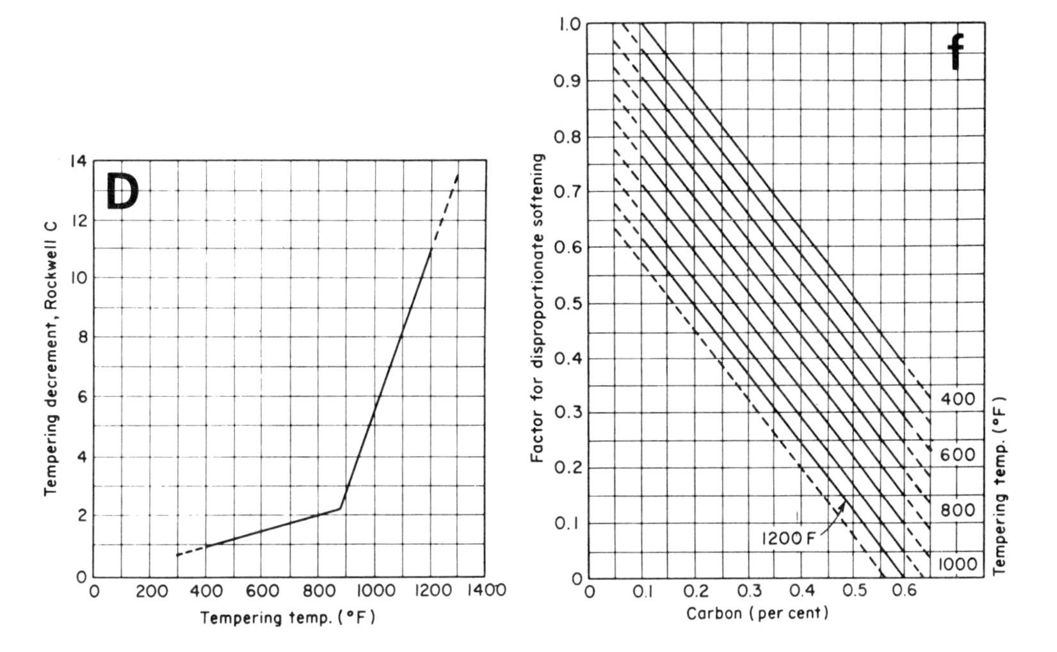

Figure 2-25 Quantities needed to calculate tempered hardness. *(From "Physical Metallurgy for Engineers," D. S. Clark and W. R. Varney, © 1952 by Litton Educational Publishing, Inc. Reprinted by permission of Van Nostrand Reinhold Company, New York.)*

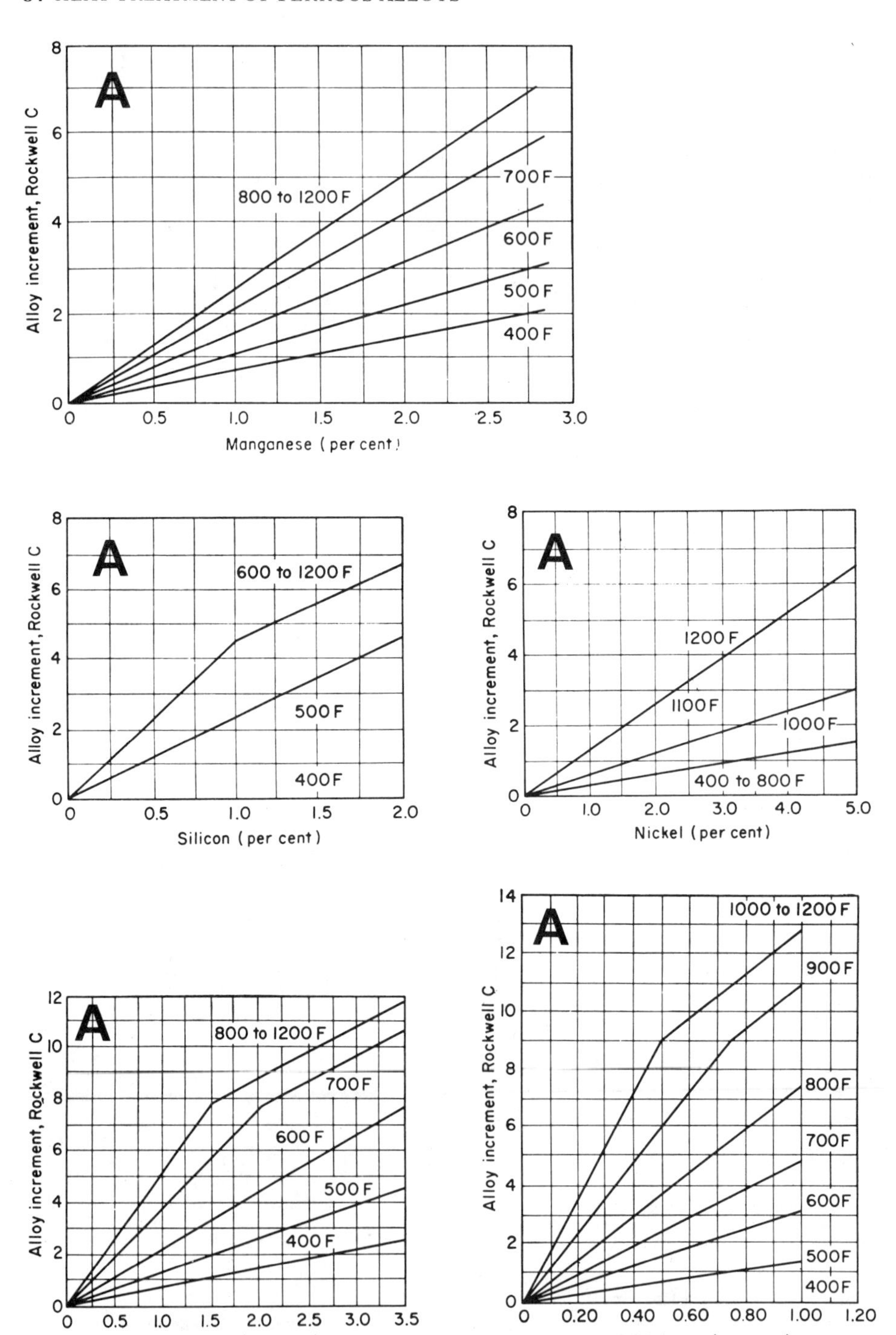

Figure 2-25 *(continued)*

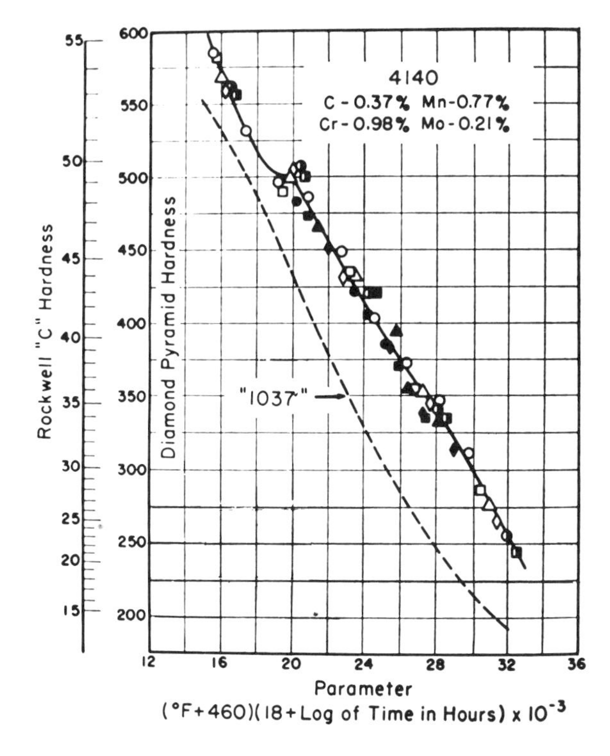

Figure 2-26 Tempering curves for a 4140 steel. *(From R.A. Grange and R.W. Baughman,* Trans. ASM, *vol. 48, p. 165, 1956.)*

quench a steel and temper it to the desired strength than to utilize an isothermal transformation treatment requiring close temperature control.

No attempt in the following sections is made to cover all the special heat treatments, nor to give all the reasons why some of them are chosen. Instead, four are discussed briefly, with a comment on the reasons for their use.

Annealing and Normalizing

Annealing is a heat treatment that involves heating a steel to the austenite region, then slowly cooling to ambient temperature. The cooling rates are usually on the order of 10^{-2} °C/s, requiring several hours cooling and usually accomplished by furnace cooling. This definition is sometimes called full annealing; however there are several specialized heat treatments in which the word annealing is involved, but we will limit our attention to the above definition. This heat treatment, of course, results in a relatively soft structure, and it is used frequently to improve machinability and also prior to cold working. Recommended austenitizing temperatures have been established and usually involve heating about 40°C above the critical temperature (austenite-ferrite boundary) for steels less than about 0.8% C. For higher carbon steels, it is customary to austenitize in the two-phase austenite and iron carbide region, about 40°C above the eutectoid temperature. Figure 2-27 illustrates the general

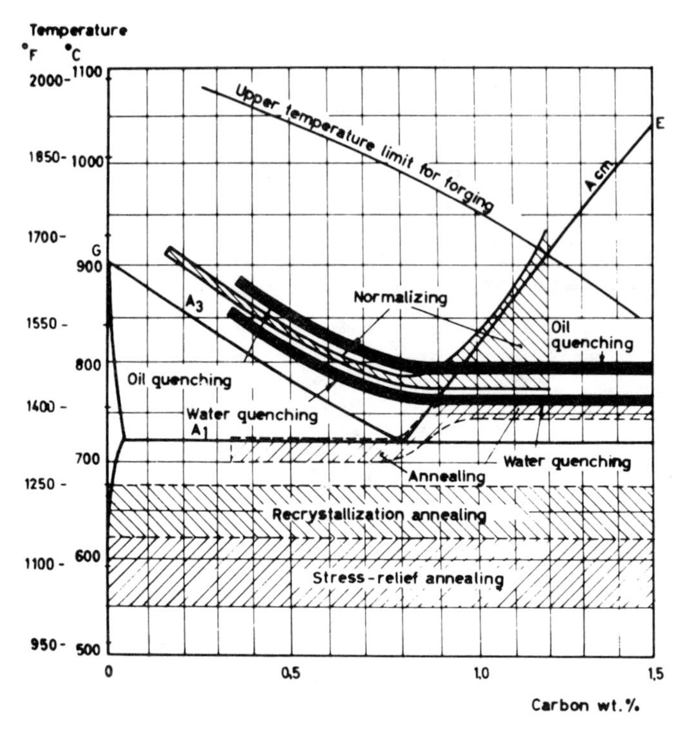

Figure 2-27 Iron-carbon phase diagram showing temperature regions for various heat-treatment operations. (Annealing in this figure is process annealing.) *(From K. E. Thelning, "Steel and Its Heat Treatment," Butterworths, Boston, 1975.)*

range of temperatures used for various heat treatments. Annealing (full annealing) would be in the same range as normalizing.

Normalizing is cooling the steel in air from the austenite region. The cooling rates are of the order of 0.1 to 1°C/s. Generally the cooling rate is such that the structure is primary ferrite and pearlite (depending upon the chemical composition). The purposes of normalizing vary; the attainment of uniform (finer) structure and improved machinability and increased strength (compared to the annealed structure) are some of the reasons for normalizing. The austenitizing temperature for normalizing is about the same as that for annealing. Normalizing is more economical than annealing as no furnaces are required to control the cooling rate.

Austempering

As will be discussed in Chap. 3, lower bainite has superior toughness to some other structures. Thus, if the steel has sufficient hardenability, isothermal transformation directly to bainite can be used as a heat treatment. In addition, because the steel is not first quenched (then tempered), there is less tendency to crack and distort. The

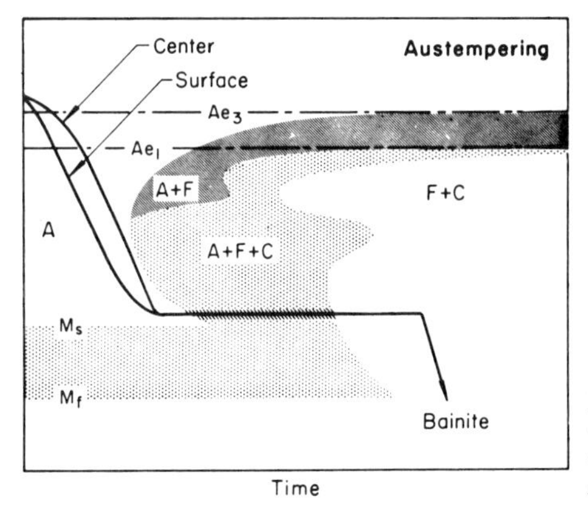

Figure 2-28 Illustration of the austempering heat treatment. *(From "Metals Handbook," 8th ed., vol. 2, American Society for Metals, Metals Park, Ohio, 1964.)*

heat treatment is illustrated in Fig. 2-28, and some properties are compared in Table 2-5. Disadvantages of this treatment, compared to quenching and tempering, are that special molten-salt baths are required for the treatment and the choice of steels may be limited.

Martempering (Marquenching)

One of the problems in hardening steels is the high distortion that develops through uneven cooling. This can be reduced by martempering. In this heat treatment (Fig. 2-29) the steel is quenched into a hot bath at a temperature just above the M_s, held at this temperature to allow the temperature of the part to equalize, then cooled relatively slowly (e.g., in air) to ambient temperature. The steel is then tempered conventionally. Not only is distortion less, but toughness may be improved (Table 2-5).

Table 2-5 Some mechanical properties (at 20°C) of a 1095 steel developed by austempering as compared to some other heat treatments

Heat treatment	Rockwell C hardness	Impact, ft-lb	Elongation in 1 in., %
Water quench and temper	53.0	12	0
Water quench and temper	52.5	14	0
Martemper and temper	53.0	28	0
Martemper and temper	52.8	24	0
Austemper	52.0	45	11
Austemper	52.5	40	8

Source: "Metals Handbook," 8th ed., vol. 2, American Society for Metals, Metals Park, Ohio, 1964.

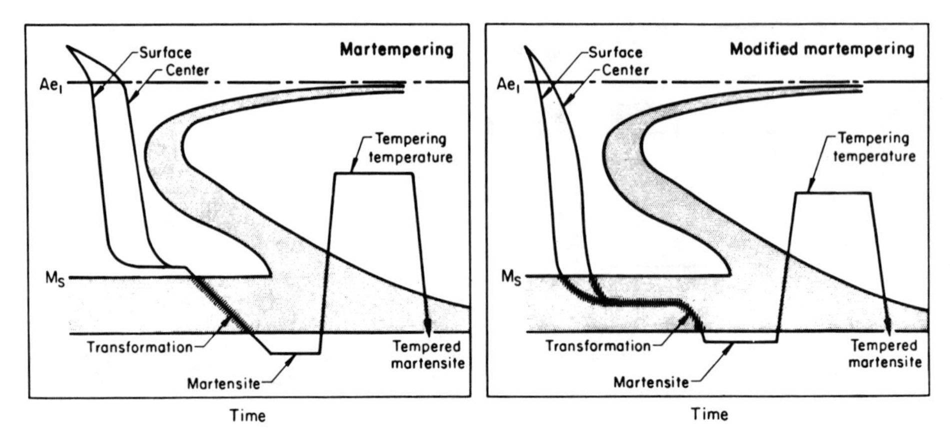

Figure 2-29 Illustration of the martempering process. *(From "Metals Handbook," 8th ed., vol. 2, American Society for Metals, Metals Park, Ohio, 1964.)*

CHAPTER
THREE
MECHANICAL PROPERTIES

In the use of steels, the mechanical properties play the central role in the choice of the steel and the heat treatment. Other properties (e.g., corrosion resistance, magnetic properties) generally play a less important role. Therefore, in this chapter the important mechanical properties of steels are reviewed and the influence of the structure on these properties is discussed. Attention is given only to carbon and low- and medium-alloy steels, in which the properties are to be developed by heat treating. The discussions for other types of steels (e.g., tool steels) are given later in the respective chapters.

The engineer is faced with applying to complex design situations property data obtained from specialized tests on specimens of specialized geometries. That is, to establish the integrity of a structure, it generally is not feasible to construct the structure, then test it to failure; then if it fails sufficiently beyond the maximum expected load, another structure just like the tested one is constructed and used. Instead, property data from small test specimens are used with theories to predict the strength in the complex situation. The theories that the design engineer uses are developed in the discipline of mechanics; these theories will be used in this chapter in only the simplest cases to illustrate their application.

Although everyone has a generalized feeling as to what is meant by "strength," in design the definition must be established rigorously. We will bypass an analysis of this problem right now, and assume for our immediate discussion that a definition has been established. Now, the strength of materials depends, of course, upon the material (e.g., its chemistry) and on its condition (e.g., microstructure). Our more immediate task, however, is first to examine factors in the test procedures which affect the numerical value of the strength. These factors can be classed generally as environment, temperature, strain rate, and stress state. The strength may depend upon whether the test is conducted in air or in vacuum; it certainly may be quite sensitive to temperature. The strength may depend upon whether the load is uniaxial or three-dimensional. It is easiest to examine these factors in terms of the specialized tests used. We then briefly examine the tensile, fatigue, and impact properties of steels and the effect of structure on these properties. Creep properties will be discussed last; they are of interest mainly at high temperature, and reference to creep is mainly in terms of stainless steels (Chap. 7).

3-1 TENSILE PROPERTIES

In the standard tensile test, a specimen of standard dimensions (e.g., a gage length of 2.000 in. and a diameter of 0.505 in.) is loaded at a constant strain rate to fracture. The load-elongation relation is recorded, from which the stress-strain diagram is obtained. If the load is divided by the original cross-sectional area, the stress is the *nominal* or *engineering stress*. For a metal, when the plastic strain in the test reaches a certain value subsequent flow is restricted to one local region, and this region undergoes continuous reduction in area (necking) to fracture. Because the supporting area is decreasing, the load decreases; thus a stress-strain diagram using the nominal stress exhibits a maximum. Although this is an artifact of the test itself, this maximum stress has been used as a design criteria; it is called the *ultimate* or *tensile strength* (or ultimate tensile). If during the test the minimum cross-sectional area is continuously measured, then the true stress can be obtained, to yield a true stress-true strain diagram. Figure 3-1 illustrates both types of stress-strain curves.

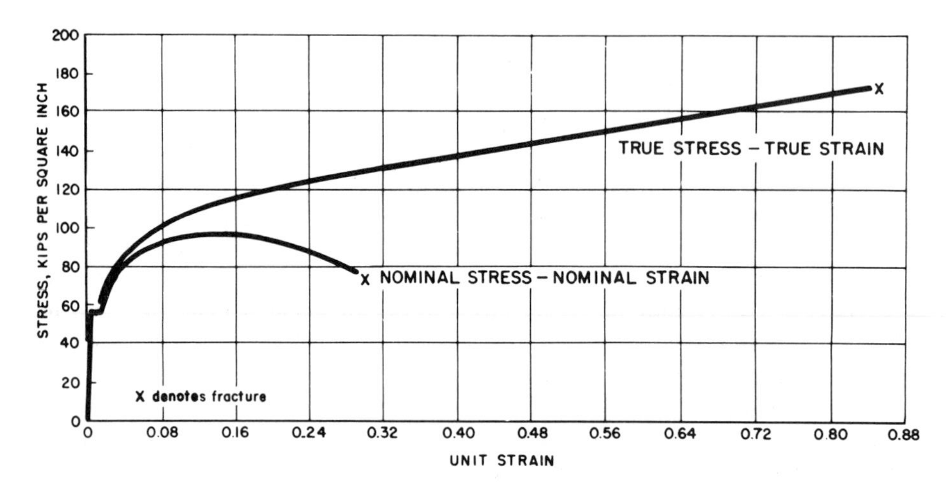

Figure 3-1 Stress-strain curves for an annealed steel. One curve is engineering stress-strain and the other is true stress-true strain. Both curves were obtained from the same set of load-elongation data. *(From H. E. McGannon (ed.), "The Making, Shaping and Treating of Steel," 9th ed., United States Steel Corporation, Pittsburgh, 1971.)*

The quantity usually desired from the test is the stress that just causes plastic deformation (yield point). The exact strain at which this occurs is difficult to establish, so in the United States a *yield strength* is defined in terms of 0.2% plastic strain. This is obtained from the intersection of a line from 0.2% strain parallel to the linear elastic region and the stress-strain curve. Unless indicated otherwise, in this book the term yield strength will mean the 0.2% yield strength.

Other properties obtained from the stress-strain diagram are the elastic (Young's) modulus, the elongation at fracture, and the reduction in cross-sectional area at fracture. These last two quantities are sometimes used as a measure of ductility. A measure of work hardening can be obtained from the slope of the stress-strain curve in the plastic region.

The mechanical properties are sensitive to the temperature of measurement. Here we are considering testing in a temperature range in a time sufficiently short that no structural changes occur during testing; thus any temperature dependence reflects the intrinsic sensitivity of the strength of the material (of a fixed structure) to temperature. This dependence is clearly seen in the stress-strain curves in Fig. 3-2. Note that the yield strength and the tensile strength increase and converge as

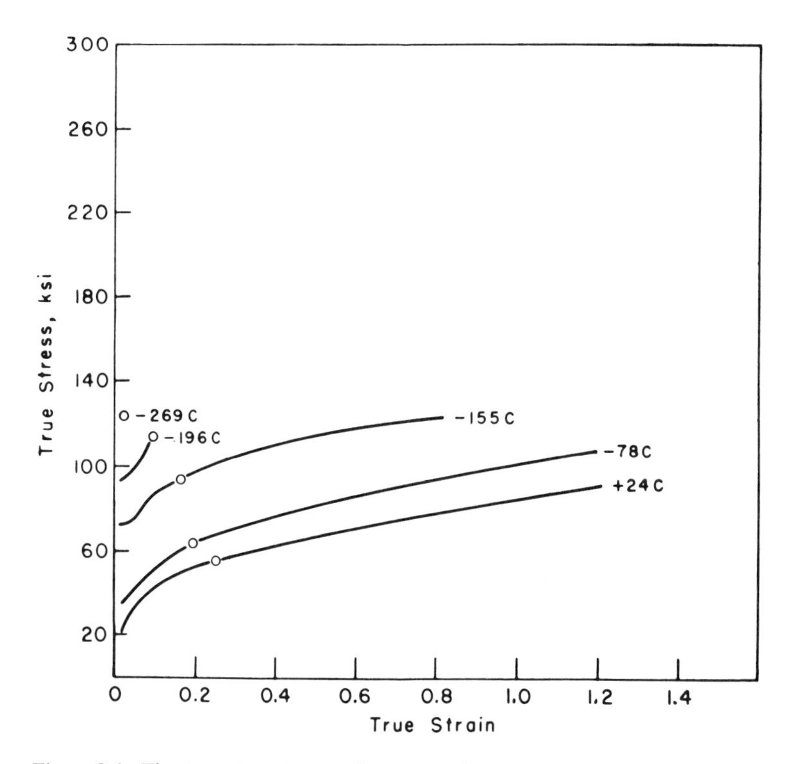

Figure 3-2 The true stress-true strain curves of commercially pure iron at different temperatures. The circles mark the location of the maximum load. *(From E. B. Kula and T. S. Desiston, in "Behavior of Materials at Cryogenic Temperatures," ASTM, Philadelphia, 1966.)*

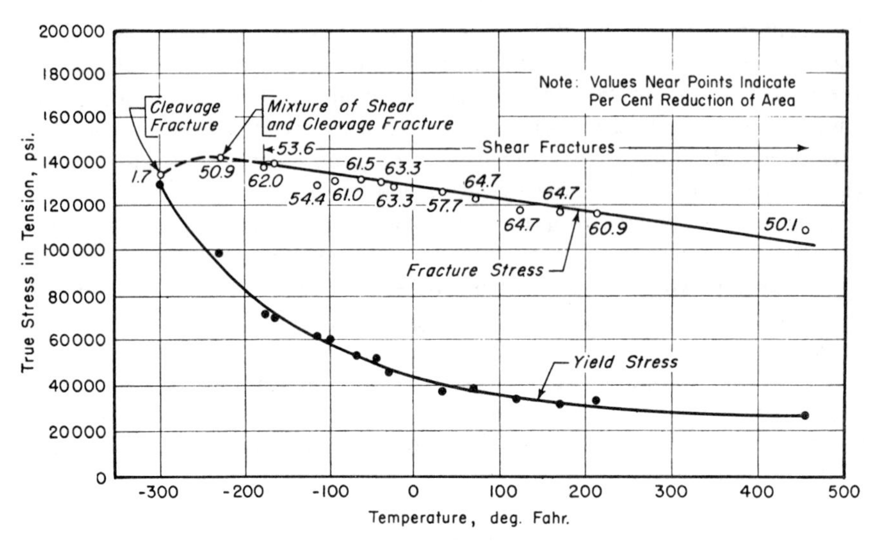

Figure 3-3 The variation of the yield and fracture stress with temperature for a 0.26% C steel in the annealed condition. The numbers give the percent reduction in area at fracture. *(From H. E. Davis, E. R. Parker, and A. Boodberg,* Proc. ASTM, *vol. 47, p. 483, 1947.)*

the temperature is lowered (Fig. 3-3), and the elongation at fracture decreases. In a general sense, the steel is becoming more brittle as the temperature is lowered. In body-centered cubic materials, as the temperature decreases the ductility decreases drastically to a very low value over a narrow temperature range. This characteristic must be considered carefully in design. One way to look at this is that if a structure is overloaded, it is best that it plastically deform, and not catastrophically fracture; low temperature favors the latter behavior.

The Yield Point Phenomena

It was pointed out that it is difficult to locate the exact stress (yield point) at which a material begins to deform plastically. This led to the definition of yield strength based on 0.2% plastic strain. However, a few classes of materials do show a sharp, well-defined yield point, and one of these is steels with certain heat treatments. This phenomena is sufficiently important industrially that we will discuss it in some detail.

The atomic mechanism of plastic deformation involves the response of dislocations to the applied force. As a brief review, the simple edge dislocation will be considered. This can be considered as an extra plane of atoms placed part of the way through a crystal, as shown in Fig. 3-4*a*. When sufficient shear stress is applied on the slip plane containing such a dislocation, a shift in the bonding occurs so that the dislocation moves until it exits at a free surface (Fig. 3-4*c*). Then permanent plastic deformation has occurred. This process must be repeated many times to generate visible offset. It is important to note that it is considerably more difficult to cause a layer of atoms to move across an adjacent layer in shear without this mechanism. Thus, crystals

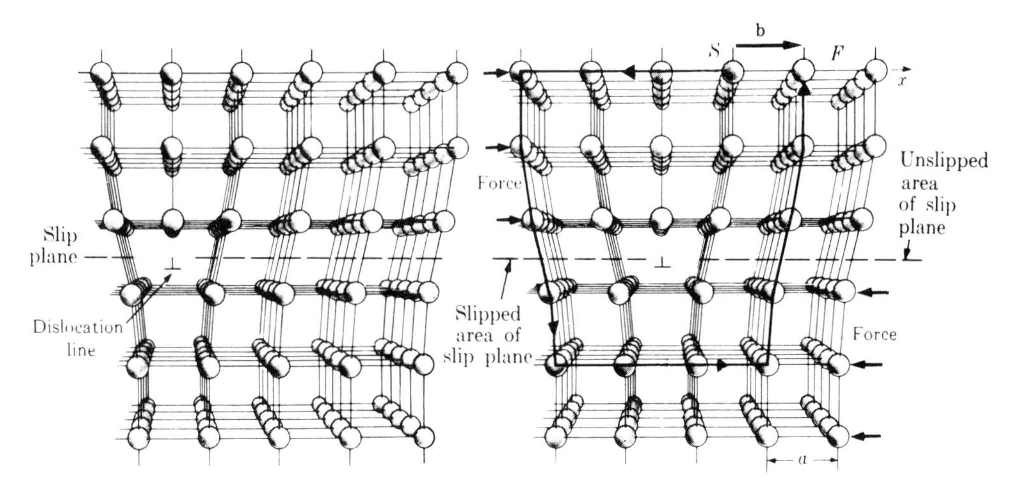

(a) An edge dislocation in a crystal structure.

(b) The dislocation has moved one lattice spacing under the action of a shear force.

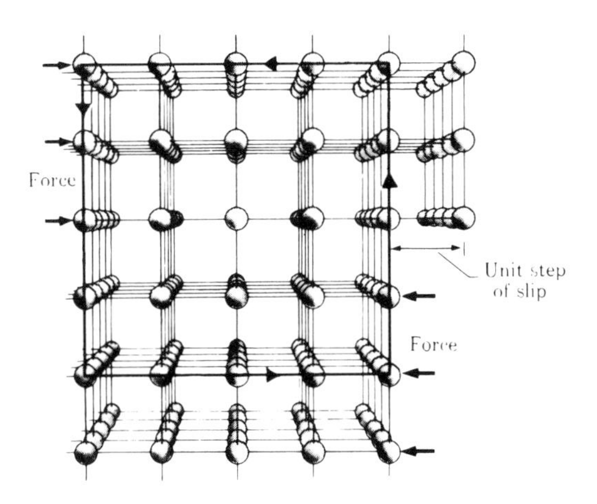

(c) The dislocation has reached the edge of the crystal and produced unit slip.

Figure 3-4 The movement of an edge dislocation, showing eventual external plastic deformation (c). *(From A. G. Guy and J. J. Hren, "Elements of Physical Metallurgy," 3d ed., Addison-Wesley, Reading, Mass., 1974.)*

free of dislocations have extremely high strength (e.g., 10^6 psi yield strength); however, such defect-free crystals are quite small (they are called whiskers) and at the present time have limited practical use. That is, most crystalline materials contain a sufficient concentration of intrinsic dislocations to establish the plastic deformation mechanism as illustrated in Fig. 3-4.

When a sample of pure iron is loaded, the shear stress increases on the slip systems, and when the critical value is reached dislocations begin to move. When a sufficient number of them move, macroscopic deformation is seen by the appearance of yielding (usually defined in terms of the 0.2% offset strain). Now consider a sample of iron with some carbon (or nitrogen) dissolved interstitially. Even though the carbon atom is relatively small, elastic strain does occur in the iron atoms in the region of the carbon atom. However, if a dislocation is present, this strain energy can be reduced if the carbon atom is located on the tension side of the dislocation (Fig. 3-5*a*). This can be accomplished by holding the iron sample at a sufficient temperature long enough so that each carbon atom eventually finds a dislocation site. The maximum solubility of carbon in body-centered cubic iron is 0.025 wt. %, and a typical dislocation density for an annealed iron sample is about 10^7 dislocations per cm^2. (That is, there are 10^7 cm of dislocation core sites per cm^3 of material.) Calculation of the number of possible carbon sites along the dislocations will show that there is a sufficient quantity of carbon atoms to saturate the dislocation sites.

Since the energy of the sample is lower with the carbon at the dislocation than if it were in the lattice elsewhere, it is more difficult to move a dislocation away from the carbon atom than if the atom were not there. Thus when the iron sample is subjected to an increasing load, and the resolved shear stress reaches the critical value for dislocation movement in pure iron, the dislocations will not move for they are pinned by the carbon atoms. A higher shear stress is required to move the dislocation. When this value is reached, and the dislocation moves away from the carbon atom (Fig. 3-5*c*), the stress is at a higher value than is required to initiate movement in the carbon-free regions; thus the dislocation continues to move, and the load required to move them decreases, if the test is being carried out such that the strain rate is constant. (This is a rather simple picture since it ignores the dynamics of dislocation movement. Plastic strain then occurs relatively easily, until the dislocations are impeded due to encounter with other carbon atoms, grain boundaries, or other dislocations. Then the load for continued extension begins to rise due to strain hardening.

This behavior is illustrated in Fig. 3-6, which is a stress-strain curve for a 0.18% C steel. The steel has been cooled slowly from the austenite region to allow the carbon atoms to saturate the dislocations. As the load was increased, the sample elongated elastically to point *a*. This was the stress level to move the dislocations away from the influence of the carbon atoms, and the load then decreased to point *b*. Extension continued at a stress corresponding to point *b* (about 34,000 psi), until point *c* was reached. Then strain hardening began to occur, and an increasing load was required to continue deformation. At point *d* the load was removed, the sample recovering elastically along line *de*. Upon reloading, the sudden elongation was not repeated, but instead plastic deformation began at point *f*, about where the sample was unloaded. The stress at point *a* is referred to as the *yield point*; it is sometime called the upper yield point, and the average stress along line *bc* is called the lower yield point.

The sudden yielding is accompanied by visible surface offset (Lueder's lines), and in fabrication of objects (e.g., automobile fenders) from steel sheet showing this behavior, surface roughness occurs in the area of the part that is deformed in the range 2 to 10%. To alleviate this problem, the sheet material is cold rolled from

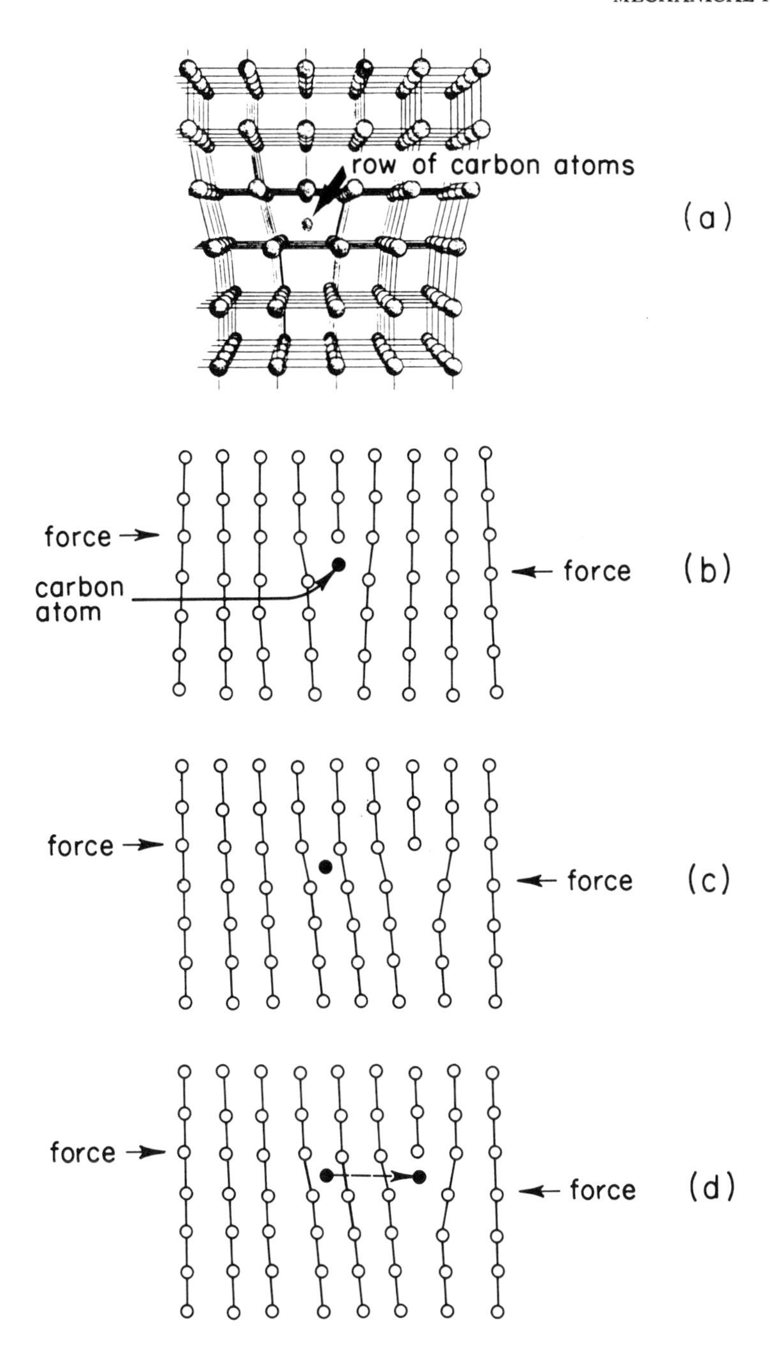

Figure 3-5 Illustration of the movement of a carbon atom to locate in the dislocation core. In (b), the shearing force will become sufficient to move the dislocation away from the carbon atom (c). However, given sufficient time, the carbon atom will migrate and eventually locate again at the dislocation (d).

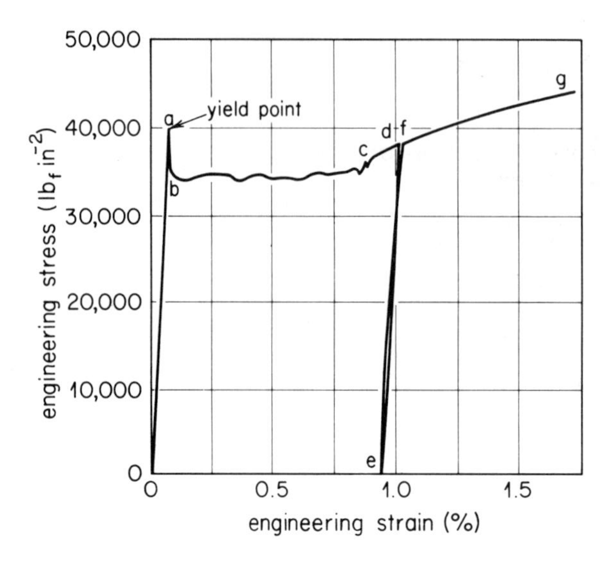

Figure 3-6 The stress-strain diagram for an annealed 0.18% C steel, showing yield point development.

1 to 10%; this moves the dislocations from the carbon atoms, and subsequent deformation exhibits no sudden yielding. The effect of this prior deformation is clearly shown in Fig. 3-7 by the stress-strain curves for various amounts of prior reduction in thickness by cold rolling. However, if the rolled sheet is allowed to age before tensile testing, the carbon atoms, which are rather randomly moving from one interstitial site to another, will eventually relocate at dislocations (Fig. 3-5), at which they are so stable that they will remain there. Then the yield point phenomena will return. This effect is illustrated in Fig. 3-8. Even at 20°C, the average residence time for a carbon atom in an interstititial site is only about one second, so that, given sufficient time, the dislocations become saturated with carbon atoms. The effect is accelerated by increasing temperature (Fig. 3-8*b*).

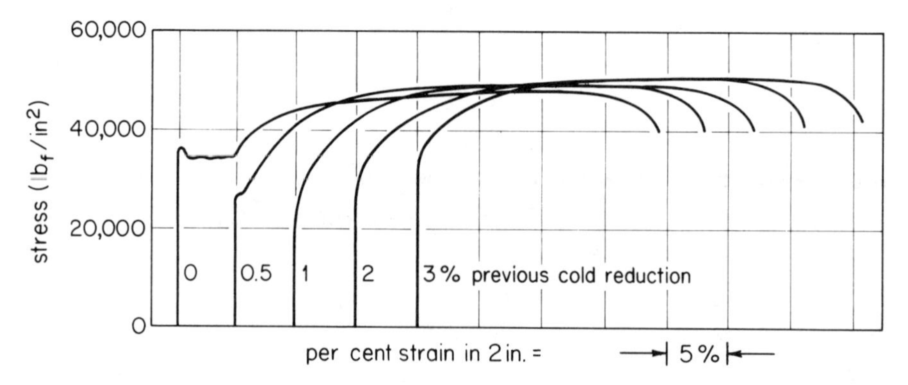

Figure 3-7 The effect of prior plastic strain on the yield point phenomena in low carbon steel sheet. *(Adapted from R. O. Griffis, R. L. Kenyon, and R. Burns, in "Yearbook of American Iron and Steel Institute for 1933," published and copyrighted by American Iron and Steel Institute in 1933.)*

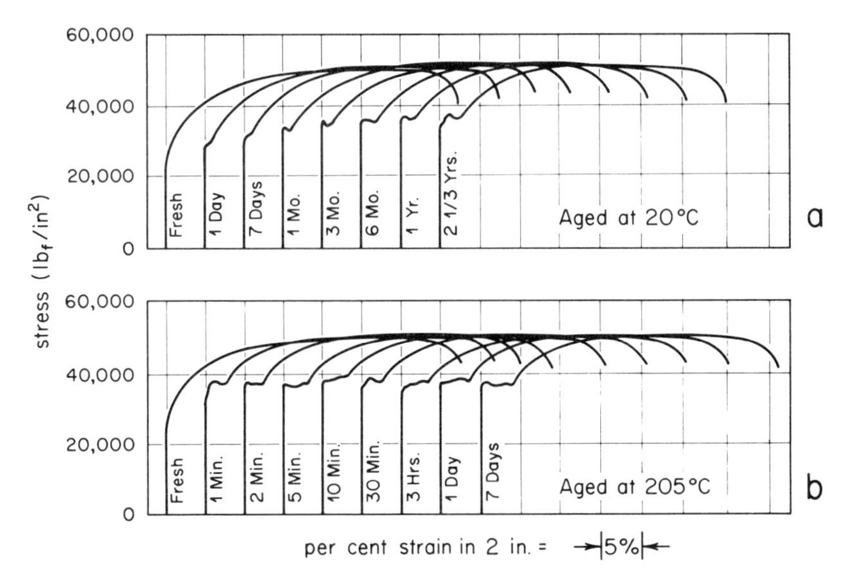

Figure 3-8 The effect of aging low carbon steel on the yield point phenomena. The steel was cold rolled 1% reduction in thickness, then aged at the temperature for the time shown prior to testing. *(Adapted from R. O. Griffis, R. L. Kenyon, and R. Burns, in "Yearbook of American Iron and Steel Institute for 1933," published and copyrighted by American Iron and Steel Institute in 1933.)*

Correlation between Yield Strength and Tensile Strength

There is a usable correlation between the yield strength and the tensile strength for steels. Figure 3-9 shows that the relation can be represented by one curve for steels that have been quenched and tempered and hence consist of small carbide particles in a ferrite matrix. The primary ferrite-pearlite structure, developed by hot rolling or by annealing, has a separate curve. The bars give the uncertainty in the correlation. For example, a steel with a yield strength of 200,000 psi has an ultimate strength of 220,000 (±15,000) psi. Neither curve here is valid for cold-worked structures.

Correlation between Yield Strength and Hardness

The relation between yield strength and hardness is shown in Fig. 3-10. Again there is one curve for the quenched and tempered steels, and one for the hot-rolled or the annealed condition.

3-2 FATIGUE PROPERTIES

Most structures are loaded with forces that are time dependent, and thus the applicability of the tensile-test data to design in such cases must be established. The fatigue

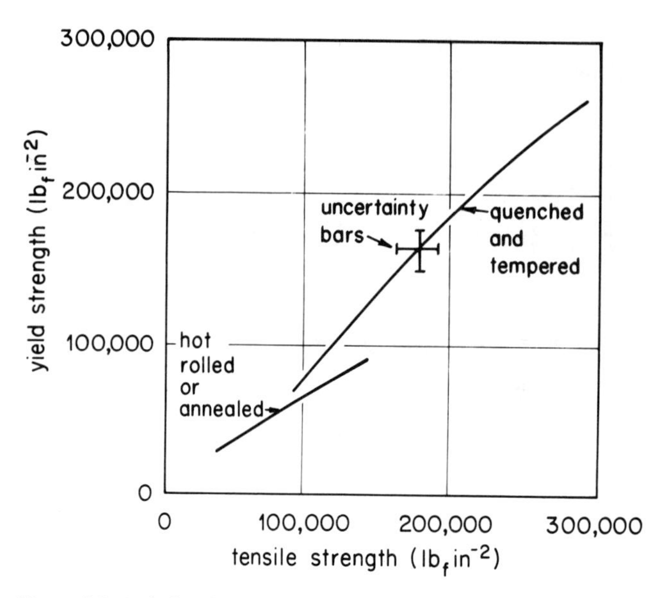

Figure 3-9 Relation between the yield strength and the tensile strength for steels. *(Adapted from "Metals Handbook," vol. 1, 8th ed., American Society for Metals, Metals Park, Ohio, 1962; W. G. Patton,* Met. Progr., *vol. 43, p. 726, 1943.)*

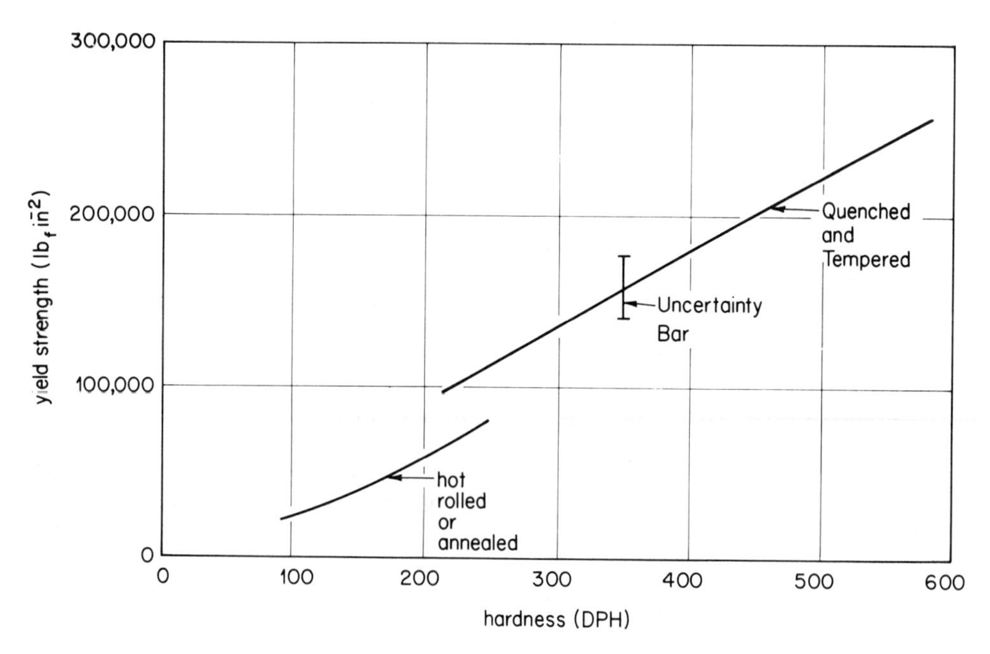

Figure 3-10 Relation between the yield strength and the hardness for steels. *(Adapted from "Metals Handbook," vol. 1, 8th ed., American Society for Metals, Metals Park, Ohio, 1962; W. G. Patton,* Met. Progr., *vol. 43, p. 726, 1943.)*

test is designed to determine the resistance to fracture under time-varying loads. We will consider here only the simple tests.

Consider a cylinder which is supported at the ends and loaded in the center at a stress level considerably below the yield strength. The cylinder is rotated, so that any element on the surface is alternately under tension and then compression. Under such loading, it is found in many cases that the material will eventually fracture even though the yield strength is not exceeded. This is an example of the dependence of the "strength" on the stress reversal. The safe limit of loading can be determined if the desired number of cycles for the "life" of the part is established. Then a failure test can be run at a series of loads, so that a plot of the number of cycles to fracture (N) versus the stress (S) can be obtained. This plot is a fatigue curve; a typical curve is illustrated in Fig. 3-11. The number of cycles to failure is quite sensitive to the specimen surface conditions, and thus the determination of the S-N curve using standardized tests requires at least 10 specimens at each stress level.

The fatigue data are usually treated statistically to obtain the S-N curve. For most steels, there appears to be a stress below which the part will not fail for very long cycles. (For example, an automobile axle will undergo about 10^8 cycles if driven 20,000 miles each year for 10 years.) This limiting stress is called the *fatigue strength* or *endurance limit*. For the data in Fig. 3-11, the fatigue strength is 85,000 psi.

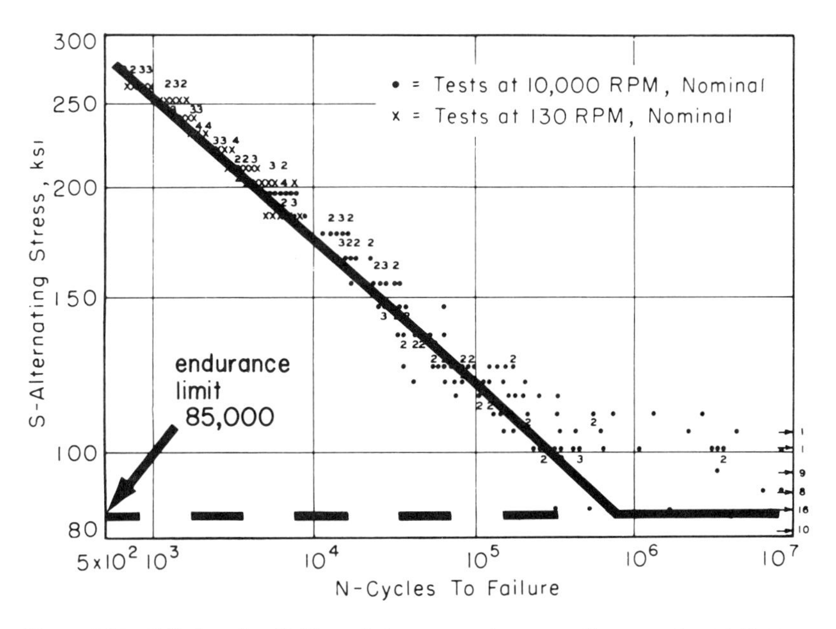

Figure 3-11 S-N data for 4340 steel, heat treated to a tensile strength of 260,000 psi (250,000 psi yield strength, hardness 52 R_c). The numbers by data points give the number of tests giving the same data point. The numbers by the arrows indicate the number of specimens at the indicated stress which did not break at 10^7 cycles. The specimens were cylinders tested in a rotating-beam configuration. Note that the results are valid for two widely different rotational speeds. *(Adapted from H. M. Cummings, F. B. Stulen, and W. C. Schulte,* Trans. ASM, *vol. 49, p. 482, 1957.)*

Correlation of Fatigue Strength with Tensile Strength, Yield Strength and Hardness

The type of correlation between fatigue strength and tensile strength usually obtained for steels is illustrated in Fig. 3-12. It is common to assume that the fatigue strength is 0.35 to 0.5 times the tensile strength. Using this curve, and the tensile strength-yield strength correlation in Fig. 3-9, the curve in Fig. 3-13 is derived. Note the uncertainty in the curve. Figure 3-14 shows the correlation between hardness and fatigue strength for several quenched and tempered steels. Note that the scatter is higher for the higher strength steels. These data, and those in the other figures (along with the correlation in Fig. 3-10), were combined to obtain the fatigue strength-hardness correlation in Fig. 3-15. Note that the scatter band increases as the strength level increases.

The uncertainty in the correlation must be considered when using this information. The scatter in the data is intrinsic in the fatigue test (Fig. 3-11) and in the uncertainty in the correlation of the strength properties and hardness (Figs. 3-9 and 3-10). There is additional uncertainty involved when applying these test data in practice. In application, the difficulty in reproducing the surface conditions, the heat treatments, and the chemistry from various lots of steels expand the problem, and all may contribute to an increase in the uncertainty.

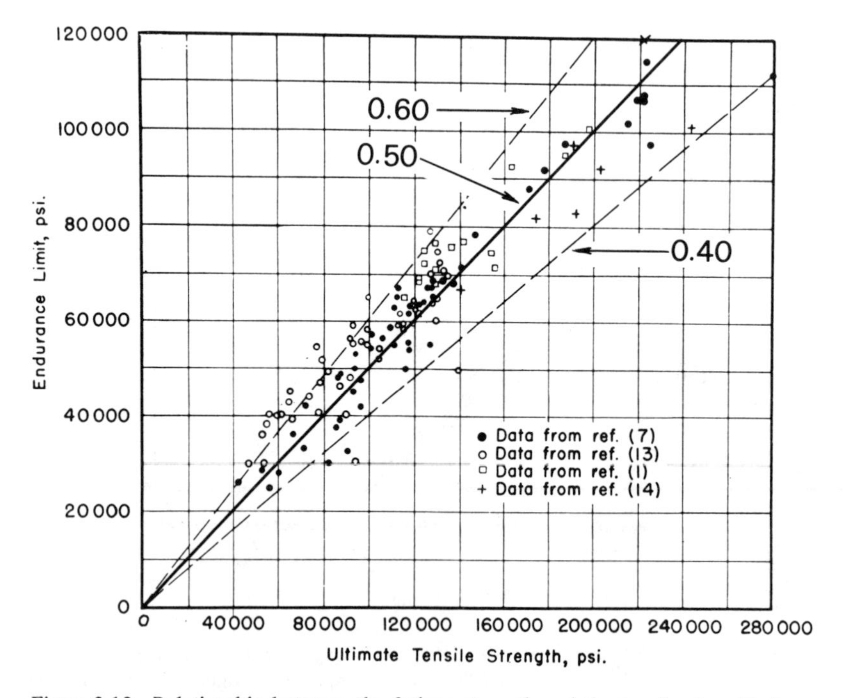

Figure 3-12 Relationship between the fatigue strength and the tensile strength for several steels. The straight lines have the slopes shown. *(Adapted from a compilation of T. J. Dolan and C. S. Yen,* Proc. ASTM, *vol. 48, p. 664, 1948. Reprinted by permission of the American Society for Testing & Materials, Copyright 1948.)*

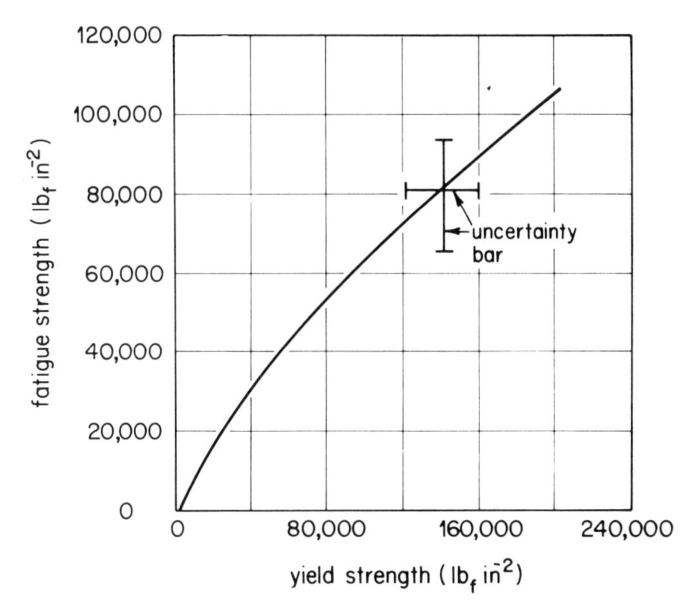

Figure 3-13 The fatigue strength as a function of yield strength for steels.

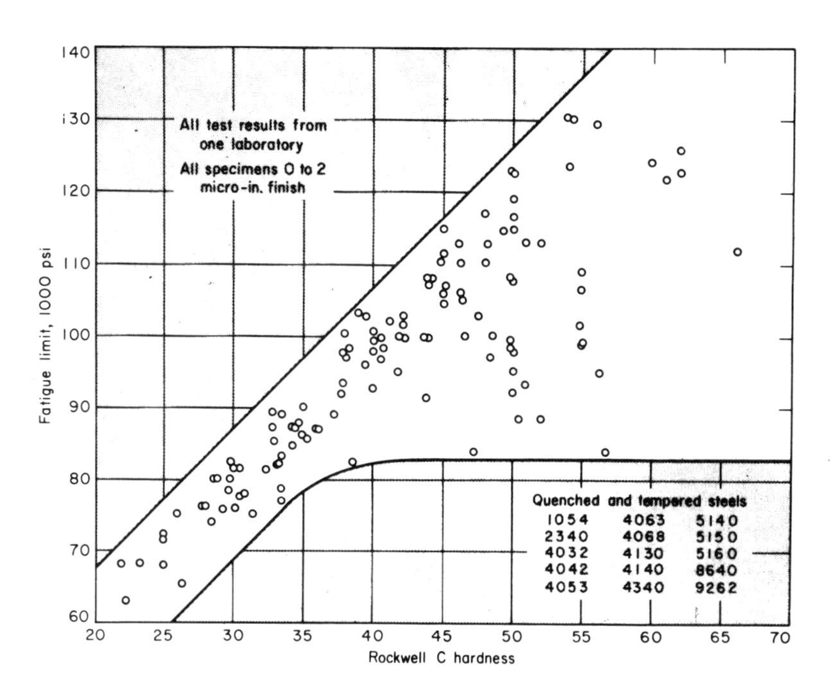

Figure 3-14 The fatigue limit as a function of hardness for several quenched and tempered steels. *(From "Metals Handbook,'. vol. 1, 8th ed., p. 217, American Society for Metals, Metals Park, Ohio, 1961.)*

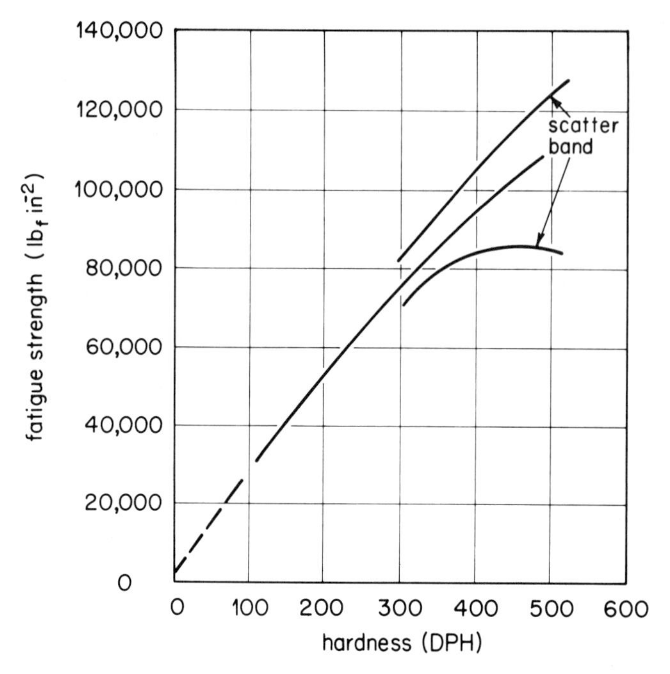

Figure 3-15 The fatigue strength as a function of hardness for steel.

3-3 IMPACT PROPERTIES

When a crystalline material is subjected to increasing stress, the shear stress on each slip system increases, and the normal stress on each cleavage plane increases. If the shear stress reaches the critical stress for significant dislocation movement before the normal stress attains the critical value for cleavage, then plastic deformation occurs. Otherwise, before significant dislocation motion occurs, a crack that forms (or pre-exists) will propogate extremely rapidly on a cleavage plane, causing sudden brittle fracture. If the temperature is sufficiently high, the slip mechanism dominates; however, at sufficiently low temperatures cleavage may dominate. In body-centered cubic materials, the temperature range below which cleavage becomes important is narrow. This behavior is easily seen in stress-strain curves obtained at low temperatures, such as those shown in Fig. 3-2 for pure iron. Note that as the temperature decreases, the yield strength increases, the tensile strength increases, the elongation at fracture decreases, and the yield strength approaches the tensile strength. These characteristics are further illustrated by the data in Fig. 3-3 for annealed 0.26% C steel. Below about −184°C (−300°F), this steel shows negligible ductility in tension.

The strength and ductility also may be quite dependent upon the strain rate. This is illustrated in Fig. 3-16 by the stress-strain curves obtained at 538°C at different strain rates. Note that with increasing strain rate the tensile strength and yield strength increase and the ductility at fracture decreases.

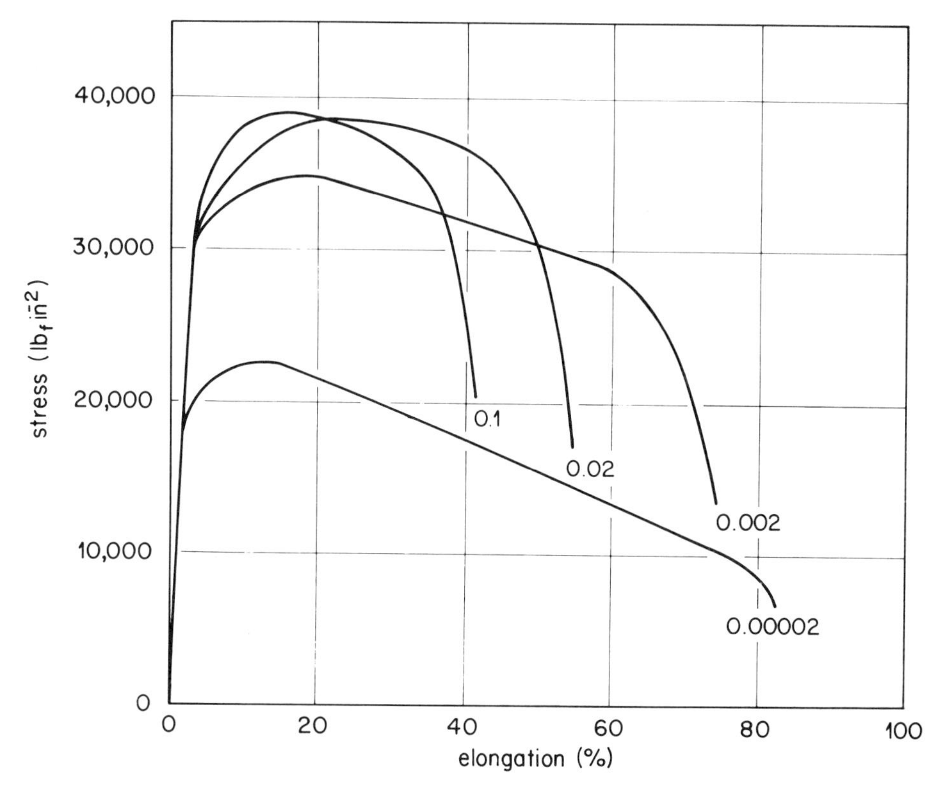

Figure 3-16 Engineering stress-engineering strain curves obtained at 538°C showing the influence of strain rate. The strain rates shown vary from 0.00002 to 0.1 min^{-1}. The steel contained 0.15% C and 0.55% Mo and had a spheroidized microstructure. *(Adapted from R. F. Miller, G. V. Smith, and G. L. Kehl,* Trans. ASM, *vol. 31, p. 817, 1943.)*

The point here is that the strength and the ability to withstand sudden brittle fracture are quite dependent upon the temperature, the strain rate, and also the stress state. The problem is to define a test that will predict this behavior. The ability of a material to absorb energy without sudden fracture is referred to in general terms as notch toughness. Thus another consideration in choosing a steel and its heat treatment is the toughness.

A useful test for toughness is not as easy to devise as resistance to static loads or cyclic loads. Since most structures contain internal flaws, either fabrication flaws or second-phase particles, the initation of cracks is of less concern than choosing a material that will resist the propagation of a crack. The discipline of fracture mechanics deals with the details of designing and using such tests. Here we will limit our attention to a test that is relatively easy to perform and that is still used. This is the impact test. A small test bar, of standard dimensions, has a notch of specified geometry in it normal to the long axis of the bar. The bar is supported such that it is broken by a sudden blow of known energy opposite the notch. The presence of the

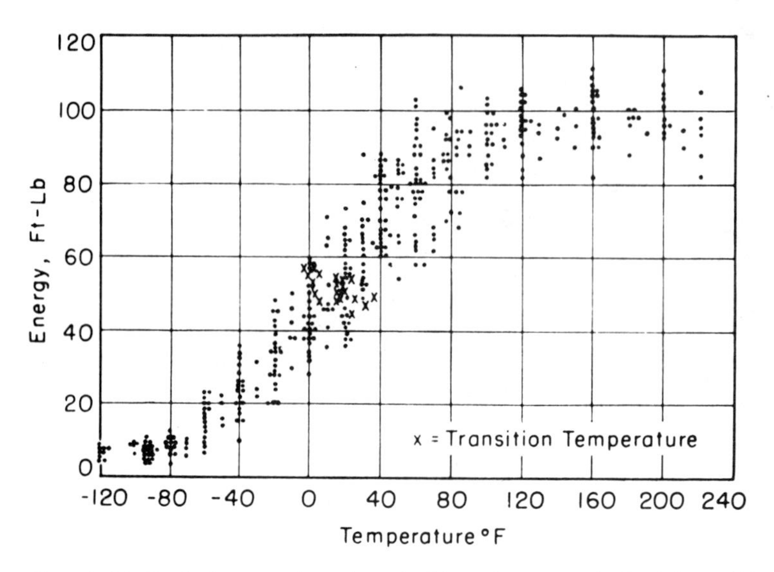

Figure 3-17 Typical impact data of steels. These data were obtained from 420 tests on steels of nearly identical chemical composition and grain size. *(From J. A. Rinebolt and W. H. Harris,* Trans. ASM, *vol. 43, p. 1175, 1951.)*

notch introduces a complex stress state, and the impact blow fractures the material under a very high strain rate. The energy to fracture the sample is measured. The test then is repeated as a function of temperature. Typical impact data are shown in Fig. 3-17.

From the impact curves in Fig. 3-18 it is seen that the energy absorbed can decrease drastically over a narrow temperature range. The temperature at which this occurs is the *transition temperature*, but this can be defined in various ways. For

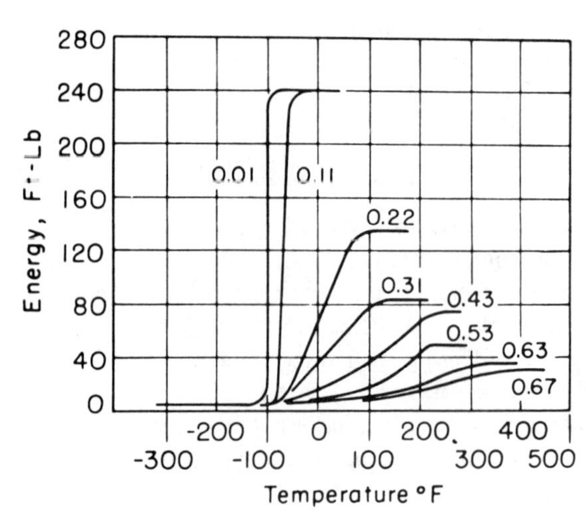

Figure 3-18 Effect of carbon content on the impact curves of annealed steels. The steels had been normalized, so that the increased carbon content is reflecting the increase in the amount of pearlite and the decrease in the amount of primary ferrite. *(From J. A. Rinebolt and W. H. Harris,* Trans. ASM, *vol. 43, p. 1175, 1951.)*

example, the temperature of the inflection point of the impact curves could be chosen. The fracture surface of the impact specimens at very low temperatures is rather flat; however, at high test temperatures the edges of the surface consist of regions in which the fracture was produced by shear. The fraction of the fracture surface covered by these shear "lips" increases with the test temperature. This is illustrated in Fig. 3-19.

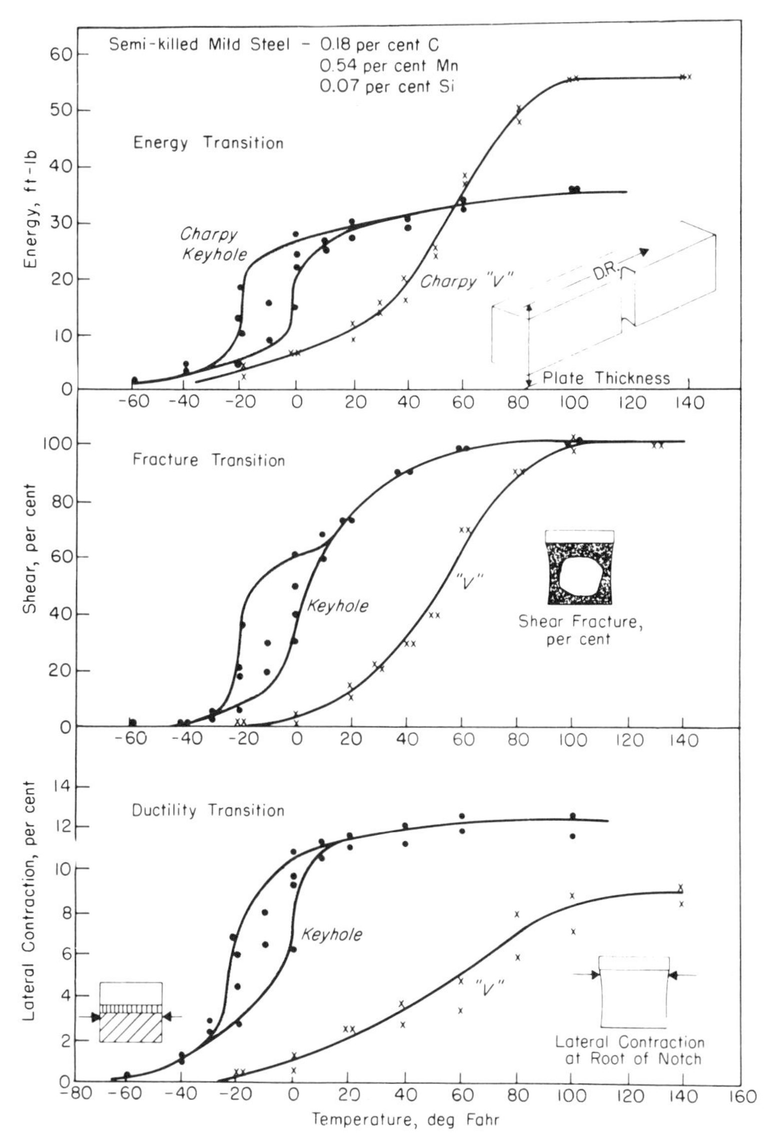

Figure 3-19 Illustration of various ways of indicating impact behavior of a steel, hence different definitions of the transition temperature. *(From W. S. Pellini, in "Symposium on Effect of Temperature on the Brittle Behavior of Metals with Particular Reference to Low Temperatures," ASTM, Philadelphia, 1954.)*

The transition temperature then could be chosen as the temperature at which this curve has an inflection point. Or the material contraction at the root of the notch can be measured as a function of test temperature (Fig. 3-19), and from this curve a transition temperature can be obtained. The point is that there are several methods of expressing the transition temperature, and each may give a different value.

A design quantity frequently used is that the minimum impact energy must be 15 ft·lbf at the lowest temperature of application. From Fig. 3-17, this temperature is ±50°F (±25°F). That this uncertainty must be kept in mind is illustrated by the curves in Fig. 3-18. Considering the 0.11% C steel, the transition temperature is about −75°F; however, an error in this value of 25°F could place the steel well within the very brittle behavior temperature region.

It is important to realize that the impact data are not used to design structures to resist impact loading, but to measure the sensitivity of the material to rapid crack propagation under the complex state of stress. Thus a steel used at a temperature where the impact energy is relatively high may still undergo brittle fracture. There are tests, however, that determine the temperature above which brittle crack propagation will not occur; this temperature is referred to as the *nil ductility transition (NDT) temperature.* In many cases, it is found that the temperature at which the impact energy is 15 ft·lbf is above the NDT, and therefore the 15 ft·lbf criteria is frequently a design criteria.

There is no general usuable correlation between hardness and strength and impact energy. In general, the impact resistance decreases with increasing strength and hardness. However, for the same strength the impact properties are quite sensitive to the structure. This will be discussed in more detail in the following section.

3-4 EFFECT OF STRUCTURE ON PROPERTIES

The yield strength of a material is related to the resistance of the dislocations to the applied force. These lattice defects control plastic deformation, and any impediment to their motion strengthens the crystal. When a moving dislocation encounters obstacles (e.g., a stronger particle), it requires an increasing stress to force the dislocation through or around the obstacle. Dislocation theory gives the required strength to be proportional to the inverse of the distance between neighboring obstacles, usually to the $\frac{1}{2}$ power. Thus in spheroidized carbide-ferrite structures, the more closely spaced the carbides (obtained by developing a finer dispersion of carbides), the greater the yield strength (Fig. 3-20). Or if the main obstacles are high-angle grain boundaries, the smaller the grain size, the higher the strength. This effect is illustrated for a steel that has been heat treated to develop a variation in the primary ferrite grain size; the remaining structure is pearlite. The carbides are quite strong, and in the pearlite the carbide-ferrite spacing is quite small compared to the primary ferrite grain size. Thus, the pearlite is strong, and yielding occurs more easily in the primary ferrite. The yield strength should, then, be related to the primary ferrite grain size. This is illustrated in Fig. 3-21. In making these comparisons other variables that affect the strength must be kept constant. For example, the yield strength is affected by substitutional elements (e.g., Cr) in solid solution in the ferrite. If, in heat treatments to vary the carbide size,

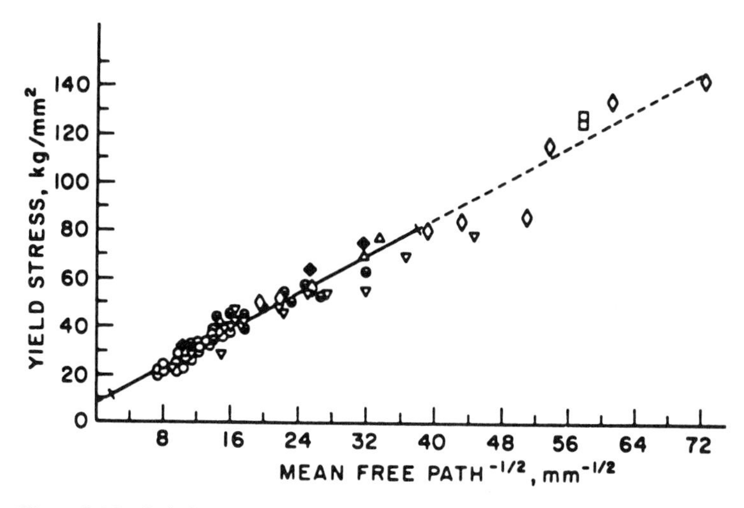

Figure 3-20 Relationship between the yield strength and average spacing between carbides in spheroidized or tempered steels. The data are from steels ranging in carbon content from about 0.1 to 1.5%. *(Adapted from J. Gurland, in "Stereology and Quantitative Metallography," ASTM, Philadelphia, 1972.)*

the alloying elements tend to be concentrated in the carbides, the solid-solution strengthening effect becomes less. These interacting effects will be discussed in more detail in the chapter dealing with structural steels.

The impact properties depend upon somewhat different factors than those for the strength. Brittle fracture is associated with a rapidly moving crack, and any obstacle

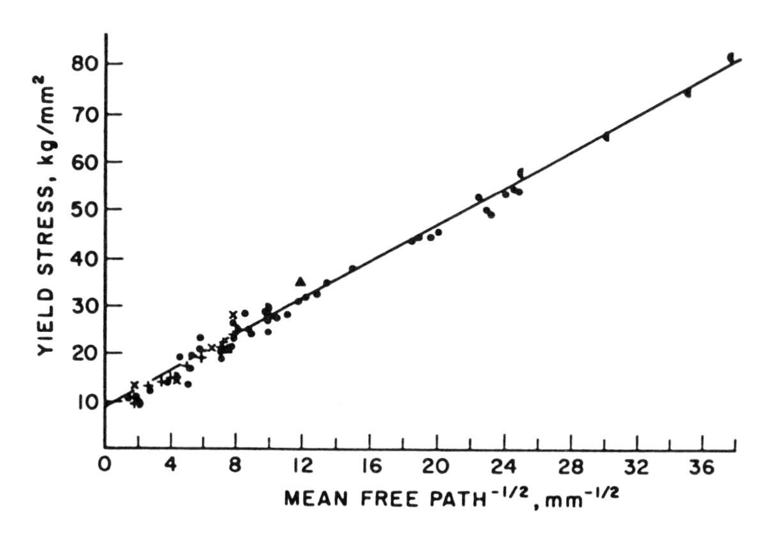

Figure 3-21 Relationship between the yield strength and the average primary ferrite grain size in steels of primary ferrite-pearlite structures. The data are from steels ranging in carbon content from about 0.1 to 0.2%. *(Adapted from J. Gurland, in "Stereology and Quantitative Metallography," ASTM, Philadelphia, 1972.)*

that impedes its motion will improve the toughness. The cleavage stress to initiate a crack in the matrix is a function of the chemical composition of the solid solution in which the crack forms. However, in multiphase structures, such as most steels, cracks may open easily in carbides or inclusions or at the particle-matrix interface, or may be present intrinsically due to fabrication techniques. Thus it is most commonly the resistance to crack propagation which controls toughness. The stress to propagate the crack depends upon the microstructure; any obstacle that impedes the propagation of the crack will improve the toughness. Thus, for example, decreasing primary ferrite grain size should decrease the NDT.

Thus, the ferrite grain size should control the toughness, assuming that the carbide distribution is held approximately constant. This effect is illustrated in Fig. 3-22 for a low-carbon alloy steel. The ferrite grain size used here is the lath size of the tempered martensite or of the bainite. As the size of the grains decreases, the transition temperature decreases, showing an improvement in toughness.

Bainite forms in two rather distinct morphologies, referred to as upper and lower bainite. Above approximately 350°C, upper bainite forms. It is comprised of ferrite plates surrounded by carbides which have precipitated along the ferrite boundaries. The dislocation density is relatively low. In upper bainite, bainite laths which differ in orientation from each other only by a few degrees form in an austenite grain. Thus, within a *prior* austenite grain the cleavage plane in the ferrite is almost continuous across the prior austenite grain. Lower bainite, forming below 350°C, has a finer ferrite grain size and a higher dislocation density than upper bainite. Further, the ferrite boundaries are relatively high angle, and the carbides are finer and distributed differently.

The finer ferrite grain size, the presence of high-angle boundaries, the high dis-

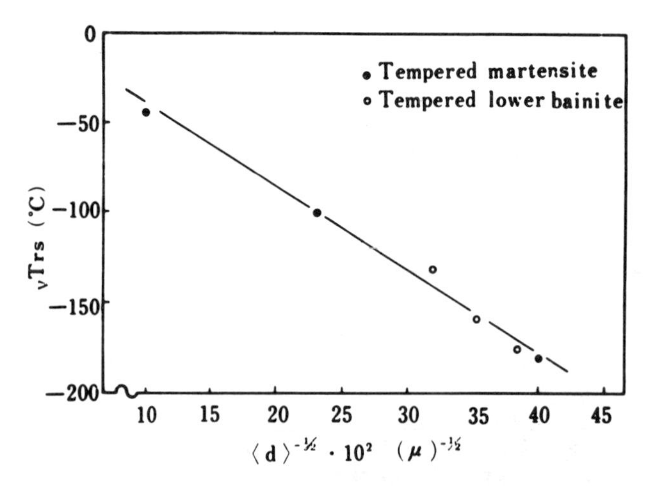

Figure 3-22 The effect of the ferrite grain size (*d*) on the impact transition temperature for a 0.12% C, 0.5% Mn, 0.2% Si, 0.4% Mo, 2.4% Ni, 1.0% Cr steel. *(From S. Matsude, T. Inoue, H. Mimura, and Y. Okamura, in "Toward Improved Ductility and Toughness," Climax Molybdenum Development Co., Japan, 1971.)*

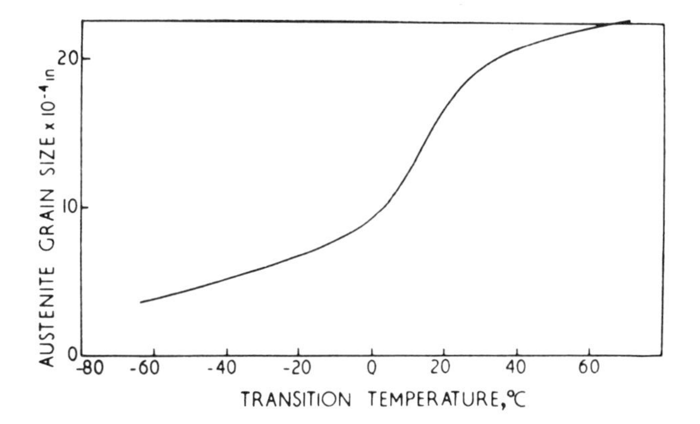

Figure 3-23 The effect of the prior austenite grain size on the impact transition temperature of upper bainite of constant tensile strength of 90,000 psi. The steel contained 0.12% C, along with Cr, Mo, and B. *(From K. J. Irvine and F. B. Pickering,* J.I.S.I., *vol. 201, p. 518, 1963.)*

location density, and the finer carbide distribution all act to improve toughness and lower the NDT in lower bainite. The impact properties thus correlate with the ferrite lath size (Fig. 3-22) and the carbide distribution. The impact properties of upper bainite are more sensitive to the prior austenite grain size, as shown in Fig. 3-23.

Generally, as the strength of the steel rises, the NDT will increase as the material becomes more sensitive to crack propagation. However, as shown in Fig. 3-24, upper bainite, even though not as strong as lower bainite, has less resistance to crack propagation. For example, forming upper bainite at the higher temperature limit of about

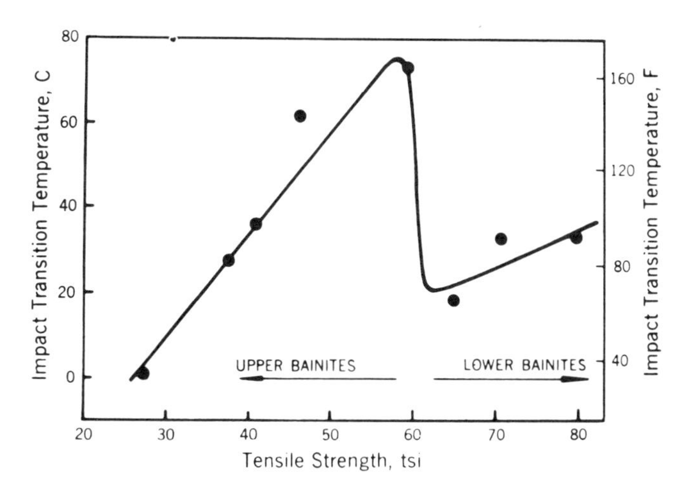

Figure 3-24 The effect of microstructure of upper and lower bainite on the tensile strength and impact transition temperature for low carbon steels. *(From F. B. Pickering, in "Transformation and Hardenability in Steels," Climax Molybdenum Co. of Michigan, Ann Arbor, Michigan, 1967.)*

650°C gives a steel with a tensile strength of 60,000 psi and a transition temperature of about 0°C. Forming upper bainite around 500°C gives a finer structure with a higher tensile strength of around 100,000 psi and a transition temperature of about 75°C. But forming lower bainite around 400°C will give a tensile strength even higher (about 130,000 psi) but with improved toughness, having a transition temperature of about 30°C.

Fresh martensite has a high strength, and crack propagation at this strength level is relatively easy. There are no carbides present in fresh martensite to impede crack growth, and the martensite plates are crystallographically related so that a crack grows easily until the high-angle, *prior* austenite boundary is encountered. Thus the impact properties of both fresh and tempered martensite are sensitive to austenite grain size. This is illustrated in Fig. 3-25 for tempered martensite.

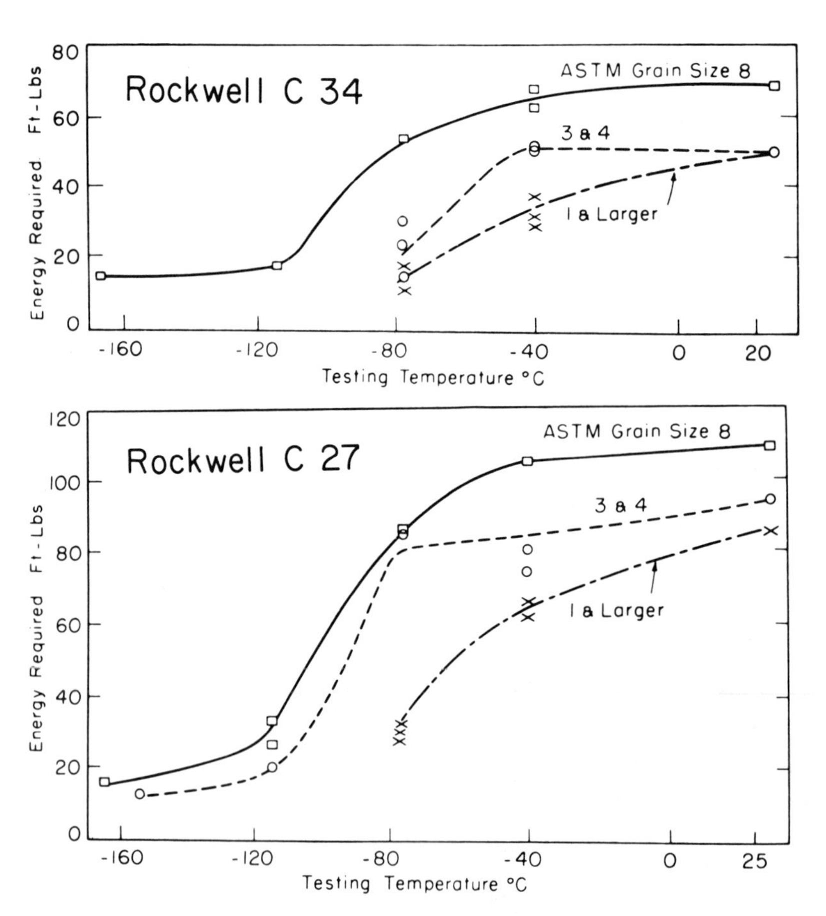

Figure 3-25 The effect of prior austenite grain size on the impact curves of NE 8640 steel. The steel samples were quenched to martensite, then tempered to give the hardness levels shown. *(Adapted from L. K. Haffe and F. F. Wallace,* Trans. ASM, *vol. 40, p. 775, 1948.)*

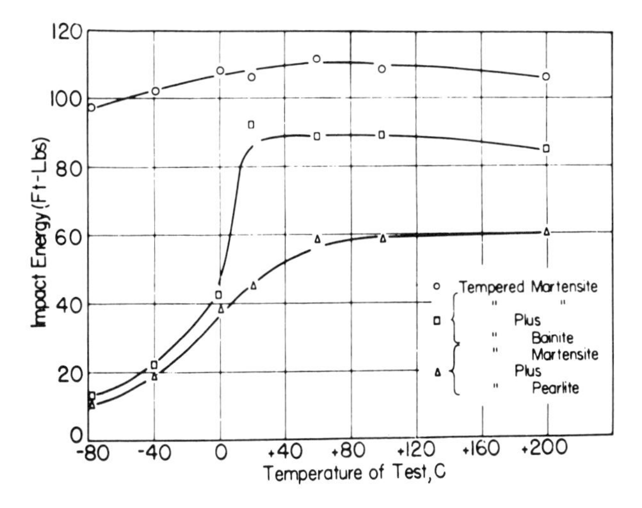

Figure 3-26 The impact curves for a NE 8735 steel of three different structures, but about the same tensile strength (125,000 psi). *(From J. H. Hollomon, L. D. Jaffe, D. E. McCarthy, and M. R. Norton,* Trans. ASM, *vol. 38, p. 807, 1947.)*

Figure 3-26 compares the impact curves for three different structures, all at approximately the same strength level. The endurance limit for all three structures also was about the same (around 70,000 psi). However, the structure has a marked effect on the impact behavior. Tempered martensite is clearly the favored structure for toughness. The difference lies in the finer carbide distribution and finer ferrite grain size for the martensite. Fresh martensite, of course, would show a much lower curve than tempered martensite.

3-5 CREEP PROPERTIES

The sensitivity of the strength of a material to the strain rate is accentuated at high temperature. To illustrate this, we will consider some tests on a typical steam-power-plant piping material (such as a steel containing 2.25% Cr, 1.0% Mo, and about 0.15% C). At 540°C (1000°F), this alloy may have a yield strength of about 28,000 psi. However, if it is loaded in tension with a stress of 10,000 psi, the sample gradually elongates, as shown in Fig. 3-27. Thus, the short-time tensile properties are misleading, for the material does deform at a much lower load. Generally elongation-time curves (creep curves) show three stages. In the initial (primary) stage, the rate of elongation is relatively rapid, but decreases with time, entering a rather linear stage (secondary creep). Near the rupture time, the creep rate may again increase. In the curves in Fig. 3-27, the primary stage is quite small, occurring within the first few hours of testing; the third stage is not shown as the samples were still in the secondary stage after 10,000 hours.

The results of the creep tests are reported in terms of two quantities. One is the minimum creep rate, obtained from the constant slope of the secondary region of the creep curve. The other is the time to rupture. Both quantities are given as a function

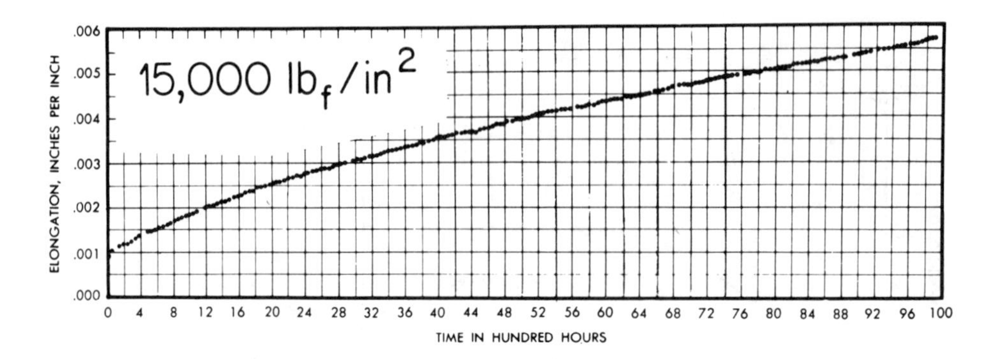

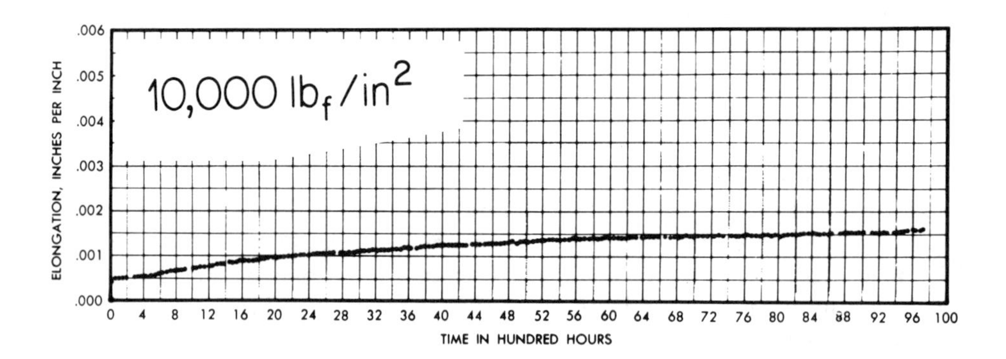

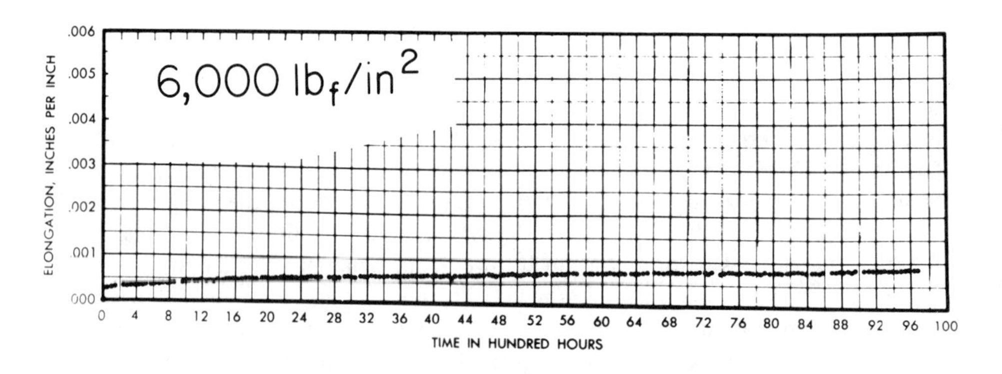

Figure 3-27 Elongation-time curves (creep curves) at 540°C for a high-temperature, medium alloy steel. *(Adapted from "Properties of Carbon and Alloy Seamless Steel Tubing for High-Temperature and High-Pressure Service," Technical Bulletin No. 5-G, The Babcock and Wilcox Co., Beaver Fall, Philadelphia, 1955.)*

of temperature, and they are illustrated in Fig. 3-28. The microstructural changes that occur involve grain growth, second-phase coarsening, and other precipitation reactions. Microstructures for this steel are illustrated in Fig. 3-29, where coarsening is observed.

The structural response of the material involves the same mechanisms that are active in annealing a cold-worked material. Recovery, perhaps recrystallization, and grain growth occur, and the material alters its shape by the movement and interaction of the dislocations, vacancies, and grain boundaries.

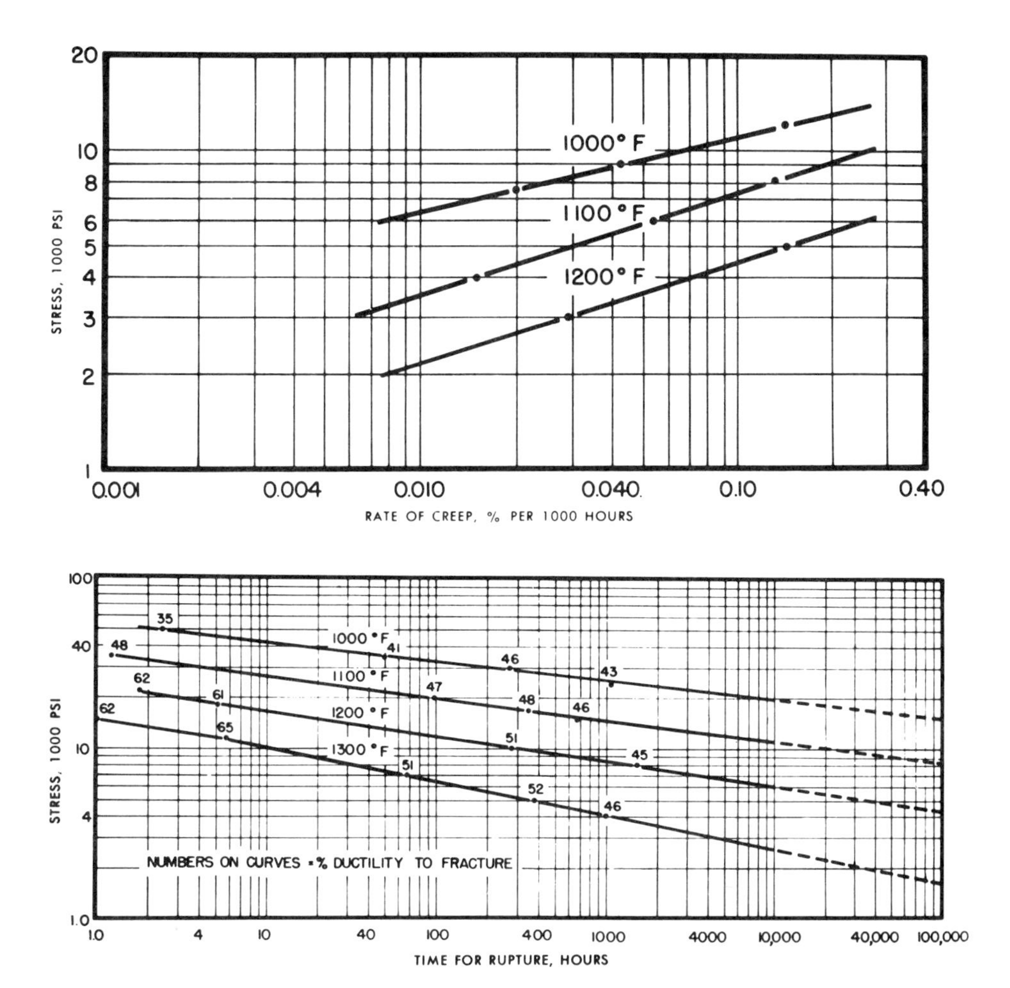

Figure 3-28 Stress-creep rate and stress-rupture curves at a 2.25% Cr, 1.0% Mo steel. *(Adapted from "Digest of Steels for High Temperature Service," The Timken Co., Canton, Ohio, 1957.)*

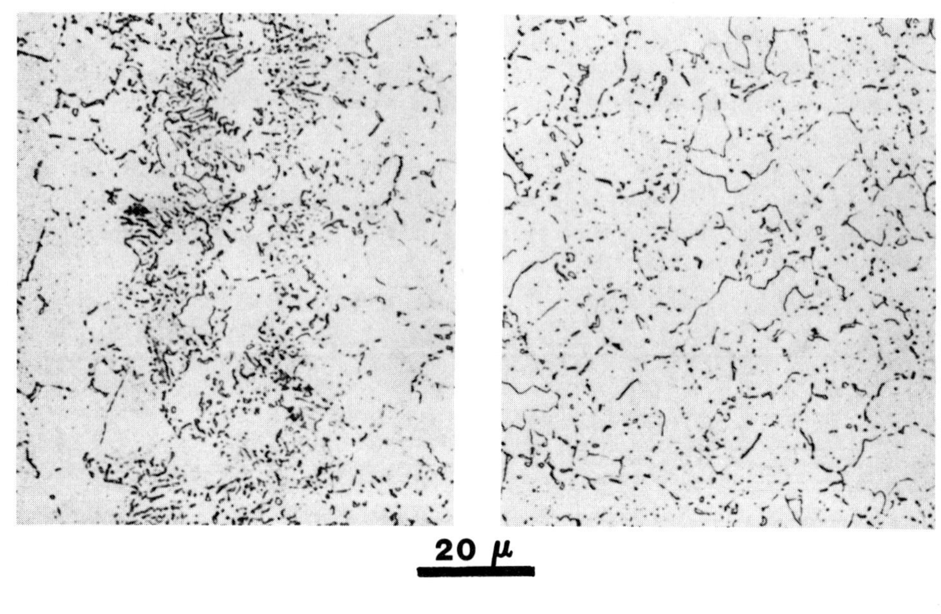

Figure 3-29 Change in microstructure upon creep testing a 2.25% Cr, 1.0% Mo steel. *(Adapted from "Digest of Steels for High Temperature Service," The Timken Co., Canton, Ohio, 1957.)*

CHAPTER

FOUR

SURFACE TREATMENTS

In many applications, it is desirable to alter the properties of the surface (or near surface) of steel parts. For example, to impart resistance to corrosion the surface may be painted, or it may be coated (e.g., zinc galvanizing). To increase the hardness, it is possible to heat just the surface layer, then quench it. Although there are a number of surface treatments having varied applications, in this chapter we will center attention only on induction and flame hardening and on carburizing.

4-1 RESIDUAL STRESSES

In heat treatments, the uneven cooling invariably introduces residual stresses. In some applications, these stresses counteract the external applied stresses, and hence some surface treatments and heat treatments are designed specifically to develop certain residual stresses. In this section, we briefly examine the origin of residual stresses; this information will be called upon in subsequent sections to illustrate the improvement of properties due to the residual stresses.

Quenching Stresses

To illustrate the formation of residual stresses, a description will be presented of the development of the stress distribution during quenching a homogeneous metal which does not undergo any phase transformation. Let the sample be a cylinder at a high temperature T_h, which is to be quenched into a medium at a lower temperature T_l. Figure 4-1 illustrates the longitudinal residual stress across the diameter at various stages in cooling. At the high temperature T_h the temperature is uniform and there are no residual stresses (Fig. 4-1*a*). The cylinder is then placed in the quenchant, which is

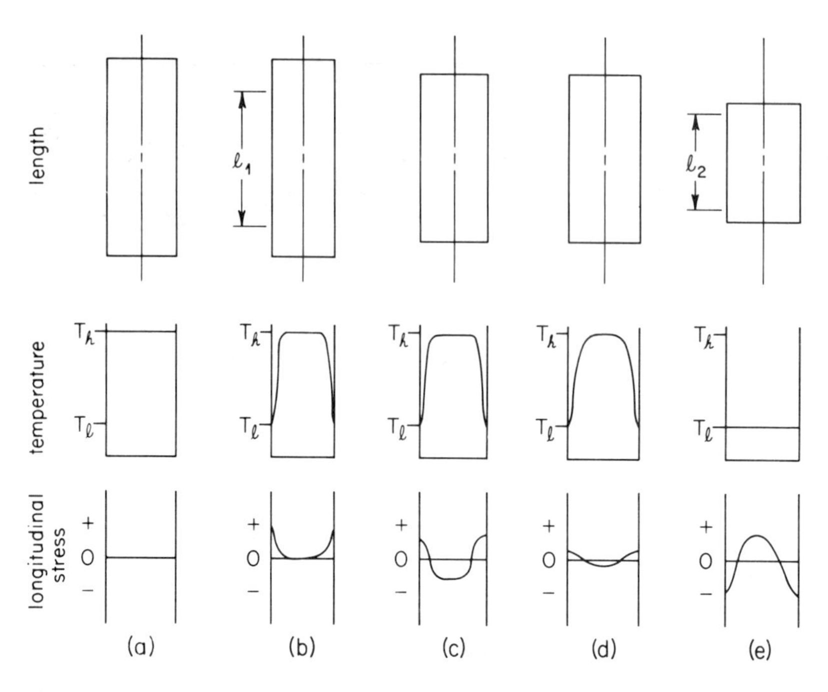

Figure 4-1 Schematic illustration of the development of longitudinal thermal stresses in a cylinder during cooling. At (a), the cylinder is at high temperature, and from (b) to (e) the cylinder is cooling.

at temperature T_l. The surface cools to T_l, but the center is still at T_h. At temperature T_l, the bar should contract to length l_1. If the center of the bar (still at temperature T_h) completely restrains the surface (at T_l) from contracting, then the surface is too long and is in tension (Fig. 4-1*b*). However, the surface will pull on the center and place it in compression. Hence the actual stress distribution at this stage, with the surface at T_l and the center still at T_h, is the surface in tension and the center in compression (Fig. 4-1*c*). If the yield strength of the material is not exceeded at any position along the diameter, then the bar will cool to T_l, with no final residual stresses. That is, the center will cool allowing the surface and the center to contract to the final length l_1 at temperature T_l. However at stage (c), if the stresses are sufficiently high, both the center and the surface begin to deform plastically. The strength of the center is less than the surface because it is at a higher temperature, but here we assume that deformation occurs equally in both sections. The cylinder then undergoes stress relaxation, and the stresses begin to vanish (Fig. 4-1*d*). Now, in this state, the relaxed length is l_1, which is the unstressed length for the surface, for it is at T_l. However, the center, upon cooling to T_l, will contract to length l_2 if unrestrained. Thus, from stage (d) on, the center cools and contracts, placing the surface in compression. The surface prevents the center from attaining the length l_2, and hence it remains in tension. The final stress distribution is the center in tension and the surface in compression (Fig. 4-1*e*).

Figure 4-2 illustrates schematically the meaning of the stress distribution. If one could take three springs, compress one, and then link it to one on each side of it, each of which had been extended, the picture would be as indicated in Fig. 4-2*a*. Here the outer layers are in tension (the spring is too long) and the center in compression (the spring is too short). To measure the residual stresses, the surface can be removed, causing the center to elongate (Fig. 4-2*b*); measurements of the dimensional changes as layers are removed (e.g., by machining) allow calculation of the residual stresses.

In actuality, the stress pattern is three-dimensional, and for cylinders one must consider the tangential and radial stresses in addition to the longitudinal stresses. Some data on a 1045 steel are shown in Fig. 4-3. Even though the structure is two phase (ferrite and carbide), in these treatments only thermal stresses are involved since the steel was quenched from below the transformation temperature. Note in Fig. 4-3*a* that the higher temperature developed the greater residual stresses, and in Fig. 4-3*b* that the faster cooling (water quench) caused larger stresses. The longitudinal stress distribution here is as predicted from the qualitative description above: compression on the surface and tension in the center. Since the residual stresses cannot exceed the yield strength, the magnitude of the stresses at the center and the surface is comparable to the yield strength for such a steel.

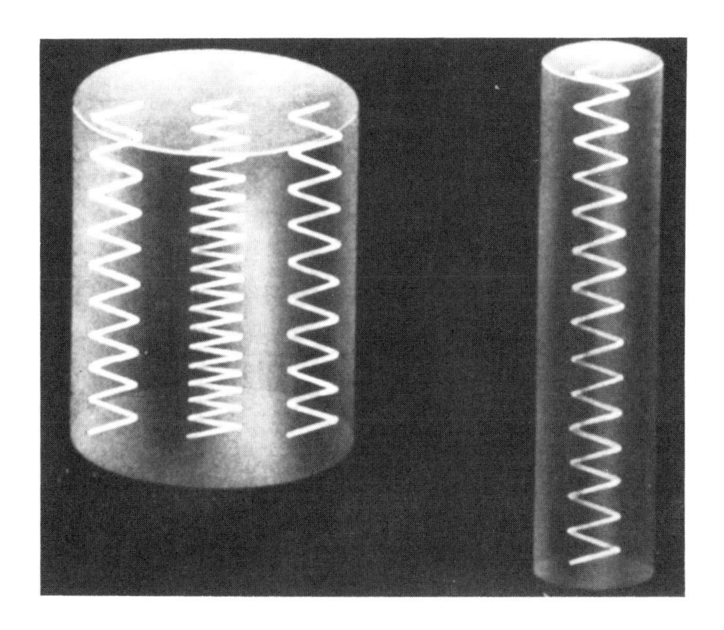

Figure 4-2 Schematic illustration of the residual stress pattern. *(From W. M. Baldwin,* Proc. ASTM, *vol. 49, p. 539, 1949.)*

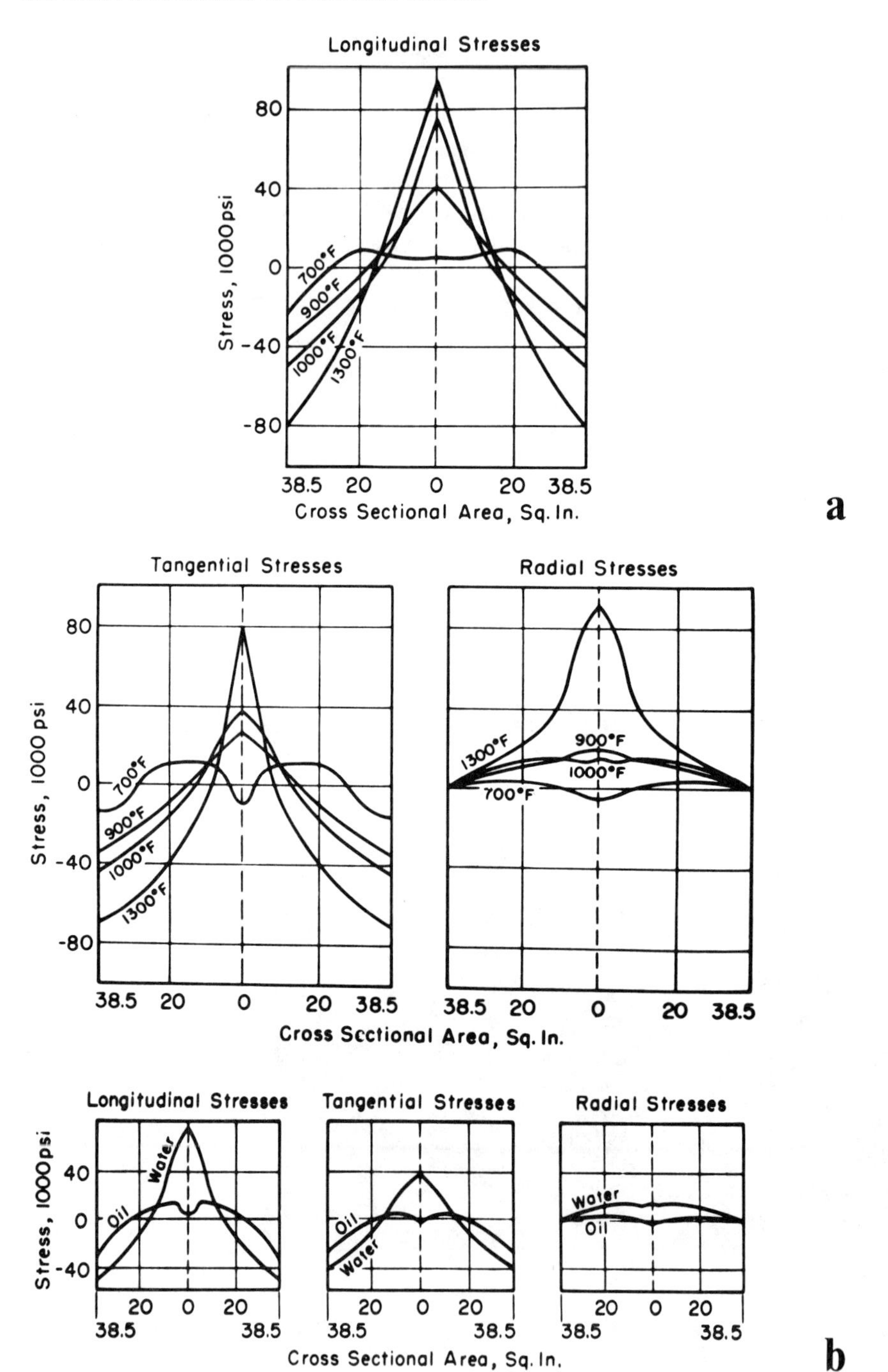

Figure 4-3 Residual thermal stress distribution in cylinders of 1045 steel for various treatments. *(From H. B. Wishart in "Residual Stress Measurements," American Society for Metals, Metals Park, Ohio, 1952.)*

Transformation Stresses

If a phase transformation occurs during the cooling process, the residual stress pattern becomes more complicated, and less predictable. The stresses are influenced by the volume change occurring, and the temperature range in which this occurs. In steels, upon rapid quenching, the surface may form martensite (with an expansion) while the remainder of the part is still austenite. However, eventually the remainder of the austenite will transform to a structure that depends upon the hardenability and the cooling rate. The expansion occurring in this region depends upon the specific structure; the formation of bainite and pearlite will occur with less volume expansion than the formation of martensite. Figure 4-4 illustrates typical stress distributions. Note that the surface longitudinal stress is compressive in the cases shown. Tempering usually reduces the residual stresses as illustrated in Fig. 4-5.

4-2 INDUCTION AND FLAME HARDENING

The development of favorable compressive residual stresses on the surface can be accomplished in some steels by selectively heating only the surface layer to the austenite region, following by rapid cooling (e.g., water spraying the surface). There are two general methods of achieving this type of thermal treatment. The surface can be heated by a flame, generated by the reaction of gases with oxygen so that the atmosphere minimizes decarburization and oxidation. This technique can be used on complex

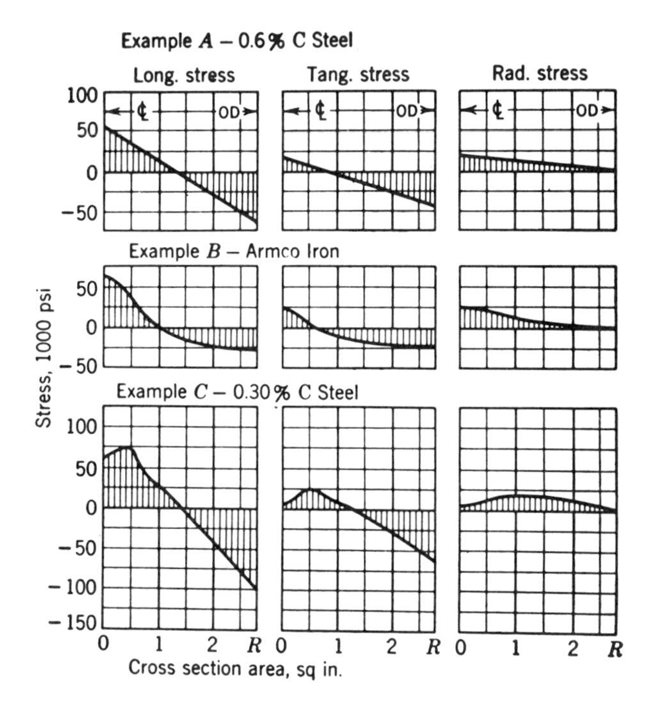

Figure 4-4 Residual stress distribution in cylinders quenched in water from 850°C. *(From H. Buhler, H. Buchholtz and E. H. Schultz,* Arch. Eisen., *vol. 5, p. 412, 1932.)*

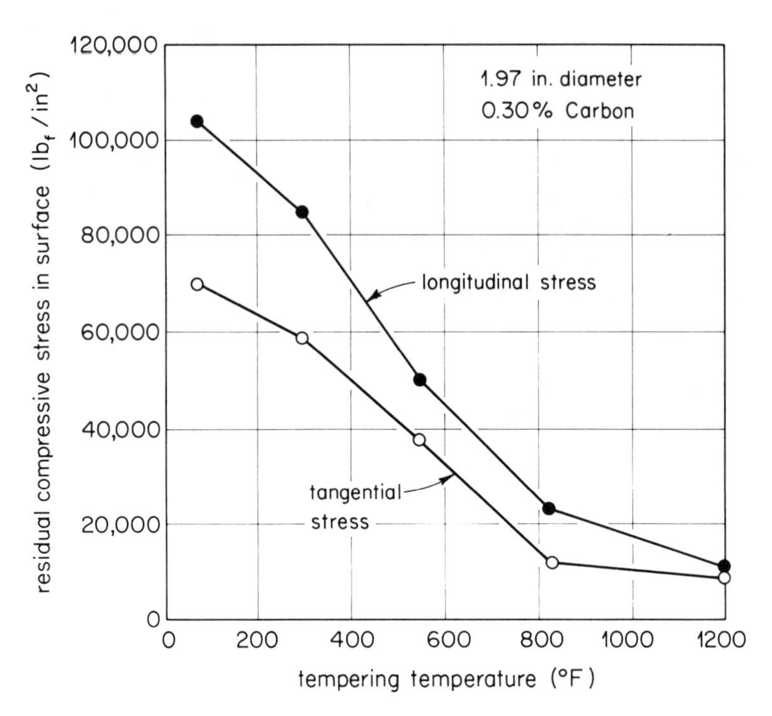

Figure 4-5 The effect of tempering on the longitudinal surface residual stress of cylinders quenched into water from 850°C. *(From O. J. Horger, in M. Hetenyi (ed.), "Handbook of Experimental Stress Analysis," John Wiley, New York, 1950.)*

geometries (e.g., gear teeth) or for localized surface hardening. The other method utilizes high-frequency electrical heating, and it is especially applicable to objects of a general cylindrical shape (e.g., shafts). A high-frequency magnetic field of the correct frequency range will induce current mainly in the surface layers of conducting parts; thus, only the surface is heated. Both of these methods of surface hardening are quite amenable to automation, and both are widely used.

In reversed-bending-fatigue loading, the fatigue cracks propagate in the surface regions when the stress is tensile. Thus, a compressive residual stress would counteract the external tensile loading, and improve fatigue behavior. If the steel is chosen properly, surface hardening will give residual compressive stresses in the surface layers. Figure 4-6 illustrates the longitudinal residual stress in cylinders that were induction hardened. Figure 4-7 shows hardness profiles for induction-hardened steels. Note that the surface hardness shows that the surface is almost completely martensitic (Fig. 1-31). The influence of the surface hardening on the fatigue behavior is illustrated in Fig. 4-8. Here the greater the case depth, the higher the fatigue strength. However, this may not always be true, especially for thick sections. Also, the exact behavior depends upon the external loading. For example, if the loading is cyclic but only tensile, then for residual stress distributions such as shown in Fig. 4-4 the center of the cylinder is the critical location.

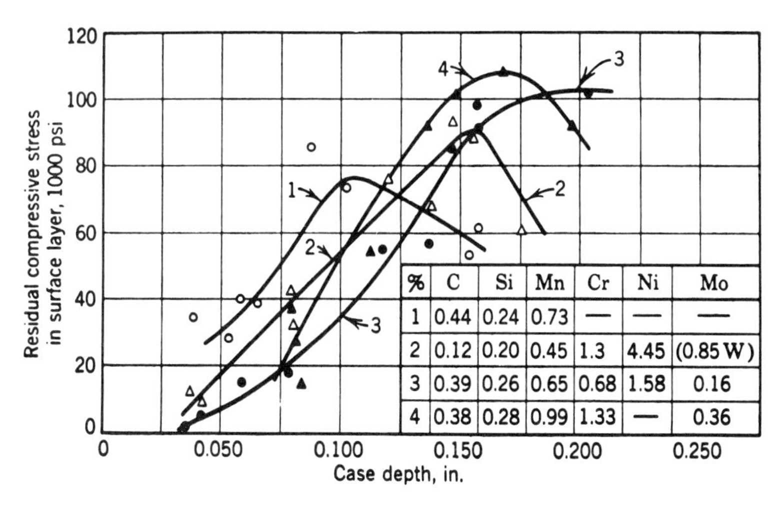

%	C	Si	Mn	Cr	Ni	Mo
1	0.44	0.24	0.73	—	—	—
2	0.12	0.20	0.45	1.3	4.45	(0.85 W)
3	0.39	0.26	0.65	0.68	1.58	0.16
4	0.38	0.28	0.99	1.33	—	0.36

Figure 4-6 The longitudinal residual stress as a function of depth in cylinders of four different steels, induction hardened. *(From I. E. Kontorocih and L. C. Livshitz,* Metallurgy, *no. 8, p. 30, 1940 as reported by O. J. Horger in "Handbook of Experimental Stress Analysis," M. Hetenyi (ed), John Wiley, New York, 1950.)*

4-3 GAS CARBURIZING

The addition of carbon preferentially to the surface layers of steels, followed by proper heat treatment, develops residual stress patterns which improve fatigue strength. In addition, the surface is hard and abrasion- and wear-resistant. There are several

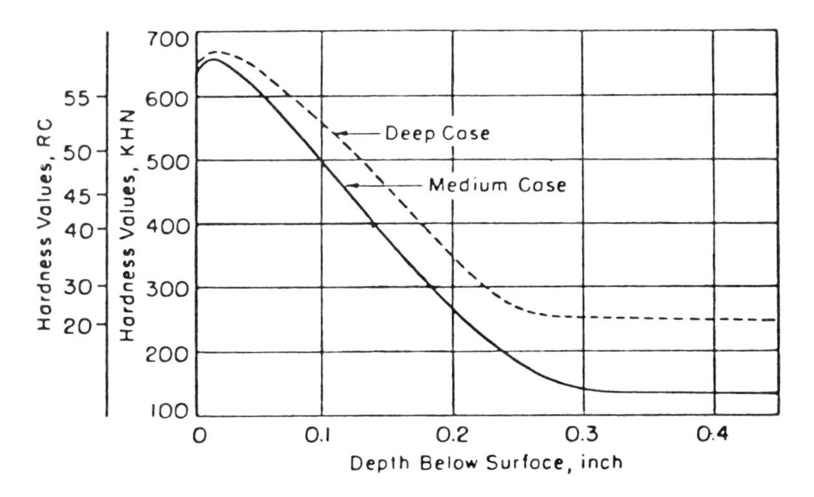

Figure 4-7 Hardness as a function of depth for an induction hardened 1045 steel. *(Adapted from S. L. Case, J. M. Berry, and H. J. Grover,* Trans. ASM, *vol. 44, p. 667, 1952.)*

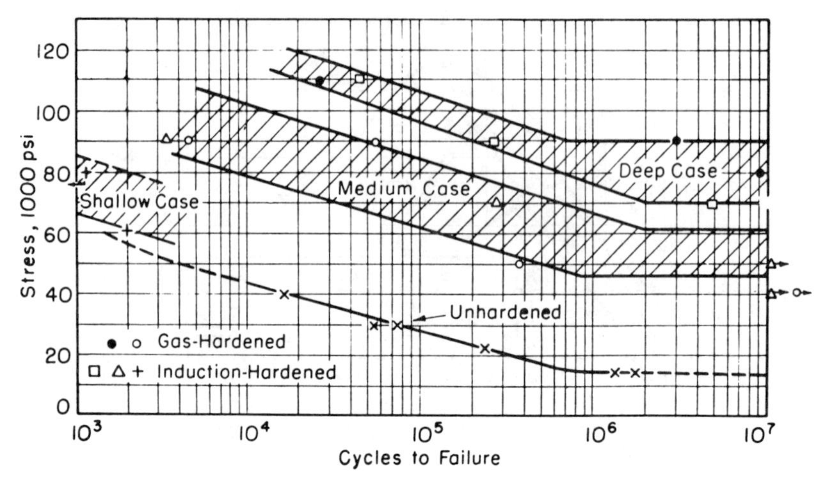

Figure 4-8 Stress-cycle curves for unhardened and surface-hardened steels of different case depth, tested in reversed bending. *(Adapted from S. L. Case, J. M. Berry, and H. J. Grover,* Trans. ASM, *vol. 44, p. 667, 1952.)*

methods for the selective addition of carbon, but in this chapter only gas carburizing will be treated.

Gas Atmospheres

When a gas mixture of CO_2 and CO is placed in contact with iron at high temperature (e.g., in the austenite range), the following reaction may occur:

$$2CO \rightarrow CO_2 + C$$

where C represents the placement of a carbon atom in the iron lattice in the surface layer. At equilibrium, there is a definite relationship (developed by chemical thermodynamics) between the concentration of the two gas species and the carbon content of the iron. It is

$$K_p = \frac{p_{CO_2}}{p_{CO}^2} a_C$$

where K_p is the equilibrium constant, p is the partial pressure of CO or of CO_2, and a_C is the "activity" of the carbon in the austenite. The partial pressure is related to the chemical composition of the gas and the total pressure by $p = XP$, where X is the volume (or mole) fraction of a given species in the gas. At a given temperature, K_p has a fixed value. For example, at 927°C (1700°F), $K_p = 0.0216$. In this case, the activity of the carbon in the austenite is related to the carbon content of the austenite by the relation $a_C = (\text{wt. \% } C)/(\text{wt. \% C})*$, where (wt. % C)* is the solubility of carbon in austenite in equilibrium with cementite at 927°C; this value is 1.25. Thus, we get

$$0.0216 = \frac{X_{CO_2}}{X_{CO}^2 P} \, \frac{\text{wt. \% } C}{1.25}$$

If the total pressure is 1 atm, the relationship between the gas composition for equilibrium with austenite containing 0.2% C is then

$$X_{CO_2} = 0.135X_{CO}^2$$

Additionally the sum of the partial pressures is equal to the total pressure, so $p_{CO} + p_{CO_2} + p_{inert} = 1$, where p_{inert} is the partial pressure of inert gases present (e.g., nitrogen). Thus, if only CO and CO_2 are present, the solution gives $X_{CO} = 0.893$ and $X_{CO_2} = 0.107$; if there is 80% inert gas present, then $X_{CO} + X_{CO_2} = 0.2$, and the solution yields $X_{CO} = 0.195$ and $X_{CO_2} = 0.0052$. The relationship between the CO and CO_2 composition at 1 atm total pressure is shown in Fig. 4-9 for several temperatures and carbon contents.

For a given total pressure and temperature, and the amount of inert species present in the gas, there is only one value of the carbon content and the CO and CO_2 contents that will represent equilibrium. If this condition is not met, the reaction will occur in one direction or the other until equilibrium is attained. Thus, if a gas

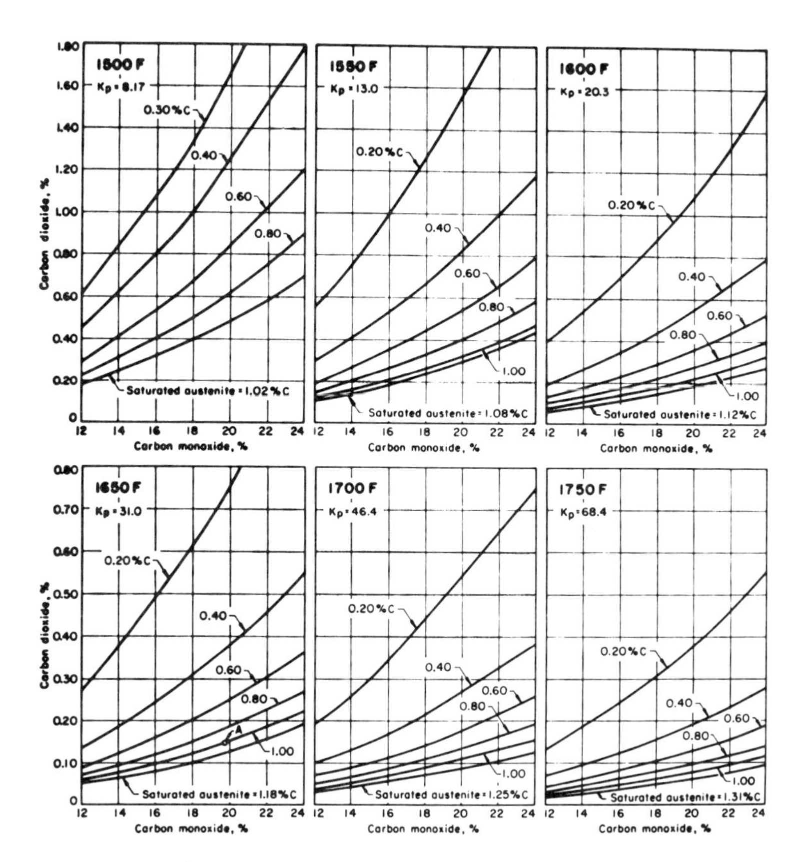

Figure 4-9 Percentages (volume %) of CO and CO_2 in equilibrium with various carbon contents of austenite at several temperatures. The total pressure is 1 atm. *(From "Metals Handbook," vol. 2, American Society for Metals, Metals Park, Ohio, 1964.)*

containing 19.48% CO (X_{CO} = 0.1948), 0.52% CO_2, and 80% N_2 is passed over a sample of pure iron at 927°C, the sample will carburize until it is homogeneously 0.2% C. If the same gas is passed over a steel sample with a carbon content higher than 0.2%, the steel will lose carbon (decarburize) until it becomes homogeneously 0.2%.

Figure 4-10 presents that data in another form. Here the temperature dependence of the equilibrium CO and CO_2 contents for various carbon contents of the austenite are shown.

Mixtures of methane and hydrogen also can be used for carburization by the reaction

$$CH_4 = 2H_2 + C$$

and the equilibrium data are shown in Fig. 4-11.

In addition to the carburizing reaction with CO and CO_2, oxidation of the iron may occur by the reaction

$$Fe + CO_2 = FeO + CO$$

The problem of whether the carburizing gas mixture chosen to carburize will also oxidize the iron must be examined. As an example, suppose it is desired to carburize a steel at 927°C so that the surface has 0.8% C. The total pressure is 1 atm, and the gas consists only of CO and CO_2. From Fig. 4-10, $p_{CO}^2/p_{CO_2} = 27$, and using $p_{CO} + p_{CO_2} = 1$, the gas composition is 95.66% CO and 4.34% CO_2. The ratio of $CO_2/CO = 0.045$ and the data in Fig. 4-12 show that this atmosphere will be reducing.

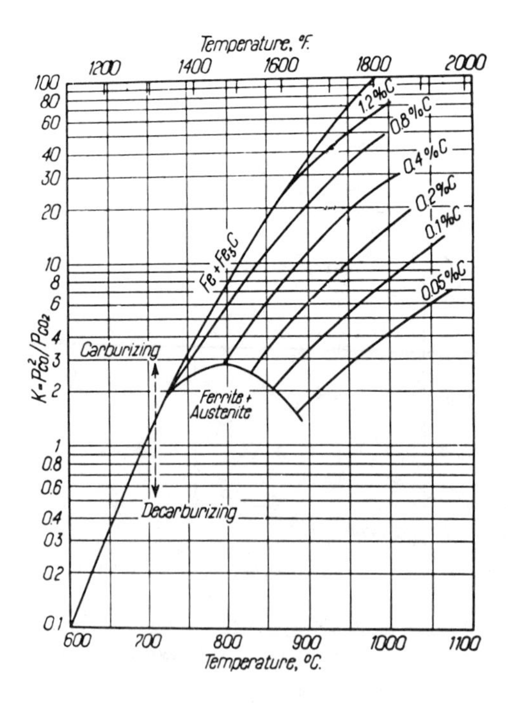

Figure 4-10 Relationships for the CO and CO_2 gas composition in equilibrium with different carbon-content irons. The total pressure is 1 atm. *(From J. B. Austin and M. J. Dary, in "Controlled Atmospheres," American Society for Metals, Metals Park, Ohio, 1942.)*

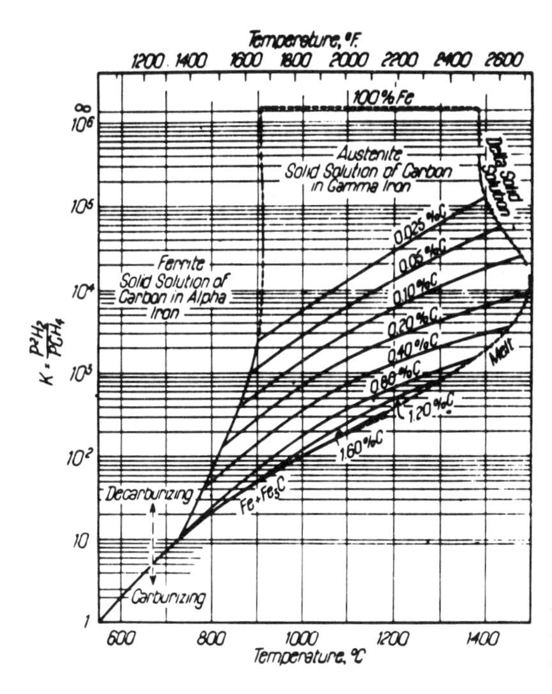

Figure 4-11 Relationships for the CH_4 and H_2 gas composition in equilibrium with different carbon-content irons. The total pressure is 1 atm. *(From J. B. Austin and M. J. Dary, in "Controlled Atmospheres," American Society for Metals, Metals Park, Ohio, 1942.)*

If water vapor is present, oxidation can also occur by the reaction

$$Fe + H_2O = FeO + H_2$$

and the equilibrium curve for this reaction is also shown in Fig. 4-12. The water vapor content usually is obtained by measurement of the "dew point," which is the temperature at which condensation occurs. Then from the solubility curve of water in the

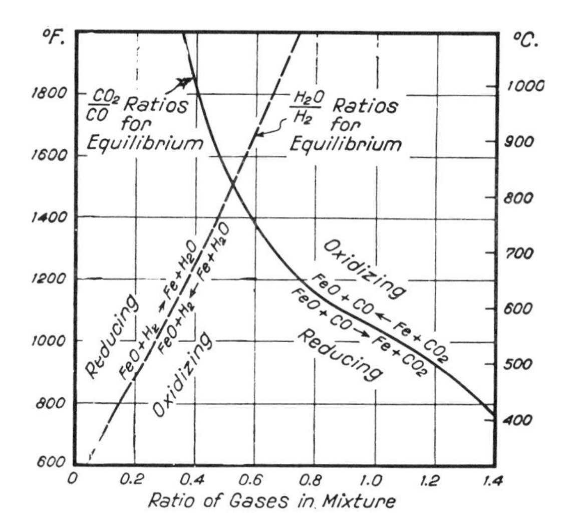

Figure 4-12 The dependence of the ratio of CO_2/CO and H_2O/H_2 on temperature for oxidation-reduction equilibrium of iron. *(Adapted from A. G. Hotchkiss,* Met. Progr., *vol. 31, p. 375, 1937.)*

particular gas as a function of temperature, the water vapor content of the gas is obtained. Because the gases being considered here are almost ideal, the solubility is not very sensitive to the type of gas. Figure 4-13 shows the solubility curve of water in such gases, given here as the water content versus dew point.

In utilizing the various data just presented to control gas atmospheres, the sensitivity of the carbon content of the austenite to variation in gas composition must be considered. An example will illustrate the problem. For a gas containing 40% H_2, 20% CO, and 40% N_2, the relationship between the equilibrium carbon content of the austenite and the amount of CO_2 or H_2O present at a given temperature can be obtained. The result is shown in Fig. 4-14. Now if we wish to control the carbon content to 0.6 ± 0.02%, then the CO_2 content must be kept to 0.16 ± 0.02%, which is below the limit of accuracy for common methods of gas analysis. However, the same equilibrium requires the dew point to be 24 ± 1°F; the dew point can easily be measured to ±1°F, so control of the carbon content is best obtained by monitoring the water vapor content by measuring the dew point.

The connection between the carbon content and the water vapor content of the gas can be seen by considering the reaction

$$CO_2 + H_2 = CO + H_2O$$

which must be in equilibrium along with the reaction

$$2CO = C + CO_2$$

For the former reaction; the equilibrium constant at 926°C is

$$1.43 = \frac{X_{H_2O}\ X_{CO}}{X_{H_2}\ X_{CO_2}}$$

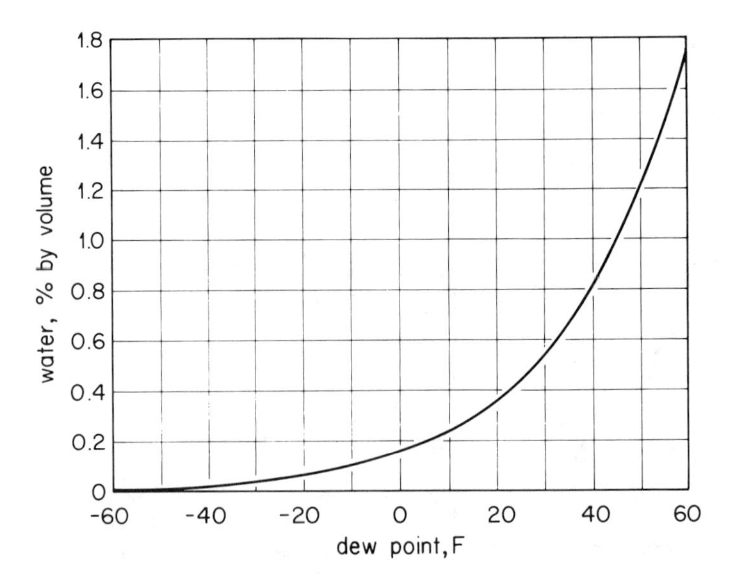

Figure 4-13 The dew point curve for water. The total pressure is 1 atm.

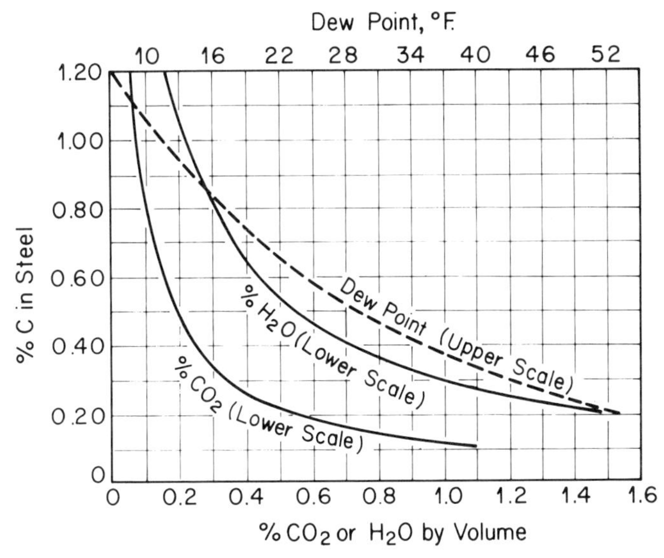

Figure 4-14 Equilibrium carbon content of austenite at 926°C as a function of the CO_2 and H_2O content for a gas containing 20% CO, 40% H_2 and 40% N_2. *(Adapted from O. E. Cullen,* Met. Progr., *vol. 66, p. 114, 1954.)*

Thus a gas containing 40% H_2, 1% H_2O, 20% (CO + CO_2), and the rest inert gas, and at a total pressure of 1 atm, will be in equilibrium with austenite containing 0.3% C. Here the water vapor content is controlled and the CO and CO_2 content will be established by the equilibrium reaction above.

Figure 4-15 shows the influence of the dew point and temperature on the carbon content. Also shown are data for some alloy steels. The alloying elements influence the equilibrium carbon content, so that data valid for plain carbon steel may not be accurate for alloy steels.

The gases used for carburizing and heat treating are usually generated by reaction of air with commercial gases. The reaction is controlled so that the approximate gas desired is obtained; however, the gas usually contains CO, CO_2, H_2, CH_4, and H_2O, in addition to the inert N_2. Figure 4-16 shows data from a specific type of gas generator. Note that the CO + CO_2 content is constant at 20%. The data in Fig. 4-15 were obtained from this gas. In the examples above, the CH_4-H_2 reaction also would have to be examined.

Case Depth

The information in the preceding section showed how the equilibrium carbon content of steel is related to the gas composition. In surface treatments, however, the steel is kept in contact with the gas only for sufficient time to develop a carbon gradient that will impart the desired properties when the steel is heat treated. This gradient depends upon the carburizing time as well as the temperature. Consider a steel part containing

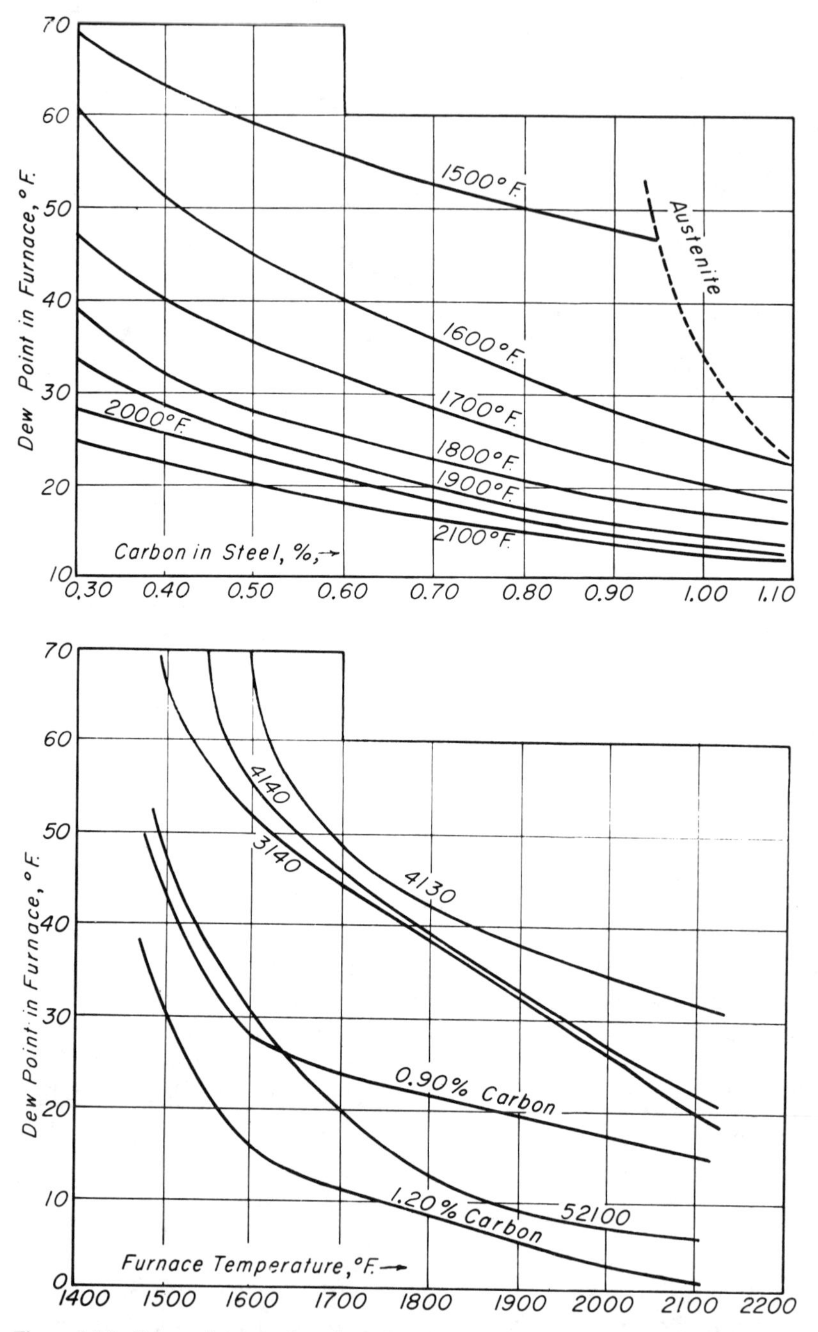

Figure 4-15 Dew point as a function of carbon content for a gas containing 40% H_2 and 20% ($CO + CO_2$). *(Adapted from "Metal Progress Data Sheets," no. 79, American Society for Metals, Metals Park, Ohio, 1954.)*

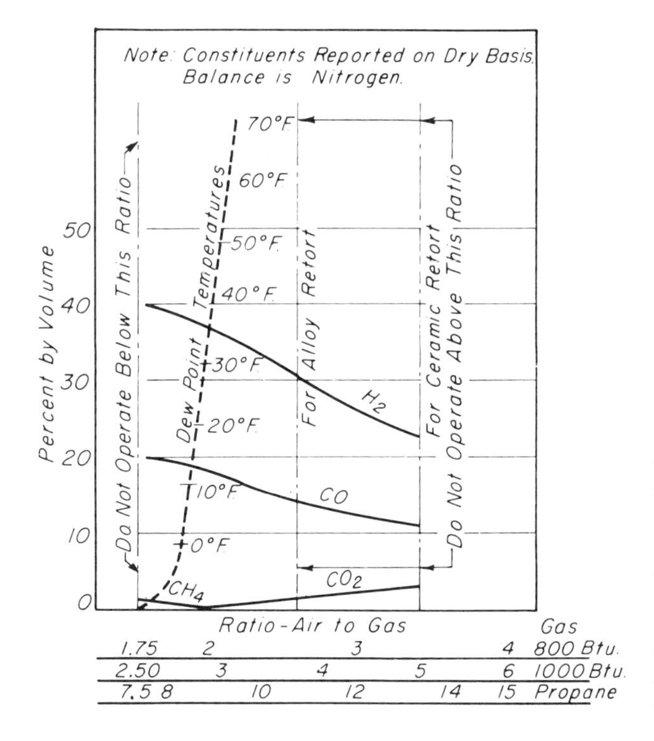

Figure 4-16 Approximate gas analysis of endothermic generator gas when cracking various gas-air mixtures. The data are applicable to natural gas having a heating value of 800 or 1000 Btu/ft^3 or for commercial propane. *(Adapted from "Metal Progress Data Sheets," no. 79, American Society for Metals, Metals Park, Ohio, 1954.)*

0.2% C initially, brought into contact with a flowing stream of gas whose composition will be in equilibrium with a carbon content of 0.8%. As the gas flows across and around the steel, the chemical reactions release carbon at the surface, where it dissolves in austenite. This will continue until the surface concentration is 0.8%. However, the surface carbon content is now higher than the interior and so carbon atoms begin to migrate toward the center, always keeping the surface carbon content slightly below the equilibrium value, so that carbon is continuously released by the gases to the surface. Thus the carbon gradient continues to decrease with time, as shown in Fig. 4-17.

The carbon gradient must be controlled to control the subsequent properties, and the *case depth* is a measure of this. It is usually defined graphically by a straight line such that the area under the line is equal to that under the actual concentration curve. This is illustrated in Fig. 4-18.

The displacement of the carbon curve with carburizing time and temperature is controlled by the rate of diffusion of carbon in the austenite. This carbon movement is governed by Fick's diffusion law, which gives the rate of carbon movement from the surface toward the center by the expression

$$J = -D \frac{\partial c}{\partial x}$$

where J is the flux of carbon atoms (the number of carbon atoms crossing an interface

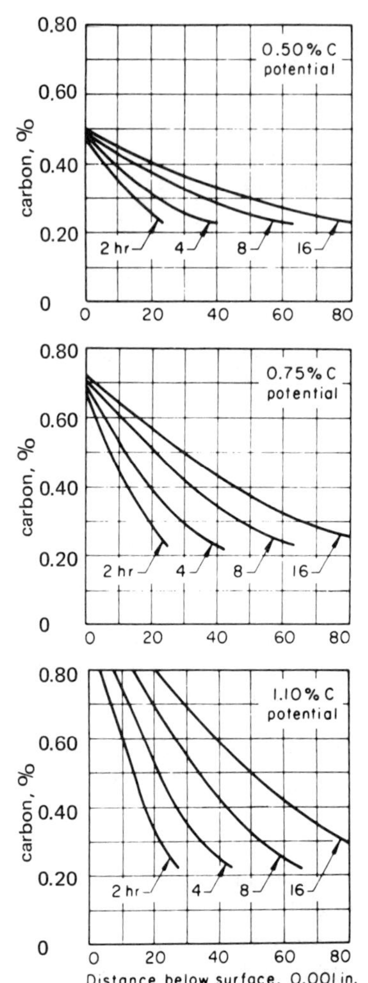

Figure 4-17 Carbon gradients on 1022 carburized at 920°C in a gas containing 20% CO and 40% H_2, with enough H_2O to be in equilibrium with the carbon content shown. *(From "Metals Handbook," vol. 2, American Society for Metals, Metals Park, Ohio, 1964.)*

per time per cross-sectional area), $\partial c/\partial x$ is the carbon gradient (c is the carbon concentration, x is the distance from the surface), and D is the diffusion coefficient. The diffusion coefficient D is quite temperature dependent, given by

$$D = D_0 e^{(-Q/RT)}$$

where D_0 is a constant, Q is the activation energy (a constant) for the diffusion of carbon in austenite, R is the ideal gas constant, and T is the absolute temperature. This is the typical form of the temperature dependence of structural activity in the solid state, and the exponential term makes the diffusion rate quite temperature sensitive. The constants are dependent upon the alloy and carbon content of the austenite.

Application of the diffusion equation to the definition of case depth leads to the relation

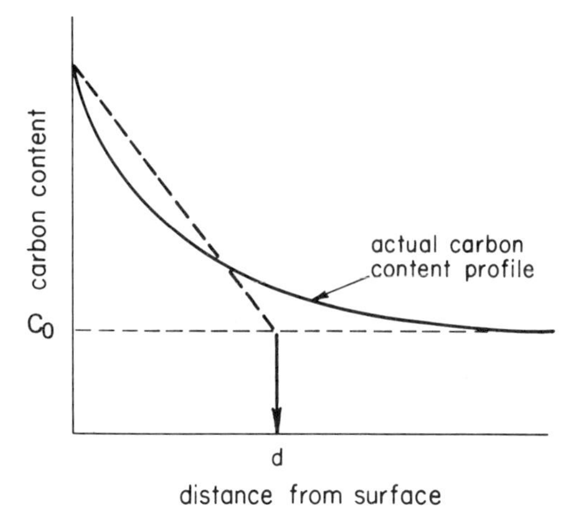

Figure 4-18 Illustration of the definition of case depth. The straight line is drawn such that the area between this line and the base carbon content (C_0) is the same as that between the actual carbon content profile and the base carbon content. Then the intercept shown defines the case depth *d*.

$$d = 80.3\sqrt{t}\, e^{-8570/T}$$

for the dependence of case depth on carburizing time and temperature. Here d is the case depth in centimeters, t is the time in hours, and T is the temperature in Kelvin.

Heat Treatments

The improvement of the fatigue properties due to using carburized and hardened steel is illustrated in Fig. 4-19. The favorable residual stress distribution (Fig. 4-20) is the

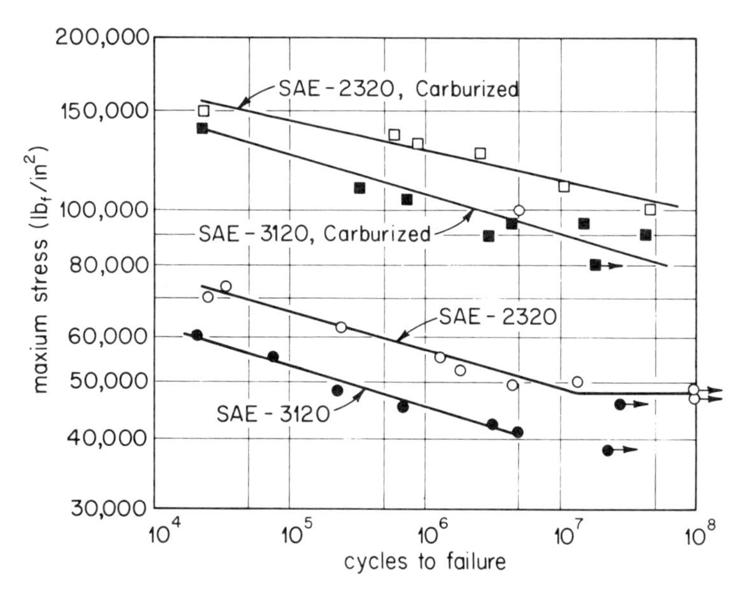

Figure 4-19 The effect of carburizing on the fatigue properties of two steels. *(Adapted from M. F. Moore and N. J. Alleman,* Trans. ASST, *vol. 13, p. 405, 1928.)*

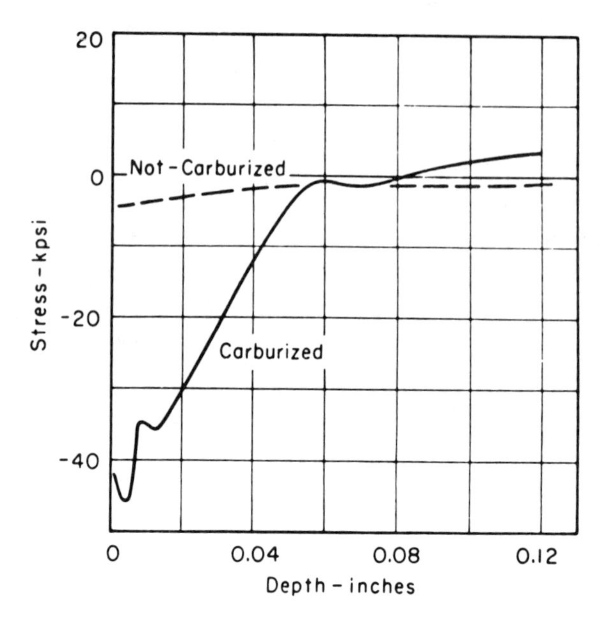

Figure 4-20 Comparison of the longitudinal residual stress distribution in 1.9-cm-diameter 8617 steel cylinders, carburized and uncarburized. The steel was carburized at 920°C to a case depth of about 0.121 cm, then oil quenched. *(From W. S. Coleman and M. Simpson, in "Fatigue Durability of Carburized Steel," American Society for Metals, Metals Park, Ohio, 1957.)*

main contributing factor to the improvement, but also the hard surface is more resistant to surface damage. The relationship between the quenched hardness and the carbon gradient is illustrated in Fig. 4-21. Since the carbon content varies with depth, the hardenability varies with depth; also the cooling rate is dependent upon depth. Thus the hardness curve depends upon both of these. For most carburizing steels, a hard core is not desired, so the alloy content is such that the surface layers form martensite and the core has a softer and less-brittle structure.

The data in Fig. 4-21 show that the hardness is a maximum not at the surface, but below it. This is due to the high carbon content on the surface lowering the M_f below 20°C, and quenching results in retained austenite. A typical microstructure is shown in Fig. 4-22. The hardness can be increased by subzero cooling to convert the retained austenite to martensite (Fig. 4-21).

The carburizing process may develop large austenite grains, and, although this gives high hardenability, the required impact properties may call for a smaller austenite grain size. This can be accomplished by cooling the steel from the carburizing temperature in air, then reheating to a lower austenitizing temperature (e.g., 850°C) for about 1 hour, then cooling in the desired quenchant.

4-4 NITRIDING

Although the development of hard surfaces and favorable residual stresses makes carburizing attractive, the control of the heat treatment and the dimensional changes induced are problems of concern. The addition of nitrogen to the surface of steels

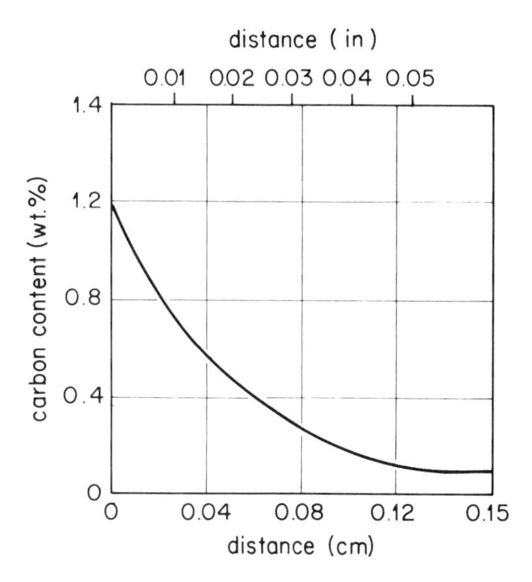

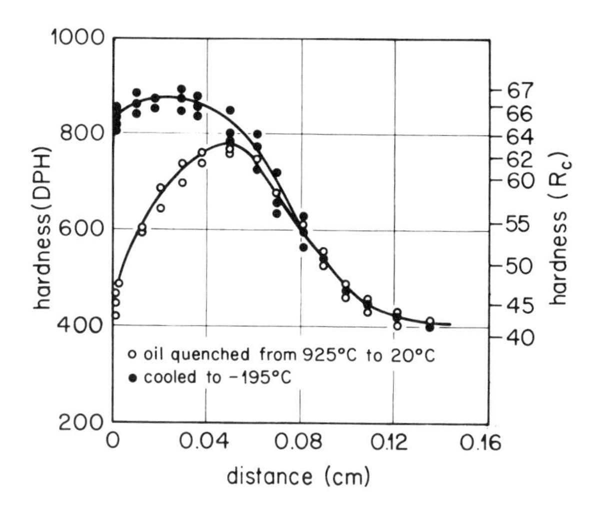

Figure 4-21 The effect of subzero cooling on the hardness gradient in a carburized and quenched 3312 steel. The initial quench to 20°C did not convert all of the austenite to martensite, since the high carbon content lowered the M_f below 20°C. Subsequent cooling to -195°C converted most of the retained austenite to martensite, raising the hardness. *(Adapted from H. Scott and J. L. Fisher, in "Controlled Atmospheres," American Society for Metals, Metals Park, Ohio, 1942.)*

produces a hard, wear-resistant surface, along with improved fatigue properties and better dimensional control. Although nitriding is not as widely used as carburizing because it is a more expensive process, it is used in certain applications (e.g., gear teeth).

Unlike carburizing, nitriding is carried out in the temperature range below austenite stability. The hardness developed depends upon the formation of hard nitrides, and not on the formation of martensite. The process is carried out in mixtures of ammonia gas (NH_3) and hydrogen, such that atomic nitrogen is released at the metal surface. These steels have specific alloying additions (e.g., aluminum) so that the nitrogen reacts with these elements to form a very fine nitride precipitate distribution. Thus the surface layer consists of a gradient of these alloy nitride precipitates.

Alloying elements are added to give the correct nitride morphology; aluminum is a common element added (from 0.85 to 1.5%). Since nitriding is carried out below the austenite range, the steel is quenched and tempered to the final desired core properties prior to nitriding. Nitriding is then performed in the range 495 to 565°C, and the times are rather long. Typical data are shown in Fig. 4-23. The improvement in the fatigue properties is illustrated in Fig. 4-24. Note also that the presence of a notch before nitriding does not alter the fatigue strength.

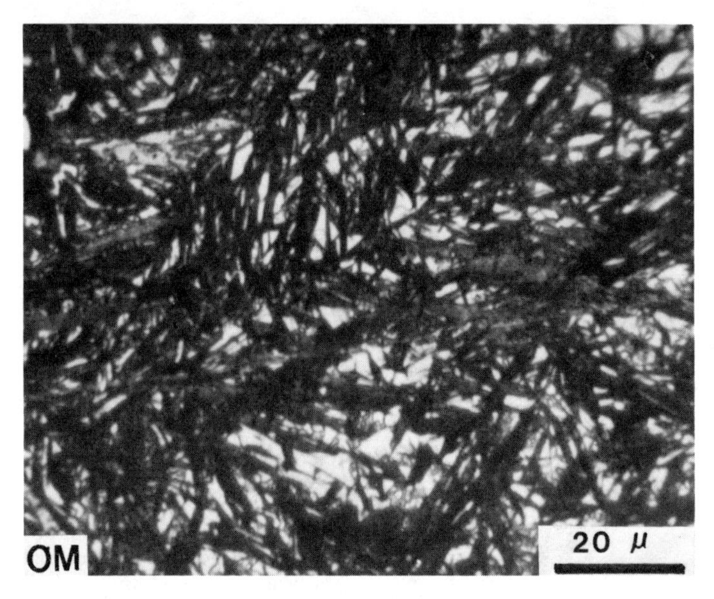

Figure 4-22 The microstructure of the surface of a carburized steel which has been water quenched to 20°C. The high carbon content has lowered the M_f below 20°C, so that retained austenite is present. The dark needles are martensite, and the white area is retained austenite.

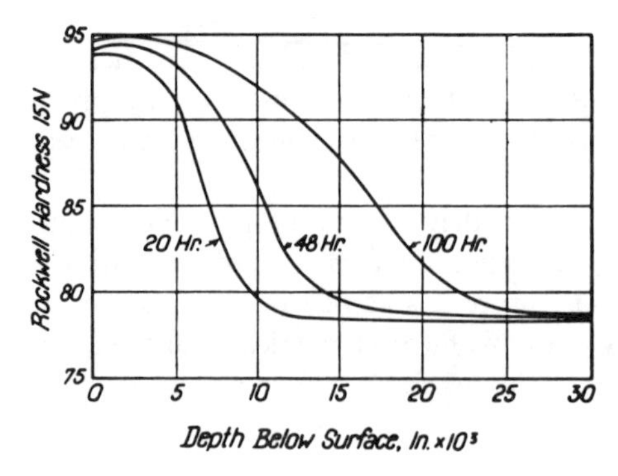

Figure 4-23 Hardness-depth curves for three nitriding times. The steel was quenched and then tempered at 680°C prior to nitriding at 525°C for the times shown. The steel contained 0.41% C, 0.055% Mn, 0.31% Si, 1.60% Cr. 1.00% Al, and 0.35% W. *(From C. F. Floe,* Met. Progr., *vol. 50, p. 1212, 1946.)*

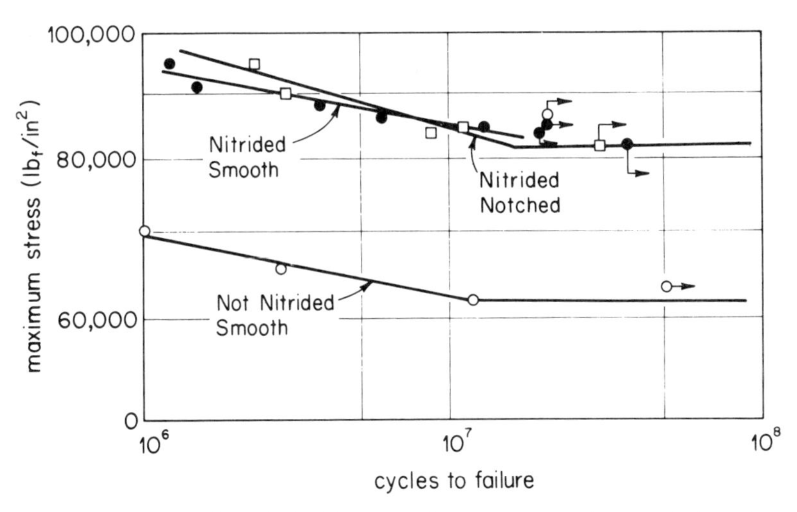

Figure 4-24 The effect of nitriding on the fatigue strength of a steel. *(Adapted from J.B. Johnson and T. T. Oberg,* Met. Alloys, *vol. 5, p. 129, June, 1934.)*

CHAPTER

FIVE

DESIGN OF HEAT TREATMENTS: SOME EXAMPLES

The choice of a steel and its heat treatment to meet required properties in a given application may not be obvious at all. In principle, the choice involves optimization of all of the variables, consistent with restrictions, to obtain the desired result most economically. A number of competing factors will be involved. For example, consideration may be given to comparing the use of existing plant equipment in a heat-treating operation to purchasing newer and more versatile equipment. The cost and availability of steei in a convenient form may dictate the use of a steel that is more expensive. The choice of heat-treating equipment may be dictated by energy conservation and environment quality considerations instead of the ease of heat treatment itself. The list of complications could be continued at some length. The point is that no simple method can be outlined which will serve as a rigorous guide to the choice of the steel and the heat treatment.

Instead of attempting to present the complete details of heat-treatment design, in this chapter examples are given within the framework of the material in the preceding chapters, to illustrate specific aspects.

5-1 HARDENABILITY CALCULATIONS

It is anticipated to manufacture a 2-in.-diameter shaft, and a 8740 steel in stock having composition 0.37% C, 0.70% Mn, 0.20% Si, 0.35% Cr, and 0.20% Mo is being considered for use. As a preliminary step in judging the suitability of this steel, and in choosing a heat treatment, the hardness distribution across the diameter of the shaft is to be calculated when it is quenched into "still" oil and "still" water.

The first step is to calculate the Jominy curve for this steel. Here a grain size will be used which will be assumed to be similar to that which would be obtained during heat treatment. The effect of error in this will be examined subsequently. From the assigned austenite grain size and the chemical composition, the ideal critical diameter can be calculated. This value can then be used to calculate the Jominy curve. The necessary data and calculations are summarized in Table 5-1. The figures from which the necessary data were obtained are indicated. The Jominy curve is plotted in Fig. 5-1. From the Jominy curve and the curves in Fig. 2-8, the hardness traverse across the diameter can be determined (Table 5-2). The results are shown in Fig. 5-2.

The increased depth of hardening due to the more severe water quench is clearly revealed by the curves in Fig. 5-2. Based only on surface hardness, a required surface value of 300 DPH or greater will not allow the use of the oil quench for this steel. In examining the suitability of this steel, the uncertainty in the calculation, and hence in the curves in Fig. 5-2, must be examined. If the grain size is taken to be ASTM 2, then $D_i = 2.6$; if it is ASTM 9, then $D_i = 1.8$. However, using the smaller austenite grain size (ASTM 9) only lowers the calculated surface hardness when quenched in water to 577 DPH, as compared to 580, which is essentially the same.

The results are more sensitive to errors in chemical composition. For example,

Table 5-1 Calculation of critical diameter and of Jominy curve

Chemical composition, wt. %	Multiplying factor (Fig. 2-9)
0.37 C	0.67 (ASTM grain size 5)
0.70 Mn	1.4
0.20 Si	1.1
0.35 Cr	1.7
0.20 Mo	1.2

$D_i = 0.67 \times 1.4 \times 1.1 \times 1.7 \times 1.2 = 2.1$ in.

Distance on Jominy bar, $\frac{1}{16}$ in.	*IH/DH* (Fig. 2-10)	Hardness, DPH
1	1.00	653 (from Fig. 2-11)
4	1.22	477
8	1.70	336
12	2.15	279
16	2.50	254
20	2.68	245
24	2.77	243
28	2.96	236
32	3.03	234

if the molybdenum content were increased to 0.30% and the chromium to 0.45%, then $D_i = 3.4$ in., and the surface hardness when quenched in oil would be calculated to be 363 DPH, a considerable increase over that shown in Fig. 5-2.

For a given hardness and geometry, hardenability can be examined to eliminate steels that just cannot attain the required hardness even for rapid quenches. For example, if the hardness at the center of a 3-in.-diameter cylinder must be 45 R_c (446 DPH), then for an agitated water quench ($H = 1.5$) the equivalent cooling rate on the Jominy bar occurs at $\frac{11}{16}$ in. (Fig. 2-8). We now could go through calculations similar to those in the above example for all possible steels, taking into account variation in chemical composition, to see which ones would not meet the hardness requirement. However, it is easier to examine the minimum Jominy curve on the hardenability band for the H steels; that is, we will base the consideration on the minimum guaranteed hardenability of the H steel. These curves are given for many steels, four of which are shown in Fig. 5-3. From these curves we find that only 4340H and 9840H meet the required conditions. If an oil quench of $H = 0.35$ is used, the equivalent cooling rate occurs at $\frac{19}{16}$ in. on the Jominy curve, and we find that only the 4340H steel is suitable for this quench.

5-2 CHOOSING A QUENCHANT AND A TEMPERING TREATMENT

We now consider designing a heat treatment to meet hardness requirements on both the surface and an interior position of a cylinder. A 2-in.-diameter shaft is to be made from a 3140 steel having the composition 0.39% C, 0.88% Mn, 0.20% Si, 1.20% Ni, and 0.70% Cr. The surface hardness must be 53 R_c and the center approximately

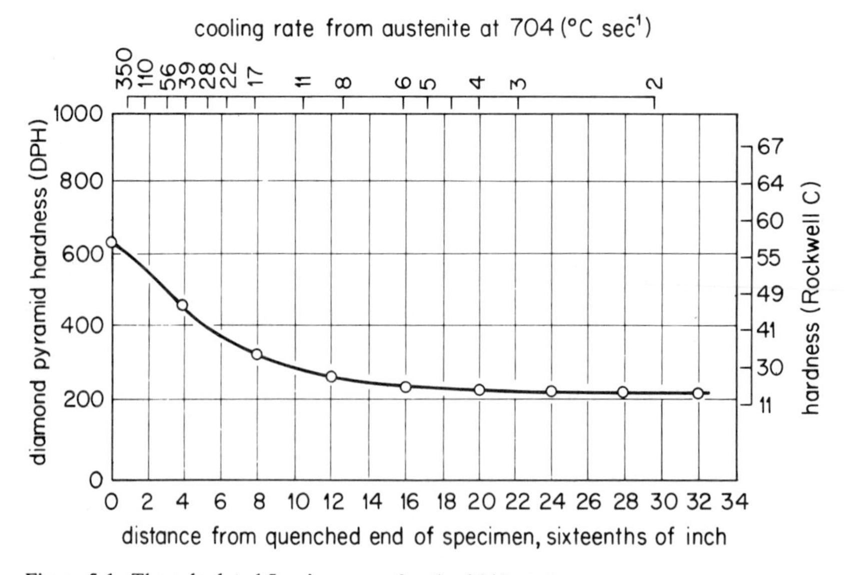

Figure 5-1 The calculated Jominy curve for the 8740 steel.

Table 5-2 Calculation of hardness distribution across the diameter of 2-in.-diameter cylinder when quenched into oil ($H = 0.20$) and water ($H = 1.0$)

$H = 0.20$ (still oil quench)

r/R (Fig. 2-8)	r, in.	Equivalent distance on Jominy curve, $\frac{1}{16}$ in.	Hardness (from Fig. 5-1)
0	0	16.0	250
0.1	0.1	16.0	250
0.2	0.2	16.0	250
0.3	0.3	16.0	250
0.4	0.4	15.5	255
0.5	0.5	15.0	260
0.6	0.6	14.5	265
0.7	0.7	14.5	265
0.8	0.8	14.0	270
0.9	0.9	14.0	270
1.0	1.0	13.0	275

$H = 1.0$ (still water quench)

r/R (Fig. 2-8)	r, in.	Equivalent distance on Jominy curve, $\frac{1}{16}$ in.	Hardness (from Fig. 5-1)
0	0	7.5	340
0.1	0.1	7.0	350
0.2	0.2	7.0	350
0.3	0.3	7.0	350
0.4	0.4	6.5	360
0.5	0.5	6.5	360
0.6	0.6	6.0	380
0.7	0.7	6.0	380
0.8	0.8	4.5	440
0.9	0.9	3.5	480
1.0	1.0	2.0	580

40 R_c. The shaft is to have a slight stress-relief temper. The geometry of the shaft changes somewhat toward each end.

Using a grain size of ASTM 7.5 and the given chemical composition, an ideal critical diameter of 3.7 in. is calculated. From this, the Jominy curve in Fig. 5-4 was constructed. Since the steel part is to have a slight stress-relief temper, the as-quenched hardness should be somewhat higher than the final required hardness. We will try for a hardness of 55 R_c (598 DPH) at the surface and 42 R_c (413 DPH) at the center. This requires a cooling rate at the surface equivalent to $\frac{3}{16}$ in. on the Jominy curve, and at the center $\frac{11}{16}$ in. (Fig. 5-4). Using the heat-transfer correlation curves (Fig. 5-5) we

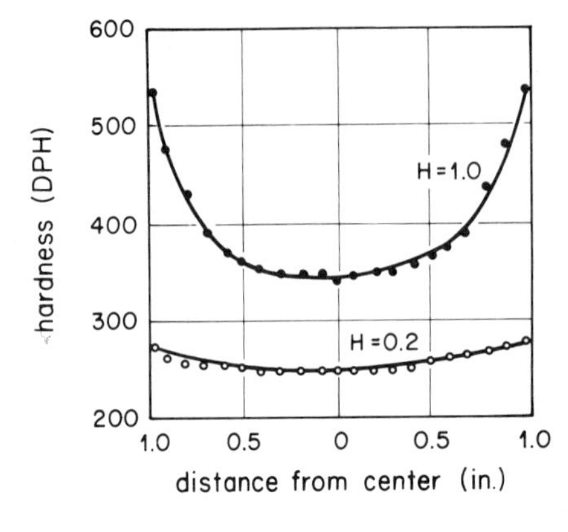

Figure 5-2 The calculated hardness distribution across the diameter of a 2-in.-diameter cylinder of 8740 steel when quenched into oil and water.

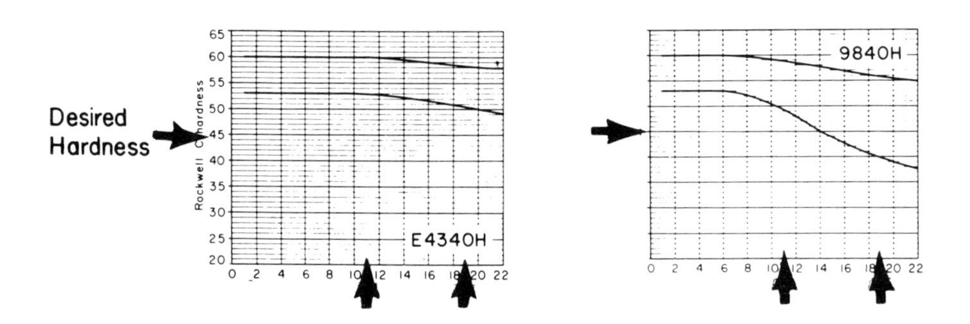

Distance from quenched end, sixteenths of an inch

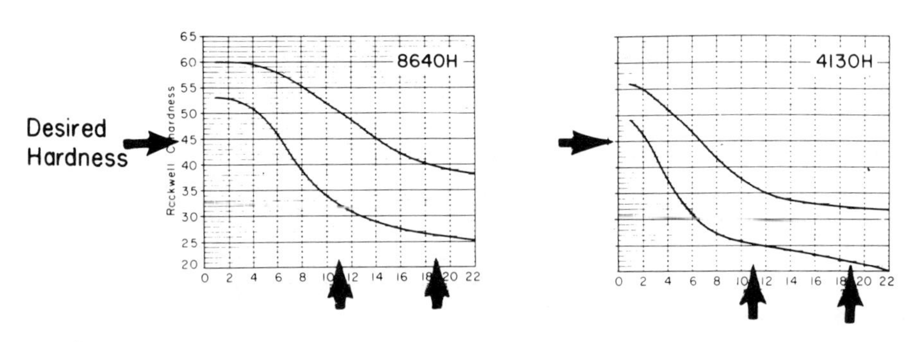

Distance from quenched end, sixteenths of an inch

Figure 5-3 Illustration of the use of the hardenability bands to determine if steels can be heat treated to meet desired hardnesses. The arrows at 11/16 and 19/16 in. are the position of equivalent rates for a 3-in.-diameter cylinder quenched in water of $H = 1.5$ and oil of $H = 0.35$, respectively. *(Curves from "Metals Handbook," vol. 1, Properties and Selection of Metals, 8th ed., American Society for Metals, Metals Park, Ohio, 1961.)*

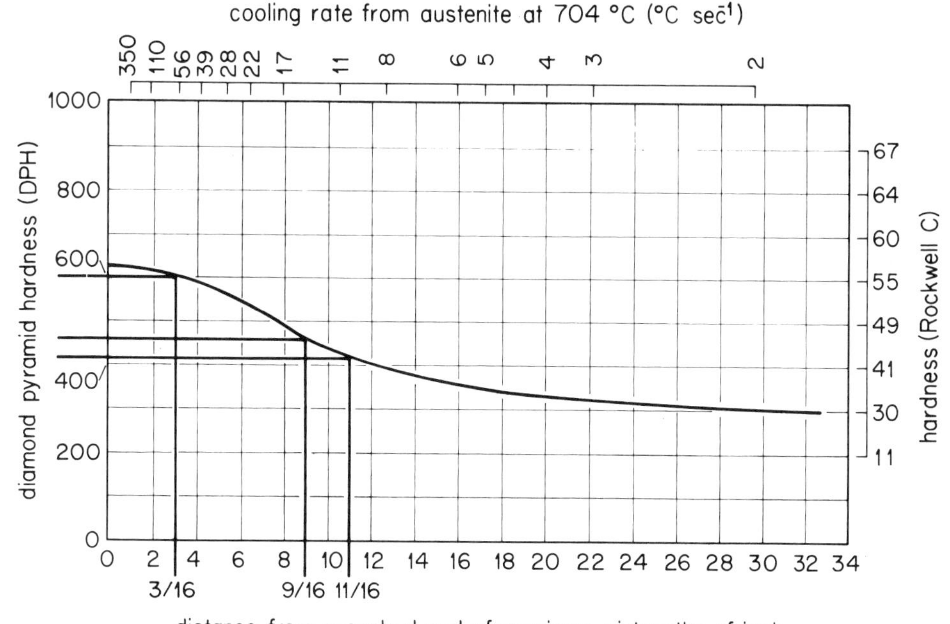

Figure 5-4 The calculated Jominy curve for the 3140 steel.

find that the surface requires a severity of quench H of about 0.7, and the center of 0.40. Obviously, both cannot be obtained simultaneously. We will see first what $H = 0.7$ gives. For this, an equivalent distance on the Jominy curve is $\frac{9}{16}$ in. for the center, and $\frac{3}{16}$ in. for the surface. From Fig. 5-4, this gives a hardness of 55 R_c at the surface and 45 R_c (446 DPH) at the center. This oil quench gives usable hardnesses and will not be a severe quench, so that cracking problems at the changing geometry on the ends of the shaft will be minimized.

We now determine the tempering temperature and time. The Crafts and Lamont correlation predicts that tempering at 205°C for 2 hours will lower the surface hardness from 55 to 53 R_c, and the center hardness will increase from 44 to 46 R_c. The increase in the center hardness is incorrect, reflecting the limitation on this correlation method. It appears, then, that tempering the steel for 2 hours at 205°C, following quenching in oil of $H = 0.7$, will give the desired surface and center hardness.

The hardenability calculations were based on a grain size of ASTM 7.5, so it is necessary to determine the austenitizing temperature and time which will give this grain size. The data of Fig. 2-14 indicate that the assumed grain size will be obtained upon austenitizing for 1 hour at 870°C.

Based on the above analysis, the following heat treatment is obtained:

1. Heat the steel for 1 hour at 870°C. This should give an austenite grain size of ASTM 7 to 8.
2. After austenitizing for 1 hour at 870°C, quench into oil of $H = 0.7$.
3. Heat the quenched steel for 2 hours at 205°C then cool in air.

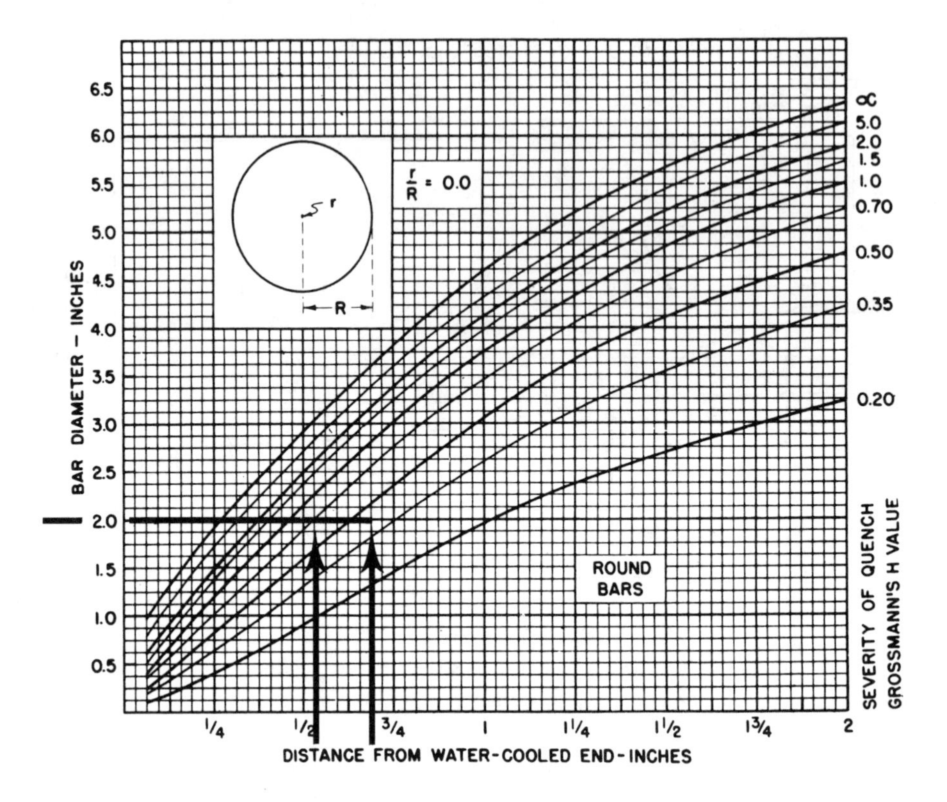

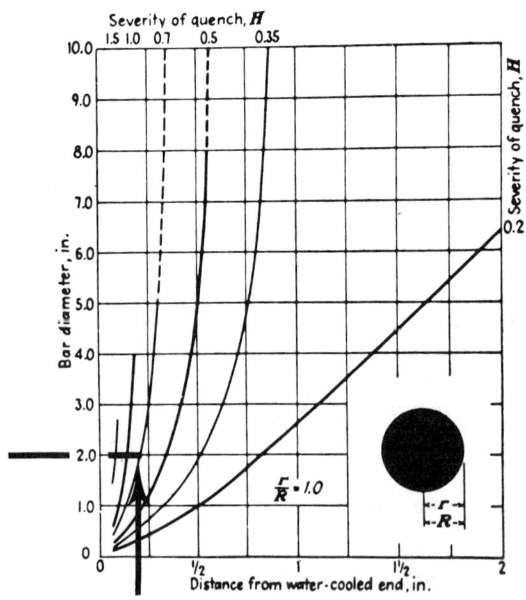

Figure 5-5 The heat-transfer correlations for a 2-in. diameter cylinder.

The uncertainties in the calculated values must be examined. If the grain size had been ASTM 5.5, then $D_i = 3.30$ in., and for $H = 0.7$ the as-quenched surface hardness is 53 R_c and the center 40 R_c. These are changed to 52 and 44 R_c, respectively, upon tempering at 205°C for 2 hours. This is a small effect. Also recall that the tempering correlation calculations have an uncertainty of about ±2 R_c. Thus, it appears that the desired hardnesses are predicted by the recommended heat treatments to within about ±2 R_c. If it is necessary to predict the hardness closer than this, then careful preliminary heat treatments must be carried out prior to production heat treatment.

The following problem illustrates some additional approaches to heat-treatment design. A 1-in.-diameter round dowel pin is to be made of a 4620H steel of the composition 0.20% C, 0.48% Mn, 0.25% Si, 1.73% Ni, and 0.24% Mo. It is to have a hardness after heat treating of 27 R_c at 3/4 radius and can be used in the untempered condition.

Using the hardenability correlations, the Jominy end-quench curves for two grain sizes were calculated and are shown in Fig. 5-6. It is seen that the required cooling rates are those equivalent to $\frac{4}{16}$ and $\frac{6}{16}$ in. on the Jominy bar. We then must examine the heat-transfer correlations. These curves are sensitive to the surface condition, and thus depend upon the degree of scale formation due to oxidation. This is illustrated in Fig. 5-7. Also note that these curves are given for the quenchant flowing at different velocities, and not in terms of a H value. If we assume that the steel is to be heated in air, then the hardness at $\frac{3}{4}$ radius will be 27 R_c for a grain size of ASTM 8, and 31 for ASTM 5, when quenched into oil flowing at 200 ft/min. Using still oil, a hardness of 22 R_c is obtained for a grain size of ASTM 8, and 27 for ASTM 5. If a nonscaling atmosphere is used, for oil flowing at 200 ft/min the hardness is 26 R_c for grain size of ASTM 8, and 30 for ASTM 5. Using still oil, these values are 24 R_c for ASTM grain size 8, and 28 for ASTM 5. Thus, in flowing oil the hardness will be between 26 and 31 R_c depending on the grain size and the surface condition. This lower figure is below the required value.

Using oil flowing at a greater velocity minimizes the effect of surface condition. Thus, for oil flowing at 750 ft/min, the hardness is 28 R_c for ASTM grain size 8, and

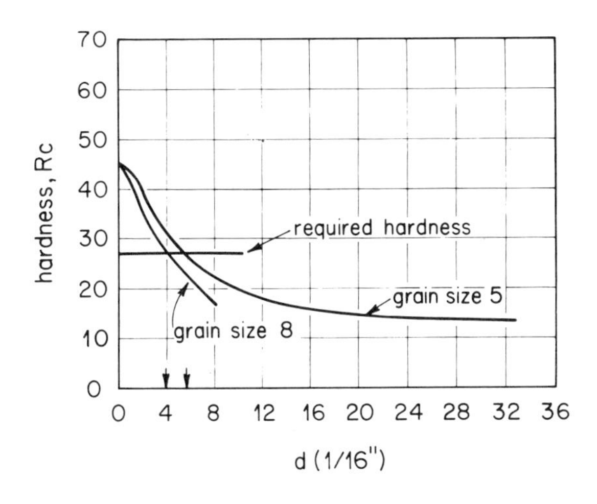

Figure 5-6 Calculated Jominy curves for a 4620H steel.

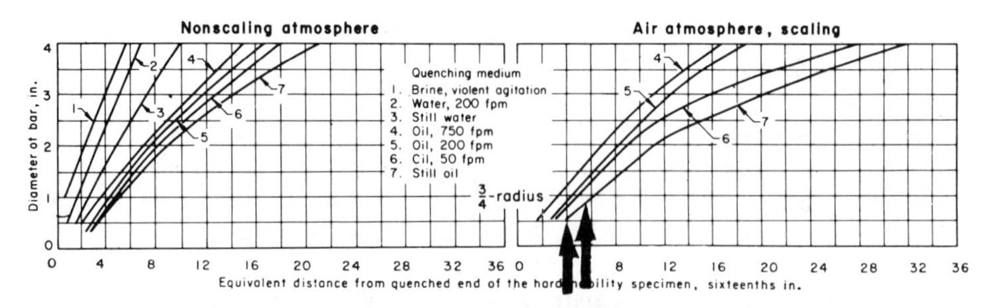

Figure 5-7 Heat transfer correlations for cooling at 3/4 radius of cylinders when austenitized in a nonscaling atmosphere and in air. *(From "Metals Handbook," vol. 1, Properties and Selection of Metals, 8th ed., American Society for Metals, Metals Park, Ohio, 1961.)*

33 for ASTM 5, whether the steel is austenitized in air or in a nonscaling atmosphere. Therefore using oil flowing at 750 ft/min will give a minimum hardness of 28 R_c, and this takes into account expected variations in the austenite grain size and surface conditions. The recommended heat treatment, then, is to heat in air to 870°C for 1 hour, then quench into oil flowing at 750 ft/min.

5-3 A CARBURIZING TREATMENT

Now we will consider a carburizing heat treatment. A 1-in.-diameter rod of 4820 steel containing 0.21% C, 0.25% Si, 3.30% Ni, and 0.26% Mo is to be gas carburized in $CO\text{-}CO_2$ to a surface carbon content of approximately 1.1%. The surface is to have a hardness of 65 to 67 R_c in the final heat-treated condition, and the center about 50 R_c. Previous tests have indicated that this amount of carbon is necessary in order to yield the correct residual stress distribution upon the formation of complete martensite at the surface layers. The desired case depth is 0.44 in. The geometry at the ends of the rod changes rather abruptly. Impact data on this steel indicate that the austenite grain size should be about ASTM 7 to 8.

The carburizing conditions must be chosen so that the carburizing time is not inordinately long. Using a temperature of 950°C in the equation (see Chap. 4)

$$d = 80.3\sqrt{t}\, e^{-8570/T}$$

(where d is in centimeters, t in hours, and T in degrees Kelvin) gives a time of 2.4 hours for a depth of 0.044 in. (0.113 cm). The required $CO\text{-}CO_2$ mixture to attain 1.1% C at the surface at 950°C can be obtained from Fig. 4-10 and is $p_{CO}^2/p_{CO_2} = 47$, yielding a CO pressure of 0.980 atm and CO_2 of 0.020 atm. This mixture is also nonoxidizing (Fig. 4-12).

We now need to choose a quench to give the desired hardness at the surface. Since the carbon content varies from the surface to about 0.1 in. below the surface, the hardenability (i.e., D_i) varies with depth in this range. Thus, to examine the effect of a quench on the hardness distribution, a Jominy curve should be calculated for each depth position. This may not be necessary, however, because of the high hardenability of this steel. Let us look at the surface first. Here the carbon content is 1.1%, but the correlation values (Fig. 2-9) only extend to 0.7% C. Extrapolating these for a grain size of ASTM 7.5, an ideal critical diameter of 5.5 in. is calculated. At the core, where the carbon content is 0.21%, D_i is calculated to be 2.6 in. For the surface layer, an oil quench of $H = 0.7$ will cool the surface at the same rate as $\frac{3}{16}$ in. on the Jominy curve, which gives a hardness of 66 R_c. This meets the surface hardness requirement. Using the $D_i = 2.6$ in. for the core, this quench gives a center hardness of 53 R_c.

It would appear that the carburizing treatment and the quench have now been specified. However, we must consider whether the rather high carburizing temperature will give too large a grain size. Figures 2-14 and 2-15 indicate that the austenite grain size should be stable around ASTM 7 to 8. Thus, the steel could be quenched in oil following the 2.4 hours in the carburizing gas.

Another point to remember is that the M_f is very sensitive to carbon content. (This is discussed in more detail in Chap. 6.) Thus, upon quenching to 25°C, although the "nose" of the TTT diagram is missed, retained austenite may be present in the high-carbon, surface layer because the M_f is below 25°C. To guarantee that any retained austenite present is converted to martensite, the quenched steel should be given a subzero treatment; cooling to −195°C (liquid nitrogen temperature) will be sufficiently low.

The final heat treatment is then the following.

1. Carburize the steel for 2.4 hours at 950°C in a CO-CO_2 atmosphere containing 2.0% CO-98.0% CO_2.
2. At the end of the carburizing time, quench the shaft into oil of $H = 0.70$.
3. Following cooling to 25°C with the oil quench, cool the part to −195°C, hold for about 30 min, then reheat in air to 25°C.

5-4 FAILURE CRITERIA

In the previous examples, the desired hardness at certain locations was given. This is usually not the case. In the design of components (such as shafts) the stress analysis will indicate the required strength, and this must be translated into a hardness for the design of the heat treatment. The example in this section deals with these problems.

A long, 3.5-in.-diameter shaft is loaded in simple torsion to a level such that the maximum normal stress at the surface is 85,000 psi. The shaft is to be made of 3140H steel containing 0.37% C, 0.60% Mn, 0.21% Si, 1.05% Ni, and 0.47% Cr. The steel is to have no tempering following the quenching operation, and the as-received condition of the shaft is hot rolled. A complete heat treatment is to be specified.

The stresses given were obtained from elastic stress analysis knowing the external loading conditions on the shaft. The condition can be seen in Fig. 5-8, which shows

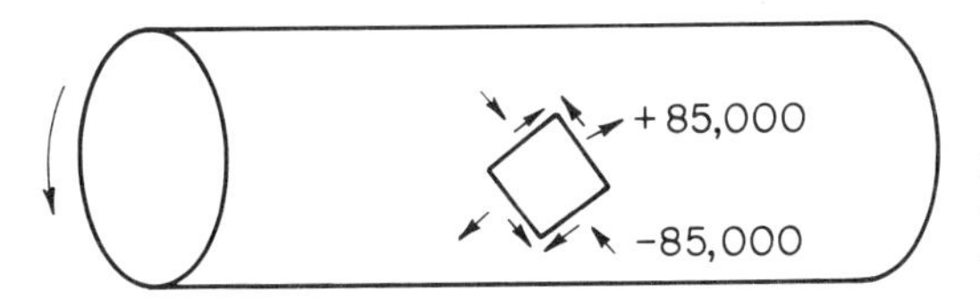

Figure 5-8 Illustration of the stresses on the surface of a shaft loaded in simple torsion.

an element on the surface. Note that this element has a tensile stress of 85,000 psi, and at ninety degrees to this a compressive stress of 85,000 psi. Now, the question is how to relate the known stresses to some property of the steel (e.g., yield strength values obtained in a simple tensile test) in order to determine whether the material is suitable. This is a most important question for engineers, and there is continuous effort to answer the question. For example, can we define failure as being attained when the maximum normal stress in the part reaches a value equal to the yield strength measured in a simple, uniaxial tensile test? We can, of course, define failure this way, but is it reasonable? Actually, several "failure criteria" have been developed, and for materials that are not brittle one can be stated as follows: A part loaded such that a complex stress state occurs is considered to have failed when, anywhere in the part, the maximum shear stress becomes equal to the maximum shear stress at the yield point in a simple tensile test. The relationship between the maximum (σ_1) and minimum (σ_2) normal stresses and the maximum shear stress is obtained from elastic theory and is

$$\tau_{\max} = \tfrac{1}{2}(\sigma_1 - \sigma_2)$$

In a simple tensile test, the maximum normal stress is the tensile stress, and the minimum is zero. Thus, at the yield point,

$$(\tau_{\max})_{yp} = \tfrac{1}{2}[\underbrace{(\sigma_1)_{yp}}_{\text{yield point stress}} - \underbrace{(\sigma_2)_{yp}}_{\text{zero}}]$$

$$= \tfrac{1}{2}\sigma_{yp}$$

For the shaft loaded in torsion, at the surface we have

$$\begin{aligned}\tau_{\max} &= \tfrac{1}{2}[(85{,}000) - (-85{,}000)] \\ &= 85{,}000 \text{ psi}\end{aligned}$$

Equating the maximum shear stresses,

$$(\tau_{\max})_{yp} = \tau_{\max}$$

or

$$\tfrac{1}{2}\sigma_{yp} = 85{,}000$$

or the required yield strength in the steel is 170,000 psi.

Now this required yield strength of 170,000 psi can be converted into a hardness value of 390 DPH (40 R_c) by use of Fig. 3-10. Using an ASTM grain size of 7.5, the calculated D_i is 2.45 in. From the calculated Jominy curve and the heat-transfer correlations, the surface hardness will be 40 R_c if quenched in oil of $H = 0.5$ and 51 R_c if quenched into oil of $H = 0.7$.

There are several uncertainties in the analysis. However, the quench of $H = 0.7$ will form a structure that is not completely martensitic, so the ductility should be sufficient to make the yield criteria valid. Another consideration is that, although the stress due to the torsional load decreases with depth, so does the hardness. Thus the surface may not be the critical location on which to base the heat treatment. The uncertainty in the hardenability associated with variable austenite grain size or chemical composition, in the hardness readings, and in the correlation of hardness and yield strength must be examined. For example, from Fig. 3-10, the hardness could actually range from 380 to 450 DPH (39 to 45 R_c). Using the $H = 0.7$ oil quench leads to a predicted hardness of 51 R_c, well above this uncertainty. These uncertainties are the basis for safety factors, so that if the maximum normal stress of 85,000 psi already includes a safety factor of, say, 2, then the usable heat treatment is to austenitize the shaft at 870°C for 1 hour and then quench into oil of $H = 0.7$.

5-5 A FATIGUE PROBLEM

A cylindrical rotating shaft 1.9 cm in diameter is loaded such that any point on the surface is alternately under longitudinal tension and compression. The load level is such that the maximum normal stress at the surface calculated from this external load varies from 150,000 psi tension to 150,000 psi compression. The shaft is to be made of 8620 steel containing 0.18% C, 0.70% Mn, 0.30% Si, 0.55% Ni, 0.52% Cr, and 0.22% Mo. To develop a favorable residual stress distribution, it is to be carburized to a case depth of 0.12 cm and to a surface carbon content of about 0.8% C. The problem is to specify a heat treatment to meet these conditions.

We first deal with the problem as if the shaft were bent by a static load, so that at the surface the maximum normal stress from the load is 150,000 psi tension. Now the shaft has been carburized, and the data in Fig. 4-20 for an 8617 carburized steel show that the longitudinal residual surface stress is about 40,000 psi compression. Thus, in bending, the actual maximum tensile stress in the surface of the shaft is 110,000 psi. In this simple bending, the minimum normal stress is zero, so that if we use the maximum shear-stress theory as a failure criteria the required yield strength at the surface must be 110,000 psi. However, the loading is cyclic, not static, and thus the fatigue strength must be 110,000 psi. By Fig. 3-13, this means that the yield strength must be 200,000 psi. This corresponds to a hardness of 450 DPH (45 R_c) which must be attained at the surface by the carburizing heat treatment.

For the surface carbon content of 0.8%, the D_i is calculated to be 5.6 in. For an austenitic grain size of ASTM 7.5, the Jominy curve is constructed, and from this and the heat-transfer correlations, an oil quench of $H = 0.35$ will give a surface hardness of 63 R_c. The uncertainties in the various correlations used should be examined, but it appears that this quench will be suitable as it gives a hardness considerably above that required. In addition, the structure for this hardness is not fully martensitic, so that sufficient ductility would be expected to make the failure criteria valid.

The residual stress assumed was based on the curves in Fig. 4-20, which are for a case depth of 0.121 cm and an oil quench. Thus, we will choose a carburizing temperature and time to give this case depth. The carburizing gas atmosphere will be chosen to maintain the surface carbon content at 0.8%, but the quenched shaft should be cooled to a low temperature to ensure complete conversion of any retained austenite to martensite. Using a carburizing temperature of 900°C leads to a calculated carburizing time of 1.9 hours. If we use a methane-hydrogen gas mixture for carburizing, by

Fig. 4-11 the gas ratio should be $H_2/CH_4 = 120$, for a total pressure of 1 atm. The recommended heat treatment, then, is the following:

1. Carburize the shaft for 1.9 hours at 950°C in a methane-hydrogen mixture having a hydrogen/methane ratio of 120.
2. Following the 1.9 hours at 950°C, quench the shaft into oil of $H = 0.35$.
3. Cool the shaft from 25°C to −196°C, then warm to 25°C.

5-6 UNCERTAINTIES AND SAFETY FACTORS

In the preceeding examples, attention has been called to the uncertainties in the results. These are caused by uncertainties in several factors, such as measuring hardness, the relationship between hardness and yield strength, the hardenability calculation, and the failure criteria. In spite of these uncertainties, the calculations are quite justified and are widely used in industry. It is best to look at these calculations as a basis for the design of heat treatments, giving information about the range of quenchants that might be usable, or clearly indicating a class of steels from which a specific steel, after closer scrutiny, will be chosen. In a production process, then, a specific steel, or several possible steels, will be examined more closely using proposed heat treatments to ensure attainment of the desired properties.

The uncertainties described above, including those associated with control of the actual industrial heat-treating process, are taken into account by the use of a safety factor. This factor effectively increases the required strength (or hardness). The exact value used is based on examination of the uncertainties mentioned above, and on the experience a given company may have had with a particular process for heat treating a particular component. The closer the quality control, the smaller the safety factor can be. In some cases the value chosen may be somewhat arbitrary and not based on a careful analysis of the situation, yet it may be sufficiently large to preclude any failures. However, if it is unnecessarily large, then excess material is being used. At least in terms of the design of heat treatments, the uncertainties mentioned above serve as a basis for the choice of a minimum safety factor.

CHAPTER

SIX

TOOL STEELS

Tool steels are a class of steels which, generally speaking, are characterized by their relatively high carbon content, high alloy content, high hardenability, and the presence of a considerable quantity of carbides in their microstructure, and hence high wear resistance. These steels are used to manufacture specific tools and to make components for fabricating objects by deformation (e.g., drawing dies). The classification of tool steels, and the range of chemical compositions of these steels, are illustrated in Table 6-1.

In this treatment of tool steels emphasis is placed on understanding the special problems associated with their heat treatment. The analysis of the heat treatment of steels in the preceding chapter will serve as a basis for extension into a treatment of these more complex steels.

6-1 TERNARY PHASE DIAGRAMS

In order to understand the behavior of tool steels it will be necessary to consider the phase equilibria in these alloys. Tool steels are multicomponent alloys, and hence the phase diagram is not the simple, two-component, binary phase diagram which can be depicted by a two-dimensional temperature-composition plot, but instead encompasses many dimensions. However, examination of the three-component, ternary-phase diagrams of the most important components of these steels (e.g., iron, carbon,

Table 6-1 Classification and nominal compositions of principal types of tool steels
AISI except for last group of steels

Steel	C	Mn	Si	W	Mo	Cr	V	Other
				Water-hardening tool steels				
W1	0.60–1.40†	-	-	-	-	-	-	-
W2	0.60–1.40†	-	-	-	-	-	0.25	-
W3*	1.00	-	-	-	-	-	0.50	-
W4	0.60–1.40†	-	-	-	-	0.25	-	-
W5	1.10	-	-	-	-	0.50	-	-
W6*	1.00	-	-	-	-	0.25	0.25	-
W7*	1.00	-	-	-	-	0.50	0.20	-
				Shock-resisting tool steels				
S1	0.50	-	-	2.50	-	1.50	-	-
S2	0.50	-	1.00	-	0.50	-	-	-
S3*	0.50	-	-	1.00	-	0.75	-	-
S4	0.55	0.80	2.00	-	-	-	-	-
S5	0.55	0.80	2.00	-	0.40	-	-	-
S6	0.45	1.40	2.25	-	0.40	1.50	-	-
S7	0.50	-	-	-	1.40	3.25	-	-
				Oil-hardening cold work tool steels				
O1	0.90	1.00	-	0.50	-	0.50	-	-
O2	0.90	1.60	-	-	-	-	-	-
O6	1.45	-	1.00	-	0.25	-	-	-
O7	1.20	-	-	1.75	-	0.75	-	-
				Medium-alloy air-hardening cold work tool steels				
A2	1.00	-	-	-	1.00	5.00	-	-
A3	1.25	-	-	-	1.00	5.00	1.00	-
A4	1.00	2.00	-	-	1.00	1.00	-	-
A5	1.00	3.00	-	-	1.00	1.00	-	-
A6	0.70	2.00	-	-	1.00	1.00	-	-
A7	2.25	-	-	1.00‡	1.00	5.25	4.75	-
A8	0.55	-	-	1.25	1.25	5.00	-	-
A9	0.50	-	-	-	1.40	5.00	1.00	1.50 Ni
A10	1.35	1.80	1.25	-	1.50	-	-	1.80 Ni

Source: "Metals Handbook," vol. 2, American Society for Metals, Metals Park, Ohio, 1964.

*In the April 1963 AISI Steel Products manual "Tool Steels," these steels were not included in the main table of compositions nor in tables of heat treating practice, because of their less common use.

†Available with various carbon contents, in increments of 0.10% within this range.

‡Optional.

and an important alloying element such as chromium) can be used as a basis for describing the phase equilibria. In this section, a discussion is presented of ternary-phase diagrams so that when presentation is made in the next section of such diagrams pertinent to tool steels, the isothermal sections and the constant-composition sections should be reasonably understood.

We first examine the simplest cases for a ternary diagram. Consider a system of

Table 6-1 *(continued)*

Steel	C	Mn	Si	W	Mo	Cr	V	Other
			High-carbon high-chromium cold work tool steels					
D1	1.00	-	-	-	1.00	12.00	-	-
D2	1.50	-	-	-	1.00	12.00	-	-
D3	2.25	-	-	-	-	12.00	-	-
D4	2.25	-	-	-	1.00	12.00	-	-
D5	1.50	-	-	-	1.00	12.00	-	3.00 Co
D6*	Now included with D3 by AISI							
D7	2.35	-	-	-	1.00	12.00	4.00	-
			Chromium hot work tool steels					
H10	0.40	-	-	-	2.50	3.25	0.40	-
H11	0.35	-	-	-	1.50	5.00	0.40	-
H12	0.35	-	-	1.50	1.50	5.00	0.40	-
H13	0.35	-	-	-	1.50	5.00	1.00	-
H14	0.40	-	-	5.00	-	5.00	-	-
H15*	0.40	-	-	-	5.00	5.00	-	-
H16	0.55	-	-	7.00	-	7.00	-	-
H19	0.40	-	-	4.25	-	4.25	2.00	4.25 Co
			Tungsten hot work steels					
H20	0.35	-	-	9.00	-	2.00	-	-
H21	0.35	-	-	9.00	-	3.50	-	-
H22	0.35	-	-	11.00	-	2.00	-	-
H23	0.30	-	-	12.00	-	12.00	-	-
H24	0.45	-	-	15.00	-	3.00	-	-
H25	0.25	-	-	15.00	-	4.00	-	-
H26	0.50	-	-	18.00	-	4.00	1.00	-
			Molybdenum hot work tool steels					
H41	0.65	-	-	1.50	8.00	4.00	1.00	-
H42	0.60	-	-	6.00	5.00	4.00	2.00	-
H43	0.55	-	-	-	8.00	4.00	2.00	-

*In the April 1963 AISI Steel Products manual "Tool Steels," these steels were not included in the main table of compositions nor in tables of heat treating practice, because of their less common use.

three components, *A*, *B*, and *C*, at constant pressure. Here we assume that the diagram is not sensitive to rather large variations in pressure, which is quite valid for most metallurgical phase diagrams. (We will not be concerned with the gaseous state, but only the liquid and solid states.) Figure 6-1 shows the assumed form of the binary diagrams of *A-B*, *B-C*, and *A-C*. In both the liquid and the solid phase, each component is completely soluble in the other.

Table 6-1 *(continued)*

Steel	C	Mn	Si	W	Mo	Cr	V	Other
			Tungsten high speed tool steels					
T1	0.70	-	-	18.00	-	4.00	1.00	-
T2	0.80	-	-	18.00	-	4.00	2.00	-
T3*	1.05	-	-	18.00	-	4.00	3.00	-
T4	0.75	-	-	18.00	-	4.00	1.00	5.00 Co
T5	0.80	-	-	18.00	-	4.00	2.00	8.00 Co
T6	0.80	-	-	20.00	-	4.50	1.50	12.00 Co
T7	0.75	-	-	14.00	-	4.00	2.00	-
T8	0.75	-	-	14.00	-	4.00	2.00	5.00 Co
T9	1.20	-	-	18.00	-	4.00	4.00	-
T15	1.50	-	-	12.00	-	4.00	5.00	5.00 Co
			Molybdenum high speed tool steels					
M1	0.80	-	-	1.50	8.00	4.00	1.00	-
M2	0.85	-	-	6.00	5.00	4.00	2.00	-
M3 Cl 1	1.05	-	-	6.00	5.00	4.00	2.40	-
M3 Cl 2	1.20	-	-	6.00	5.00	4.00	3.00	-
M4	1.30	-	-	5.50	4.50	4.00	4.00	-
M6	0.80	-	-	4.00	5.00	4.00	1.50	12.00 Co
M7	1.00	-	-	1.75	8.75	4.00	2.00	-
M8*	0.80	-	-	5.00	5.00	4.00	1.50	1.25 Cb
M10	0.90	-	-	-	8.00	4.00	2.00	-
M15	1.50	-	-	6.50	3.50	4.00	5.00	5.00 Co
M30	0.80	-	-	2.00	8.00	4.00	1.25	5.00 Co
M33	0.90	-	-	1.50	9.50	4.00	1.15	8.00 Co
M34	0.90	-	-	2.00	8.00	4.00	2.00	8.00 Co
M35	0.80	-	-	6.00	5.00	4.00	2.00	5.00 Co
M36	0.80	-	-	6.00	5.00	4.00	2.00	8.00 Co
M41	1.10	-	-	6.75	3.75	4.25	2.00	5.00 Co
M42	1.10	-	-	1.50	9.50	3.75	1.15	8.00 Co
M43	1.25	-	-	1.75	8.75	3.75	2.00	8.25 Co
M44	1.15	-	-	5.25	6.25	4.25	2.25	12.00 Co

*In the April 1963 AISI Steel Products manual "Tool Steels," these steels were not included in the main table of compositions nor in tables of heat treating practice, because of their less common use.

The phase relations can be presented in a Cartesian coordinate system. For example, temperature, and the percent (say by mass) of A and the percent of B, can be used; the percent of C is obtained by subtracting the sum of the percent of A and B from 100. However, it is more convenient visually to use a triangular coordinate system to represent the composition of all three components, with temperature represented vertically and normal to the triangle. This is illustrated in Fig. 6-2 for the A-B-C system. In this figure the upper surface and the lower surface contain between

Table 6-1 *(continued)*

Steel	C	Mn	Si	W	Mo	Cr	V	Other
			Low-alloy special-purpose tool steels					
L1	1.00	-	-	-	-	1.25	-	-
L2	0.50–1.10†	-	-	-	-	1.00	0.20	-
L3	1.00	-	-	-	-	1.50	0.20	-
L4*	1.00	0.60	-	-	-	1.50	0.25	-
L5*	1.00	1.00	-	-	0.25	1.00	-	-
L6	0.70	-	-	-	0.25‡	0.75	-	1.50 Ni
L7	1.00	0.35	-	-	0.40	1.40	-	-
			Carbon-tungsten special-purpose tool steels					
F1	1.00	-	-	1.25	-	-	-	-
F2	1.25	-	-	3.50	-	-	-	-
F3	1.25	-	-	3.50	-	0.75	-	-
				Mold steels				
P1	0.10	-	-	-	-	-	-	-
P2	0.07	-	-	-	0.20	2.00	-	0.50 Ni
P3	0.10	-	-	-	-	0.60	-	1.25 Ni
P4	0.07	-	-	-	0.75	5.00	-	-
P5	0.10	-	-	-	-	2.25	-	-
P6	0.10	-	-	-	-	1.50	-	3.50 Ni
P20	0.35	-	-	-	0.40	1.25	-	-
P21	0.20	-	-	-	-	-	-	4.00 Ni, 1.20 Al
				Other alloy tool steels				
6G	0.55	0.80	0.25	-	0.45	1.00	0.10	-
6F2	0.55	0.75	0.25	-	0.30	1.00	0.10‡	1.00 Ni
6F3	0.55	0.60	0.85	-	0.75	1.00	0.10‡	1.80 Ni
6F4	0.20	0.70	0.25	-	3.35	-	-	3.00 Ni
6F5	0.55	1.00	1.00	-	0.50	0.50	0.10	2.70 Ni
6F6	0.50	-	1.50	-	0.20	1.50	-	-
6F7	0.40	0.35	-	-	0.75	1.50	-	4.25 Ni
6H1	0.55	-	-	-	0.45	4.00	0.85	-
6H2	0.55	0.40	1.10	-	1.50	5.00	1.00	-

*In the April 1963 AISI Steel Products manual "Tool Steels," these steels were not included in the main table of compositions nor in tables of heat treating practice, because of their less common use.

†Available with various carbon contents, in increments of 0.10% within this range.

‡Optional.

them the two-phase α + liquid region. For any temperature above the upper surface, the alloy is all liquid; for any below the lower surface, it is all solid solution α.

Isothermal Sections

We now examine what constant-temperature sections (isotherms) in this ternary-phase diagram look like. In Fig. 6-3 are indicated five temperatures on the binary

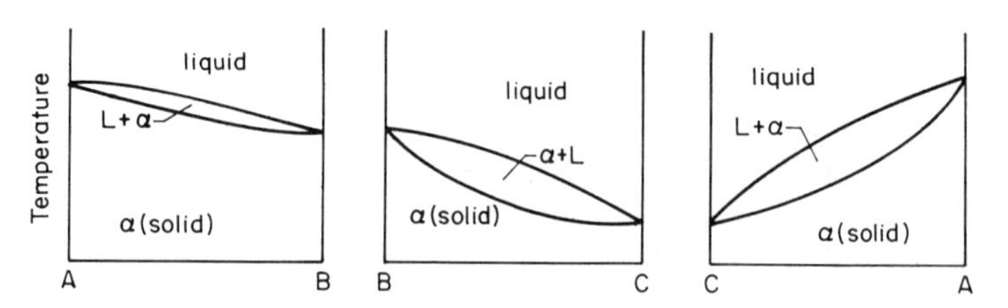

Figure 6-1 Binary phase diagrams for the isomorphous system *A-B, B-C,* and *C-A.*

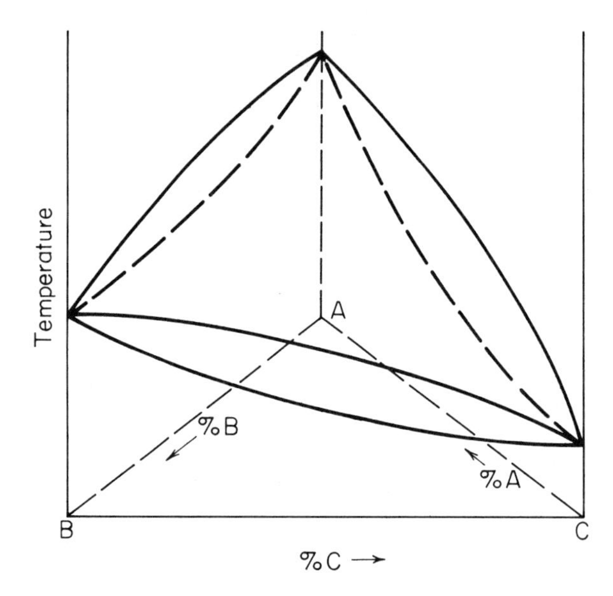

Figure 6-2 Temperature-composition plot of the ternary phase diagram based on the three binary systems shown in Fig. 6-1.

diagrams. Shown also are what the traces of the intersection of a plane parallel to the triangular base at a given temperature with the surface gives. For the temperature T_3 this is shown in more detail in Fig. 6-4. Note that for the binary system *B-C* at this temperature the solubility of *B* in liquid *C* is given by X_2, and the solubility of *C* in solid α is given by X_1. That is, X_1 and X_2 are the chemical composition (percent *C*) of the solid and liquid phases, respectively, when they are at equilibrium at temperature T_3. A similar meaning for X_3 and X_4 applies to the system *A-C*. For ternary compositions of *A*, *B*, and *C*, the two-phase region is bounded by a line connecting X_1 and X_4 and a line connecting X_2 and X_3. This gives the isothermal section shown for T_3 in Fig. 6-3. The shape of the bounding lines depends, of course, on the shape of bounding surfaces in Fig. 6-2, and on the temperature under consideration.

We will now consider how to read the chemical composition (percentages of *A*, *B*, and *C*) on the triangular plots. Consider the isotherm at T_3, as shown in Fig. 6-5. If we had a binary alloy of 20% *A*-80% *C*, we would be located at point *a*. Likewise, if we had a binary alloy of 20% *A*-80% *B*, we would be at point *b*. If we take the 20% *A*-80% *C* alloy (point *a*) and substitute enough *B* atoms for *C* atoms so that

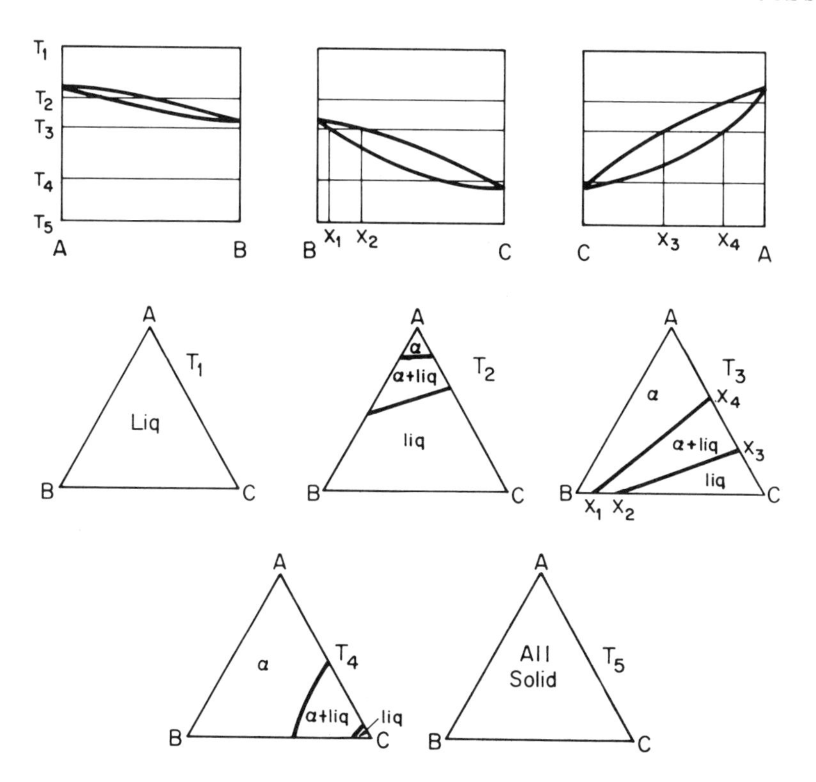

Figure 6-3 Isothermal sections from the ternary system *A-B-C* at five temperatures, showing the relation to the binary phase diagrams at these temperatures.

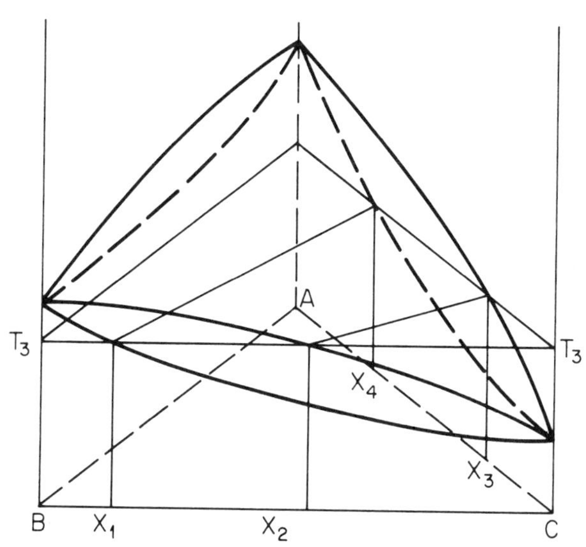

Figure 6-4 The ternary phase diagram for the system *A-B-C* showing the isotherm at the temperature T_3.

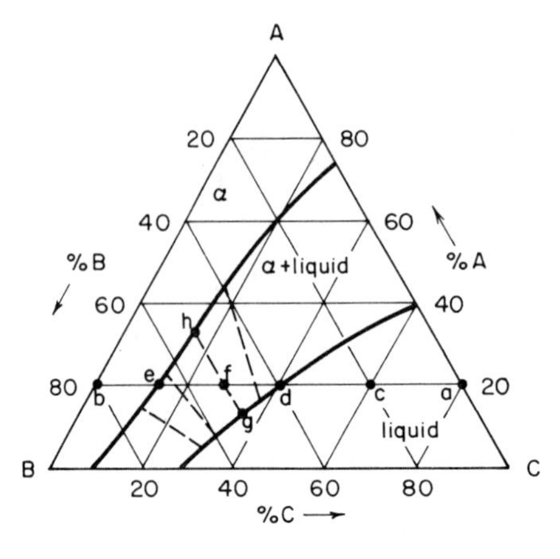

Figure 6-5 An isothermal section from the ternary system *A-B-C*.

the composition is changed to 20% *B*–20% *A*–60% *C*, we will have moved from point *a* along the line *ab*, to point *c*. We are removing *C* atoms and replacing them with *B* atoms, so that the percentage of *A* remains at 20%. Thus for this fixed percentage of *A*, when sufficient *B* is substituted to attain a composition of 20% *A*–40% *B*–40% *C*, point *d* has been reached. Substitution of more *B* for *C* will cause precipitation of α from the liquid. Although it is not obvious, the composition of the α will not be that of point *e*, but of some other point which must be determined experimentally.

Now, when more *B* is added, precipitation of α from the liquid occurs. However, the composition of the α that is precipitated (and is in equilibrium with the liquid) is not apparent from the phase diagram. These values must be determined experimentally. In the situation discussed here, consider sufficient *B* added to attain an overall percentage so that the composition of the liquid is 38% *C*-49% *B*-13% *A* (point *g*) and that of the α in equilibrium with it is 16% *C*-49% *B*-35% *A* (point *h*). It is assumed that the overall amount of *A* in the mixture is still 20%.

We now calculate the composition of the mixture. We already required that the percentage of *A* be 20%. Using the symbols for the mass of the liquid, α, and mixture, respectively, m_l, m_α, m_0 we can write a material balance around each component. Thus:

$$X_C\, m_0 = 0.16\, m_\alpha + 0.38\, m_l$$

$$X_B\, m_0 = 0.49\, m_\alpha + 0.49\, m_l$$

$$X_A\, m_0 = 0.20\, m_0 = 0.35\, m_\alpha + 0.13\, m_l$$

Also, we have

$$m_0 = m_\alpha + m_l$$

$$100 = X_A + X_B + X_C$$

and four of these five equations can be solved for the composition and for the amounts of liquid (m_l/m_0) and α (m_α/m_0) present. The composition of the mixture is found to be 31% *C*-49% *B*-20% *A*. This is point *f* in Fig. 6-5. Note that it lies on a straight line between points *g* and *h*. The amount of α (m_α/m_0) is 31.8% and that of liquid is 68.2%.

Thus if a mixture of *A, B*, and *C* having the composition of point *f* attains equilibrium at the temperature for which the phase diagram in Fig. 6-5 is valid, the composition of the liquid is found experimentally to be that of point *g*, and the composition of the α that of point *h*. The line *gfh* is referred to as a tie line. Other typical tie lines are shown in Fig. 6-5.

As more *B* is added to the mixture, but with the percentage of *A* maintained at 20%, the composition of the mixture moves along the line *ab*. The composition of the α and liquid phases is given by the appropriate tie line, with the amount of liquid decreasing and vanishing when point *e* is reached. The α then has the chemical composition given by point *e*. Continued addition of *B*, maintaining the amount of *A* at 20%, moves the alloy along line *ab*, until a binary alloy of 80%-20% is attained (point *b*).

Isopleths*

Another convenient method of presenting ternary phase equilibrium is to plot the phase diagram with the percentage of one component constant, the other two varying. The type of diagram obtained, for a given fixed percentage of one component, is seen by examining the trace of a plane parallel to the temperature axis along the lines corresponding to the fixed percentage. This is shown in Fig. 6-6 for the diagram under consideration, with the plane inserted at a constant percentage of 20% *A*. If the intersection of the phase boundaries and the plane is displayed on a separate graph, the diagram in Fig. 6-7 at 20% *A* is obtained. Also shown in this figure is the variation in the isopleth as the amount of *A* is increased from 0 to 80%.

Although not obvious from the isopleths, an isotherm intersecting the phase boundaries does *not* give the chemical composition of the phases. These values must be read from the isothermal section.

6-2 THE IRON-CARBON-CHROMIUM PHASE DIAGRAM

As a basis for examining in detail the heat treatment of tool steels, the iron-carbon-chromium phase equilibria will be discussed. The binary diagrams are shown in Fig. 6-8. The ternary diagram is quite complicated, mainly because chromium is a strong carbide former. In the chromium-carbon binary system, several carbides exist, and others may exist in the ternary system. These changes are best seen by examining isopleths for increasing chromium content. In Fig. 6-9 isopleths are shown at 0, 5, 13, and 17% chromium. Note that increasing chromium content causes the austenite region to narrow because chromium is body-centered cubic and favors the body-

* Isopleth: A line on a map connecting points at which a given variable has a specified constant value.

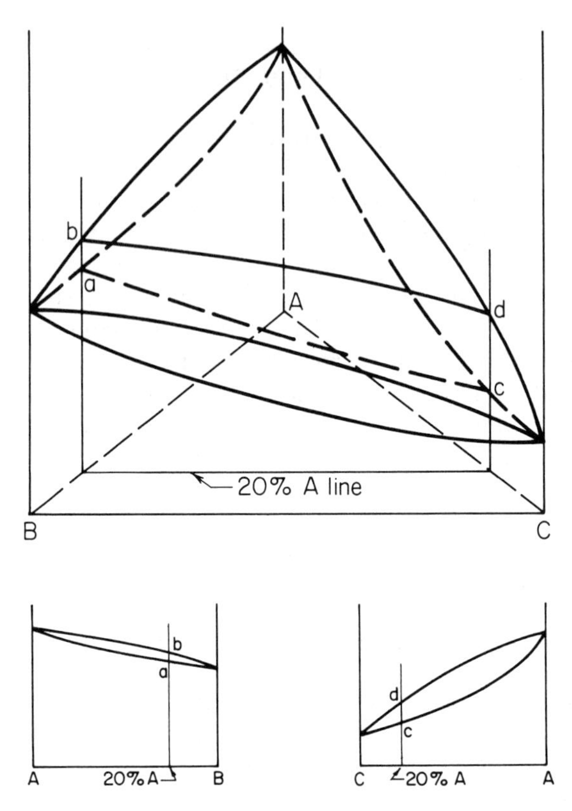

Figure 6-6 Ternary diagram showing an isopleth at 20% A.

centered cubic form of iron-carbon alloys, the ferrite and δ phases. For 17% chromium, this form of the alloy is stable from the melting point to room temperature for a carbon content below about 0.1%. Note that the carbon content of the eutectoid point is lowered from about 0.8% for no chromium to about 0.2% carbon for 13% chromium. Also the eutectoid temperature is raised from about 723°C for no chromium to about 1000°C for 17% chromium. The boundary of the austenite region is projected onto a common plane for comparison in Fig. 6-10.

Now consider a specific alloy: 13% Cr and 1.5% C. We want to examine the phase relations as a function of temperature for this alloy. Figure 6-9*b* shows that this alloy will not obtain the all-austenite region at any temperature. At a given temperature the isopleth of Fig. 6-9*b* cannot be used to obtain the chemical composition of the austenite and the carbide. Instead, these values must be obtained from the isothermal sections.

Consider the temperature 1150°C; the isothermal section in Fig. 6-11 shows the chemical composition of the alloy at point *a*. The tie line is shown dashed, with the chemical composition of the austenite given by the point *b* to be 9% Cr and 0.8% C. The composition of the carbide is given by point *c*, 51% Cr and 7.8 C. From these values, a material balance shows that the alloy at 1150°C consists of 9.4% carbide and 90.6% austenite.

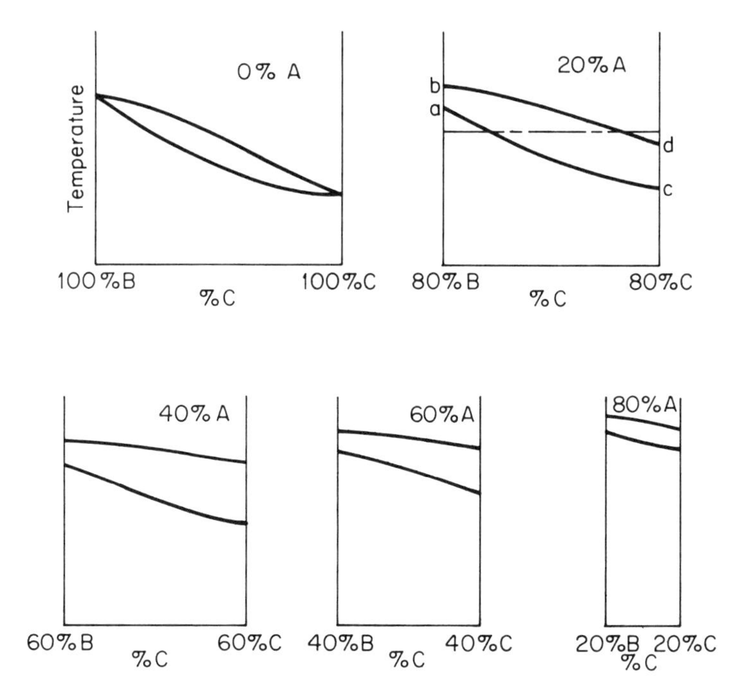

Figure 6-7 A series of isopleths from the ternary system *A-B-C.*

We now make the same type of analysis at 850°C using the isotherm as shown in Fig. 6-11. The chemical composition of the alloy (1.5% C and 13% Cr) is located at point *a*, and that of the two phases at *b* and *c*. A material balance shows that at this temperature the alloy contains 17.0% carbide and 87% austenite. Figure 6-12 shows the variation in the chemical composition of the austenite with austenitizing temperature, as well as the amount of carbides present at each temperature.

This steel has such a high carbon and alloy content that its hardenability is very high. An isothermal TTT diagram is shown in Fig. 6-13 for a commercial steel similar to the 1.5% C-13% Cr alloy. It is relatively easy to cool this steel from the austenite plus carbide region (e.g., 1000°C) without the austenite decomposing to ferrite and carbide. The austenite should transform to martensite. However, the martensite start and finish temperatures depend strongly on the carbon and alloy content *of the austenite*; these values may not be the same as those of the steel because the chemical composition of the austenite is not the same as that of the steel.

6-3 THE EFFECT OF AUSTENITIZING TEMPERATURE ON THE AS-QUENCHED HARDNESS

Since the carbon and chromium content vary with austenitizing temperature, the amount of martensite at room temperature after quenching to room temperature will

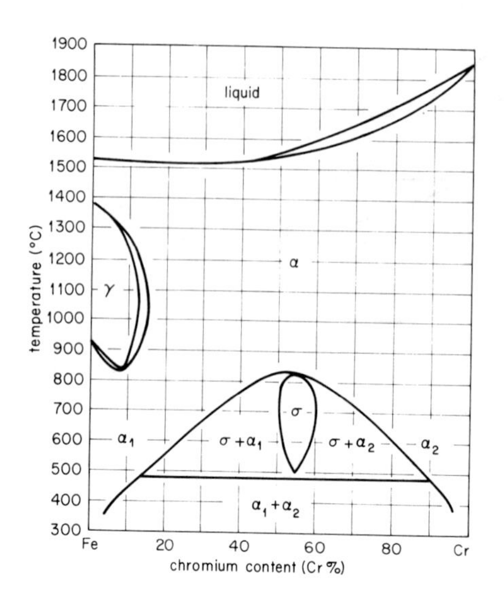

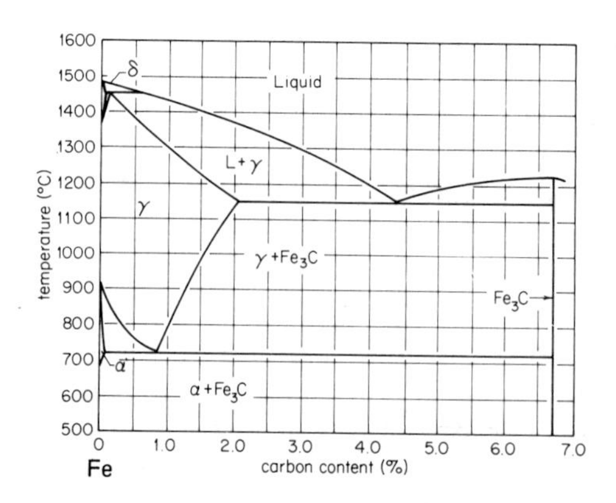

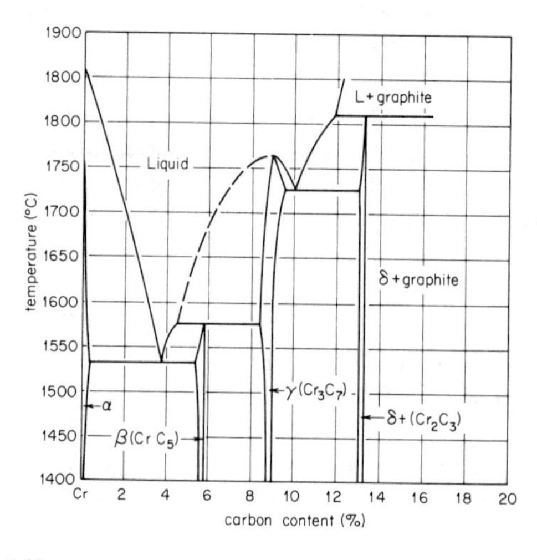

Figure 6-8 The binary phase diagrams iron-chromium, iron-carbon, and chromium-carbon. *(Adapted from "Metals Handbook," vol. 8, American Society for Metals, Metals Park, Ohio, 1973.)*

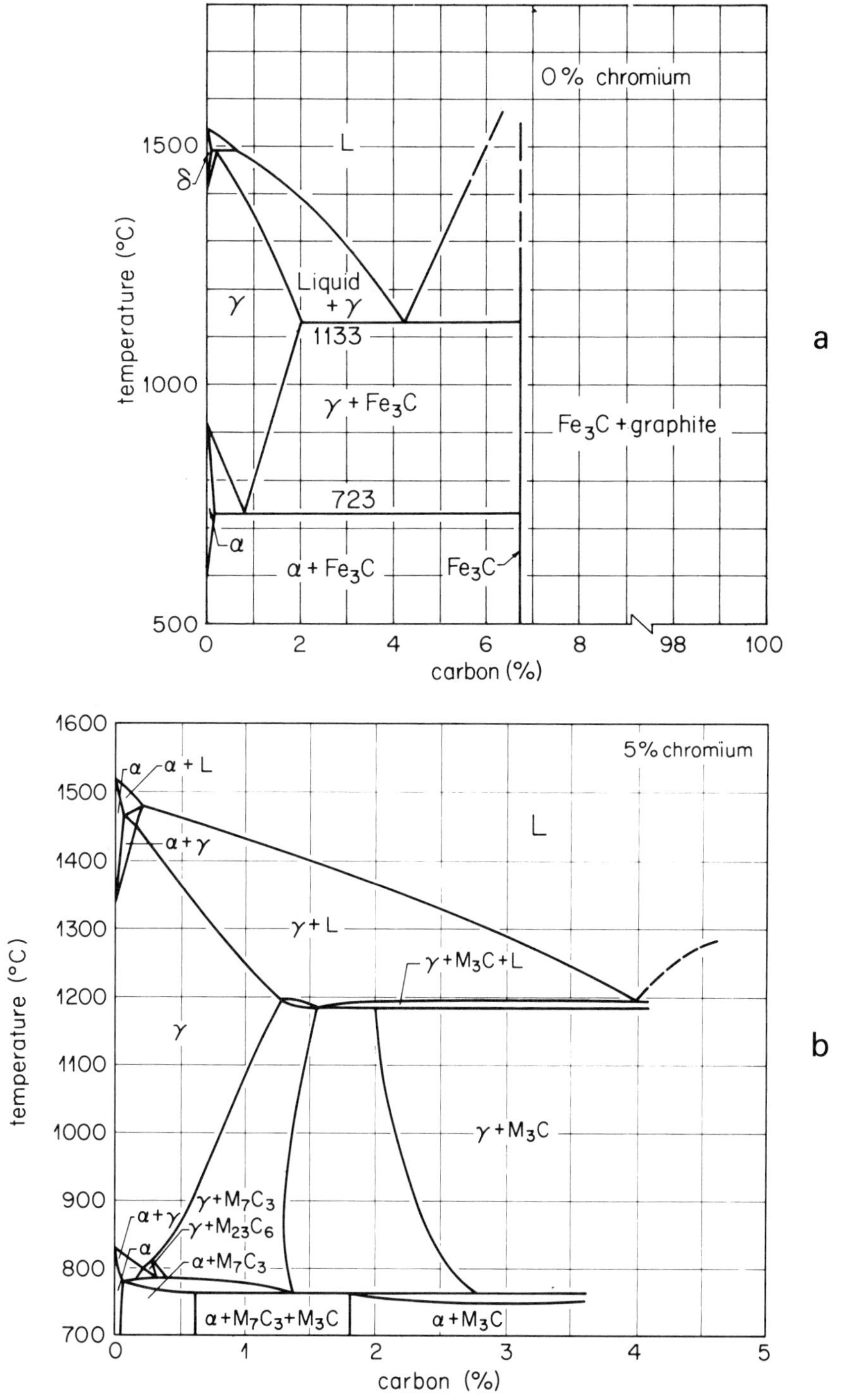

Figure 6-9 Isopleth at increasing chromium content. *(Ternary sections adapted from K. Bungardt, E. Kunze, and E. Horn,* Arch. Eisen., *vol. 29, p. 193, 1958.)*

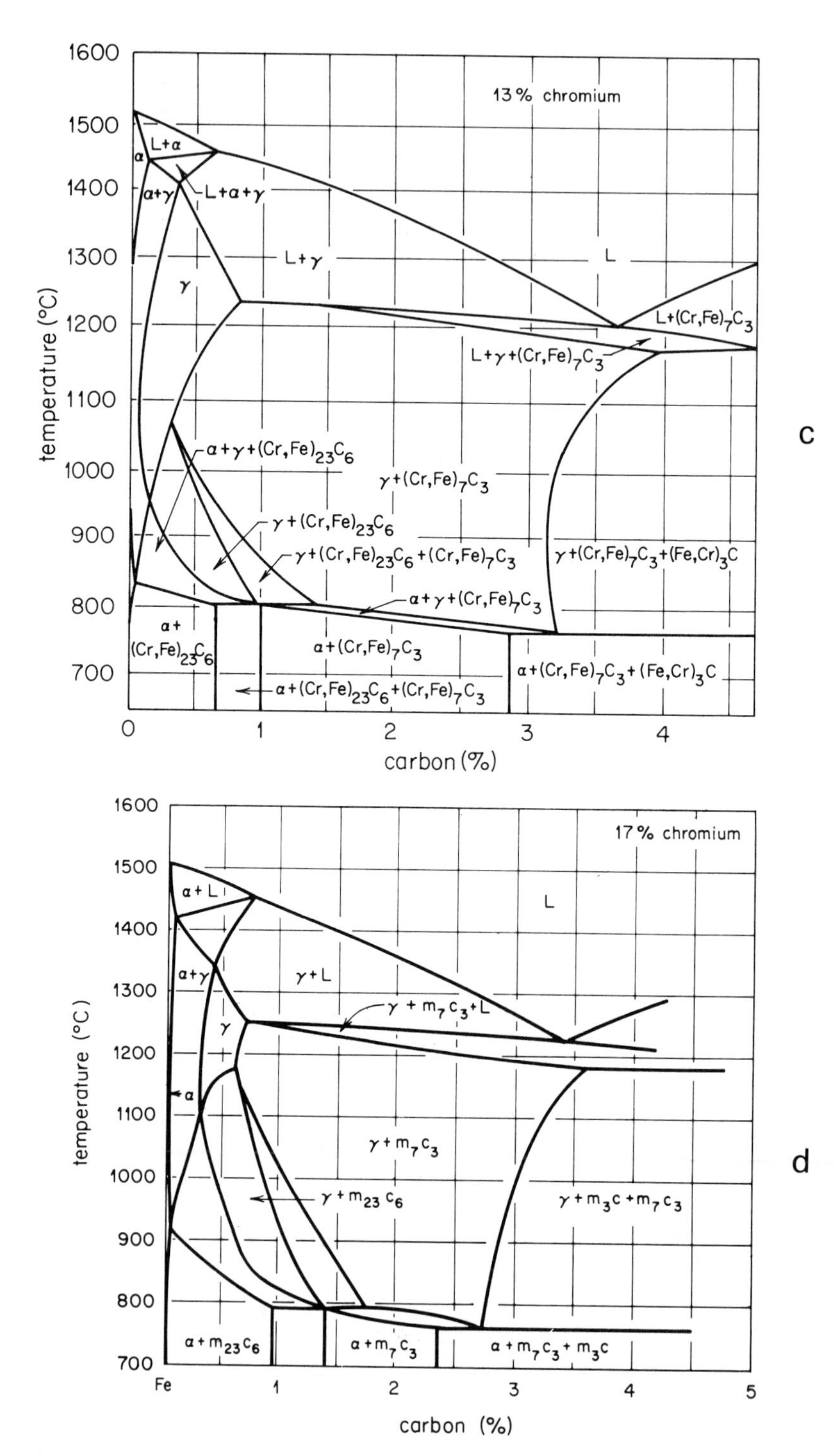

Figure 6-9 *(continued)*

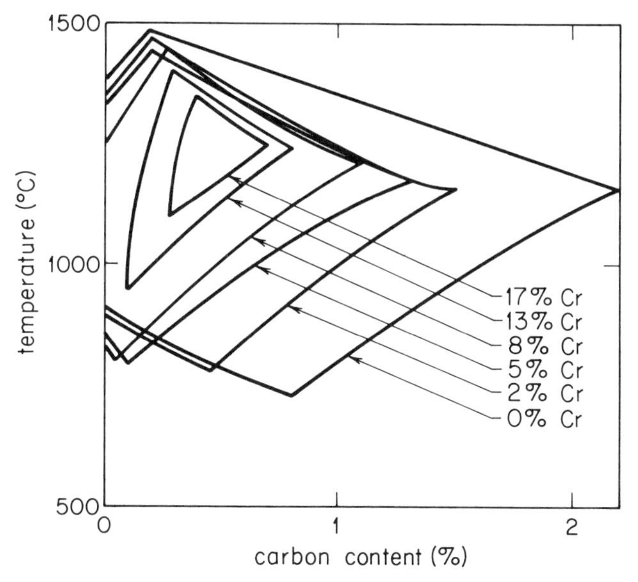

Figure 6-10 Boundary of the austenite region for various chromium contents. *(Adapted from K. Bungardt, E. Kunze, and E. Horn,* Arch. Eisen., *vol. 29, p. 193, 1958.)*

vary with austenitizing temperature. We now examine an estimation of the amount of retained austenite, and estimate the as-quenched, room-temperature hardness for the 1.5% C-13% Cr steel.

The effect of the carbon content on the amount of retained austenite in plain carbon steels is shown in Fig. 6-14. Alloying additions will increase the amount of retained austenite further (cobalt is an exception). Although the effect of chromium additions up to 13% has not been determined, we will use an approximation that the amount of retained austenite increases 11% per 1% of chromium addition. Using with this value the data in Fig. 6-12 and Fig. 6-14, the curve shown in Fig. 6-15 is obtained. An assumption here is that the steel has been cooled from the austenite sufficiently rapidly to avoid decomposition of the austenite prior to the M_s temperature; the TTT diagram of Fig. 6-13 shows that the cooling rate can be relatively slow yet still allow this condition to be attained. For the steel being considered here, the values of the amount of austenite at the austenitizing temperature (Fig. 6-12*c*) must be multiplied by the values in Fig. 6-15 to obtain the amount of retained austenite and martensite at 25°C after quenching; the remainder of the structure is the carbides which were present at the austenitizing temperature. The results of this calculation are shown in Fig. 6-16.

We can now estimate the as-quenched hardness of the steel at 25°C as a function of the austenitizing temperature. We make the assumption that the hardness will be given by the sum of the amount of each phase multiplied by the hardness of the phase. The hardness of the martensite depends only on the amount of the carbon in the

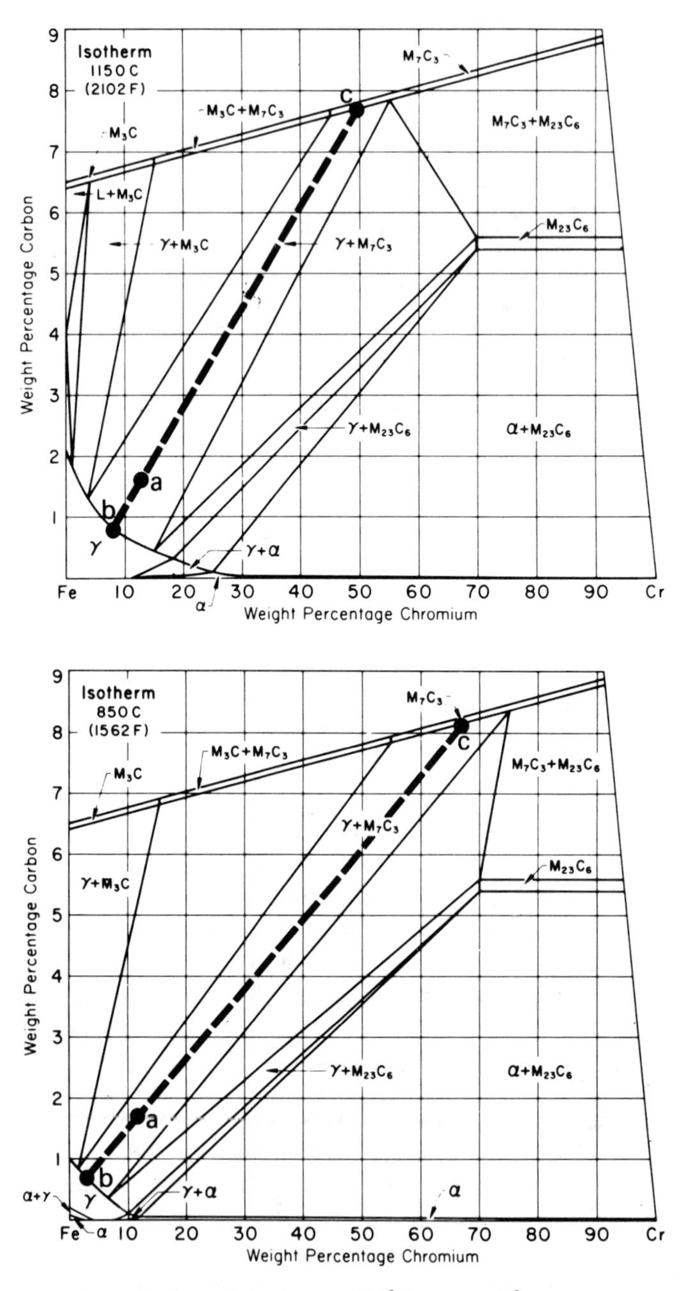

Figure 6-11 Isothermal sections at 850°C and 1150°C for the iron-chromium-carbon system. The tie lines are shown for the 1.5% C–12% Cr alloy (a). From the intersection of the tie line with the phase boundaries, the chemical composition of the carbide (c) and the austenite (b) is obtained. *(Adapted from K. Bungardt, E. Kinze, and E. Horn,* Arch. Eisen., *vol. 29, p. 193, 1958.)*

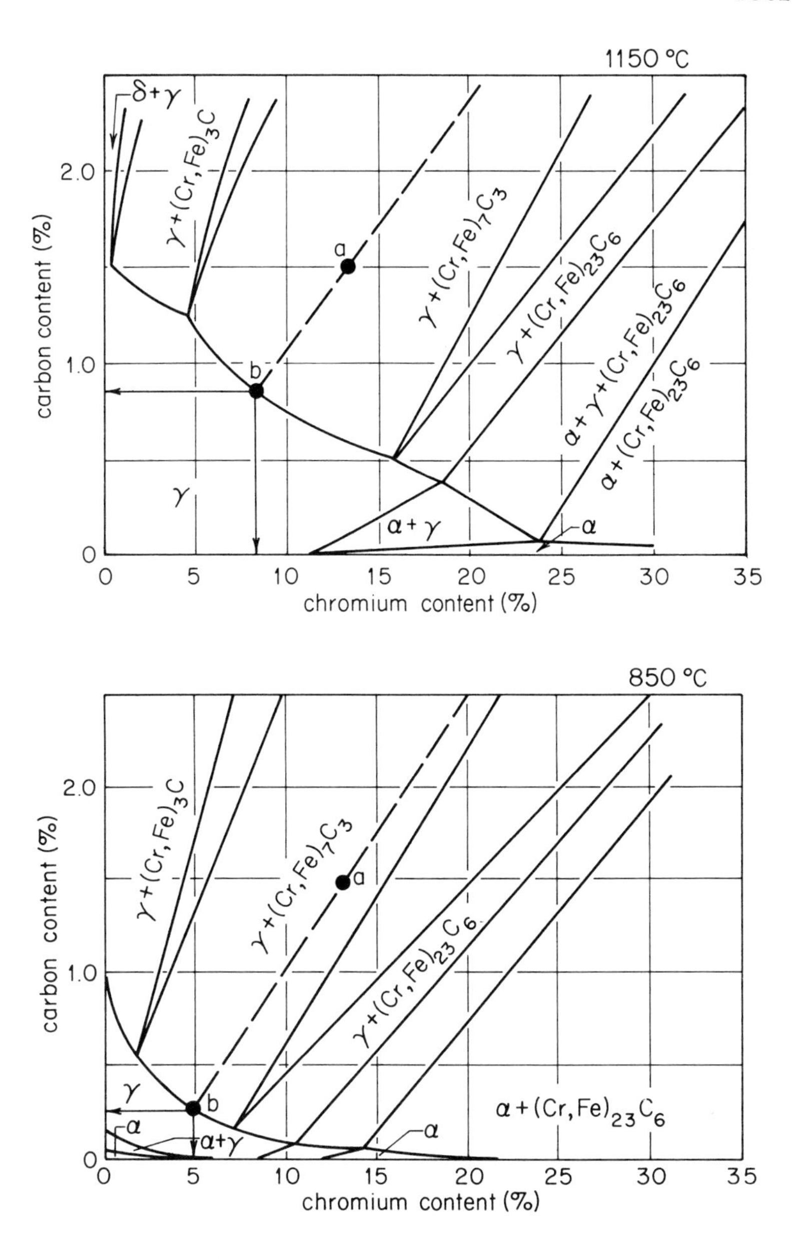

Figure 6-11 *(continued)* These show the iron-rich corner enlarged, with the carbon and chromium content of the austenite shown by the intersection.

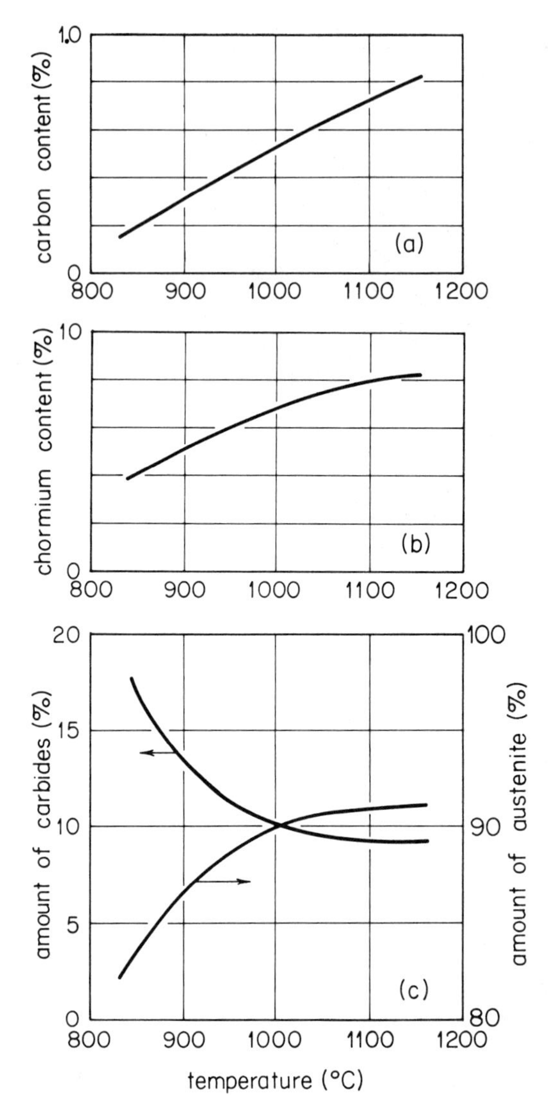

Figure 6-12 The variation in the carbon content (a), chromium content (b), and carbides and γ present at the austenitizing temperature (c) as a function of austenitizing temperatures.

austenite. The contribution of the martensite to the hardness can be calculated using Fig. 6-16 and Fig. 1-31. The hardness of carbides ranges from about 62 to 72 R_c, depending upon the type of carbide. For the chromium carbides formed in this steel, the hardness will be about 70 R_c.

The hardness of the austenite depends upon its carbon and alloy content (solid-solution hardening). We will use a value of 10 R_c (90 R_b) for our calculations. The final hardness of the as-quenched steel is shown in Fig. 6-17.

Therefore, it is seen that the hardness after quenching to 25°C is a sensitive function of the austenitizing temperature used. Too low a temperature results in the

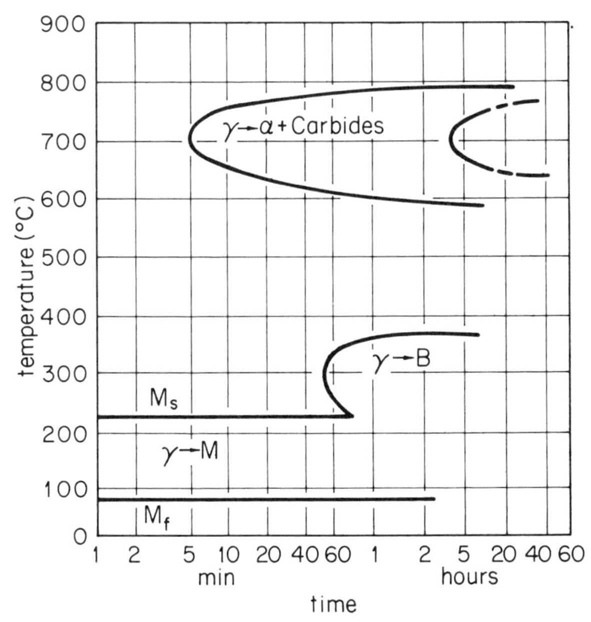

Figure 6-13 Isothermal time-temperature-transformation diagram for a commercial steel similar in composition to that of the 1.5% C–13% Cr steel. *(Adapted from P. Payson and J. Klein,* Trans. ASM, *vol. 31, p. 218, 1943.)*

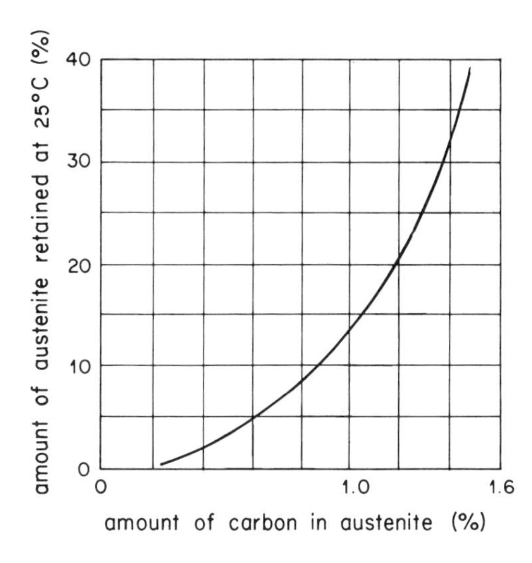

Figure 6-14 The variation of the percent retained austenite as a function of carbon content of the austenite for plain carbon steels. *(Adapted from R.T. Howard and M. Cohen,* Trans. AIME, *vol. 176,, p. 384, 1948.)*

austenite formed having such a low carbon content that the martensite formed upon quenching has a low hardness. Increasing the austenitizing temperature increases the carbon content of the austenite, and when this is converted to martensite upon quenching the higher carbon content causes the martensite to have a higher hardness. However, too high an austenitizing temperature increases the carbon and chromium content of the austenite sufficiently that the martensite finish temperature is below 25°C; upon quenching to this temperature not all of the austenite is converted to martensite,

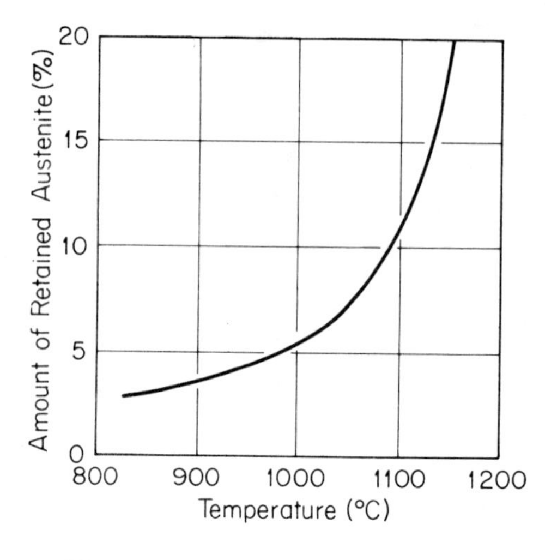

Figure 6-15 Variation with the austenitizing temperature of the amount of austenite which is retained (not converted to martensite) upon cooling to 25°C.

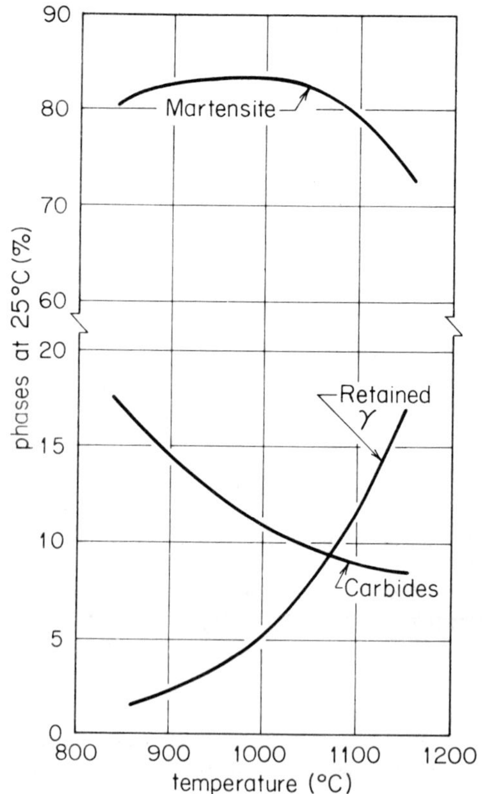

Figure 6-16 The variation with austenitizing temperature of the amount of the phases carbide, martensite, and retained austenite present after quenching to 25°C.

and soft retained austenite is present. This causes the hardness of the steel to decrease. (The hardness can be increased by cooling the steel to below 25°C to convert the retained austenite to martensite.) Thus there is an intermediate temperature which will yield a maximum as-quenched hardness. For tool steels, this is generally around

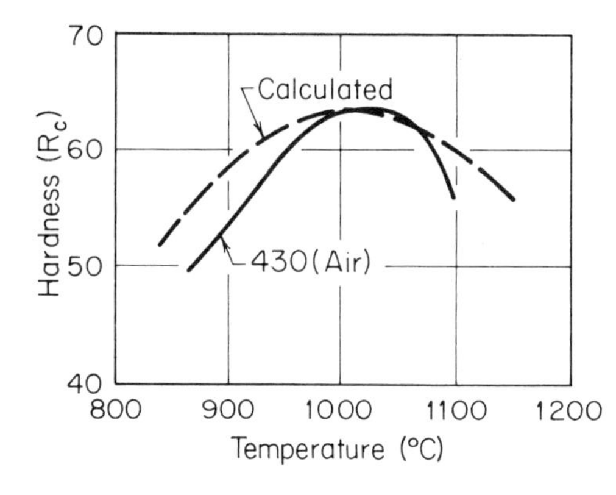

Figure 6-17 Variation with austenitizing temperature of the hardness upon quenching to 25°C for a 1.5% C–13% Cr steel. The data for the commercial tool steel 430 (D2), which contains about 12% Cr and 1.5% C, are also shown. *(Experimental curve adapted from G. A. Roberts, J. C. Hamaker, and A. R. Johnson, "Tool Steels," 3d ed., p. 505, American Society for Metals, Metals Park, Ohio, 1962.)*

980°C, whereas the usual austenitizing temperature for low-alloy and plain carbon steel is around 870°C.

A microstructure developed by quenching from a temperature that causes austenite to be retained is shown in Fig. 6-18. The long, dark needle-shaped regions are martensite. The small, round spheroids are the carbides which were present at the austenitizing temperature. The remaining matrix is the retained austenite, which is about 70% of the structure in this case. Note that the austenite grain size is quite large, due to the high austenitizing temperature.

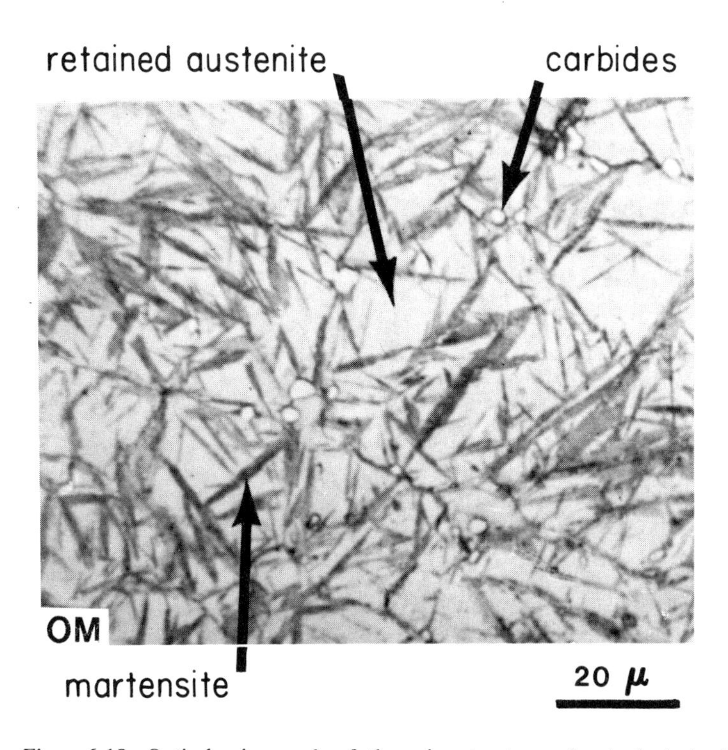

Figure 6-18 Optical micrograph of the microstructure of a tool steel with retained austenite present.

6-4 STABILIZATION

When very high carbon- and alloy-content steels of the tool steel class are cooled, the amount of austenite converted to martensite depends upon the delay encountered in cooling. For example, if a sufficiently high austenitizing temperature is used, retained austenite will be present at 25°C after quenching. It would appear that cooling to a sufficiently low temperature below 25°C will convert the retained austenite to martensite. However, the time delay at 25°C following the quench has in some steels a potent effect on the amount of the remaining austenite converted to martensite. This phenomena is referred to as stabilization, as the low-temperature aging causes structural changes that stabilize the austenite so that it is more difficult to convert to martensite.

Such behavior is illustrated in Fig. 6-19 for a 1.1% C-1.5% Cr tool steel. The curve *abcd* is obtained if the steel is cooled continuously to 25°C, then immediately to 0°C. However, if the steel is quenched to 44°C (point *c*) and then held at this temperature for 30 min, upon subsequent cooling the start of further martensite formation is retarded, as shown by the curve *cef*. The steel must be cooled to 28°C (point *e*) to cause more martensite to form, and the amount of martensite formed upon continued cooling is less than that obtained by continuous cooling (curve *abcd*).

If a tool steel has austenite retained upon cooling to 25°C, complete conversion to martensite can be attained by cooling to a sufficiently low temperature (sub-zero quench). However, the possibility of stabilization of the retained austenite requires that the sub-zero quench be conducted without excessive delay at 25°C. For example, austenitizing a 2% C-12% Cr tool steel at 1040°C and then oil quenching will result in a hardness of 62 R_c, with about 30% retained austenite present. If the quenched steel is cooled within an hour to about −85°C, the subsequent hardness will rise to 67 R_c. However, if a week passes prior to the sub-zero quench, the hardness will only increase to 63 R_c. Thus, the consequence of an excessive delay prior to a sub-zero cooling can be important.

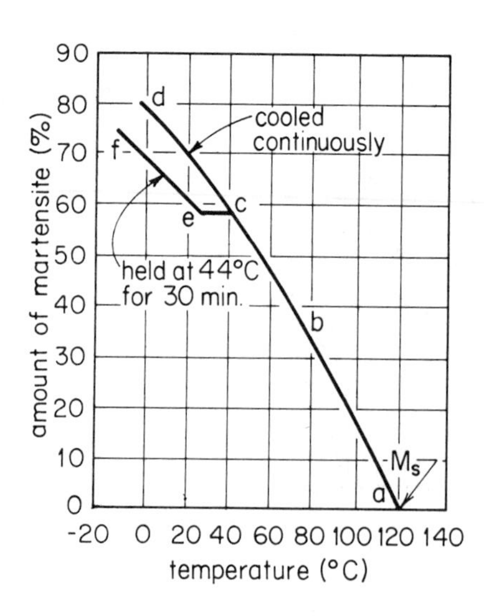

Figure 6-19 The effect of time of holding above the *Mf* prior to continued cooling on the amount of retained austenite. *(Adapted from M. Cohen,* Trans. ASM, *vol. 41, p. 65, 1949.)*

Stabilization also makes the amount of retained austenite at 25°C following the initial cooling dependent on the cooling rate. The high-hardenability tool steel can be cooled relatively slowly without falling below the critical cooling rate. Thus the air-hardening tool steels (e.g., 1% C-5% Cr-1% Mo-0.2% V) can be cooled as slowly as approximately 2°C/s.

Thus one would normally expect that the amount of retained austenite at 25°C following either water quenching or air cooling would be identical. However, the slower air cooling allows stabilization to occur, since the austenite is held in the stabilization range longer than with the faster water quench, and so the air-cooled specimen will have more retained austenite at 25°C. Such an effect is illustrated in Fig. 6-20*a*. For all austenitizing temperatures there is more retained austenite for air cooling as compared to oil quenching. The difference in the amount of retained austenite remains at about 10%, and the hardness difference remains small and approximately constant. In some alloys, though, this effect can be significant.

6-5 TEMPERING

Tempering of steels is done for several reasons, depending on the type of steel and its applications. In many applications, it is important to reduce the residual stresses

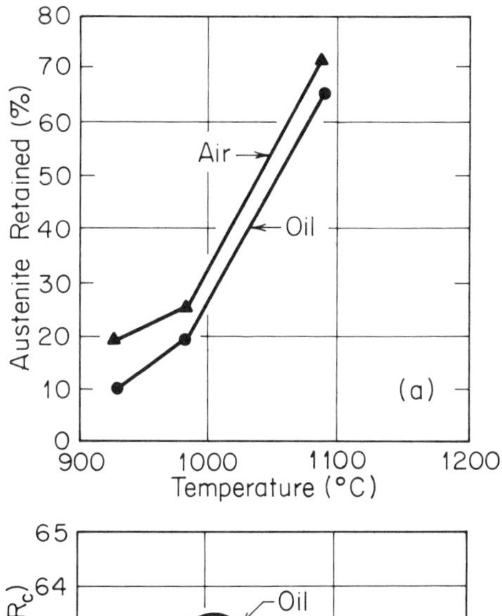

Figure 6-20 Effect of cooling rate on the amount of retained austenite at 25°C. The steel was a high-carbon, high-chromium tool steel (1.60% C–11.95% Cr). *(Adapted from O. Zmeskal and M. Cohen,* Trans. ASM, *vol. 31, p. 380, 1943.)* (b) Effect of cooling rate on the as-quenched hardness of a 1.5% C high-carbon, high-chromium tool steel. *(Adapted from G. A. Roberts, J. C. Hamaker, and A. R. Johnson, "Tool Steels," 3d ed., p. 505, American Society for Metals, Metals Park, Ohio, 1962.)*

induced by uneven cooling and martensite formation, and frequently this is the main reason for tempering. In addition, the toughness is increased by tempering, although in some steels at a sacrifice of strength. For plain carbon steels, the tempering heat treatment converts the martensite to carbide and ferrite. However, in high carbon, high alloy steels, the transformation is considerably more complex. In this section we examine the details of the phase transformations that occur upon tempering such steels.

The tempering curves in Fig. 6-21 illustrate the behavior of many tool steels. The hardness depends upon both the tempering temperature (and, of course, time) and the austenitizing temperature. In many steels a maximum, referred to as *secondary hardening*, is developed around 550°C, and its occurrence is important in the use of tool steels.

Because the as-quenched structure depends upon the austenitizing temperature, allowing various amounts of martensite and retained austenite to be present, the decomposition of both of these phases must be considered in tempering.

For the decomposition of martensite, the phase diagram will indicate the form of the equilibrium carbide. In plain carbon steels, this will be Fe_3C; in alloy steels more complex carbides will be present. However, non-equilibrium carbides may form as an intermediate step to the final structure. Even in alloy steels we might expect that Fe_3C would form first, since its formation would involve only the movement

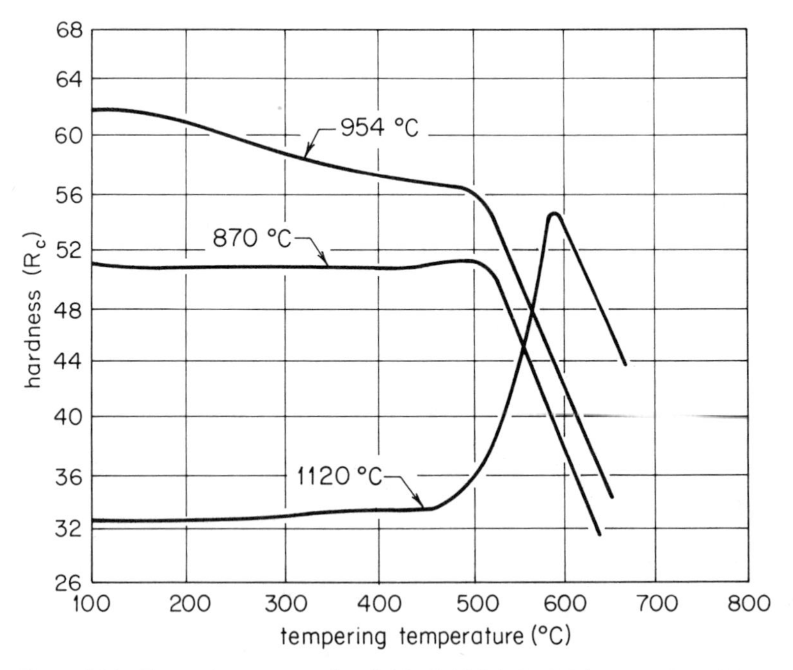

Figure 6-21 Tempering curves of a 1.4% C–12.2% Cr–3.5% Co tool steel, austenitized at the temperatures shown for about 1 hour, oil quenched, and tempered for 30 min. *(Adapted from J. P. Gill,* Trans. ASST, *vol. 15, p. 387, 1929.)*

of the carbon atoms, whereas the formation of a complex carbide (e.g., Cr_7C_3) would involve the movement of the substitutional elements, a slower process generally than for carbon atoms. The metastable Fe_3C would eventually disappear as the complex carbide formed.

It might be expected that the progress of the decomposition of the retained austenite can be predicted from the isothermal TTT diagram. This diagram can serve as a guide, but the austenite present after quenching and then reheating to the tempering temperature does not necessarily decompose at the same rate as austenite quenched directly from the austenitizing temperature to the transformation temperature. The retained austenite has a more complex dislocation structure from the residual stresses developed during the quench and these defects, along with the carbon segregation to these defects, cause the austenite to decompose at a different rate.

The exact details of the decomposition of the retained austenite and the martensite to the equilibrium phases differ, sometimes greatly, in steels, so that it is impossible to describe one exact process for all steels. However, the detailed research of the tempering processes in steels has revealed common features and made clear distinctions among various steels, and it has led to the separation of the tempering process into four categories. It is easiest to do this in terms of the effect of tempering temperature for a fixed tempering time (say 2 hours). Some of the tempering stages overlap in temperature; and, of course, the details of the tempering process, even for a given steel, vary depending on the tempering time chosen for comparison.

Stage 1: Tempering in the range 25 to 200°C causes the martensite to decompose to a metastable carbide, designated ϵ carbide, which is approximately $(Fe, M)_{2.3}C$, where M stands for other substitutional alloying elements in the carbide with the iron (e.g., chromium). This fine precipitate will frequently cause a slight increase in hardness. An example is shown in Fig. 6-22 for a plain carbon steel. The same effect in this stage of a tool steel of approximately 2% C-12% Cr is also shown. Here, comparison must be made for the lowest austenitizing temperature because the structure will then be mainly martensite and carbides, with little retained austenite.

Stage 3: From 200°C to higher temperatures, $(Fe, M)_3C$ precipitates from the martensite. In the lower temperature range, the carbides are needle-like; at higher temperatures they are spherical. The relatively coarse carbide formed in this stage causes the hardness to decrease, as shown in Fig. 6-22. Note that the presence of alloying elements such as chromium retards the rate of softening when compared to the iron-carbon alloy.

Stage 4: The presence of strong carbide-forming elements (e.g., molybdenum) results in the replacement of the coarse $(Fe, M)_3C$ carbides, which caused softening, with a finer dispersion of alloy carbides. Because these fine carbides coarsen slowly the hardness in this stage will increase, but at the upper extreme of this temperature range sufficient coarsening can occur to cause softening. The magnitude of this effect depends upon the specific steel. For example, the chromium steel shown in Fig. 6-22, austenitized to have in the as-quenched condition little retained austenite, does not show a peak in this range. In this alloy the chromium-rich carbide that forms, $(Cr, Fe)_7C_3$, coarsens too rapidly to cause the hardness to increase, but clearly the rate of softening (compared to the iron-carbon steel) is retarded. Tool steels with alloying

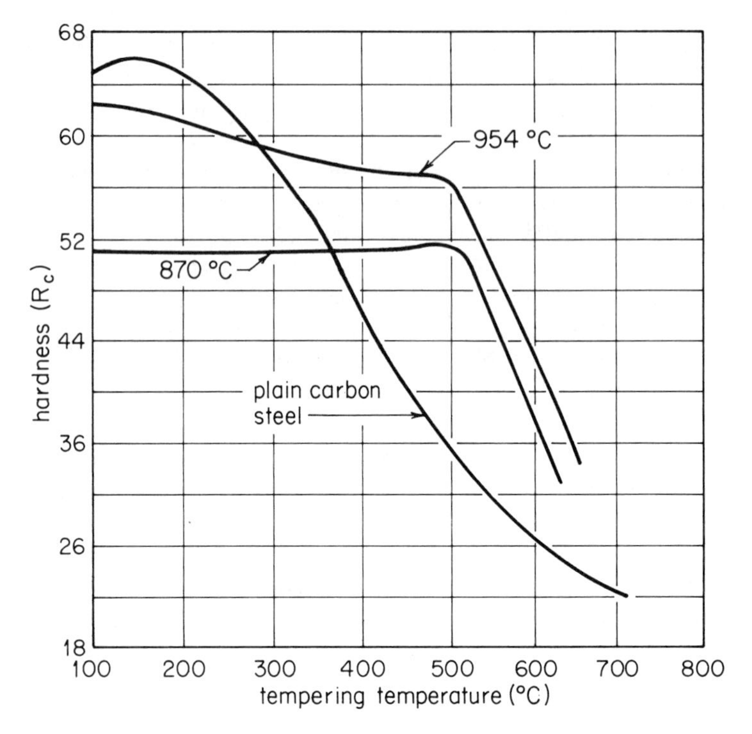

Figure 6-22 Comparison of tempering curves for a plain carbon steel (1.4% C, austenitized at 1200°C, water quenched and subzero cooled, tempered 1 hour) to those of a commercial 1.4% C–12.2% Cr–3.5% Co steel (austenitized at temperature indicated, quenched in oil to 25°C, tempered for 30 min). *(Alloy steels: adapted from J. P. Gill,* Trans. ASST, *vol. 15, p. 387, 1929; plain carbon steels: adapted from B. S. Lement, B. L. Averbach, and M. Cohen,* Trans. ASM, *vol. 47, p. 291, 1955.)*

elements that cause the precipitation of the more stable, finer carbides show a marked maximum in this stage (secondary hardening), and this is illustrated in Fig. 6-23. The steel contained approximately 18% W-4% Cr-1% V-0.7% C; it was austenitized for 2 min at 1280°C, then oil quenched, followed by tempering for 2.5 hours at each temperature, then air cooling. The variation in the hardness with tempering time for three temperatures is also shown, and it is clear that maximum hardness is achieved for tempering at 500°C for about 5 hours. The austenitizing temperature chosen results in little retained austenite, so that the secondary hardening in this case is associated with changes in the martensite. Examination of the type of carbides precipitated in this specific steel indicates that the secondary hardening is due to the formation of a very fine (e.g., 30 Angstroms) precipitate of $M_{23}C_6$ carbide.

Stage 2: This stage of tempering refers to the decomposition of the retained austenite. The decomposition can occur over the whole temperature range if the time is sufficiently long, transforming the austenite to primary carbides and to bainite in the lower temperature range, or to primary carbides and pearlite (or a ferrite-carbide aggregate which may not be lamellar) in the higher temperature range. Figure 6-24

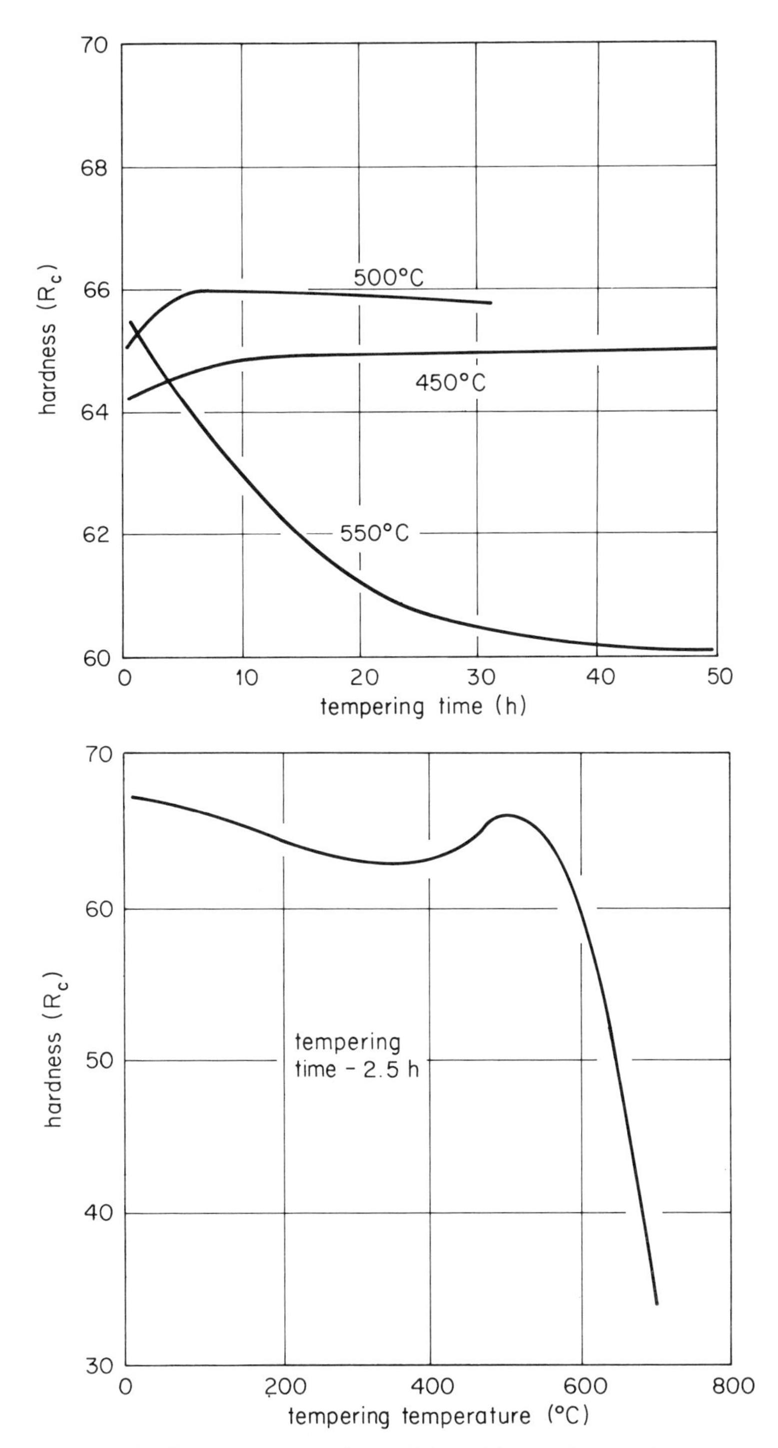

Figure 6-23 Tempering curves for a 18% W–4% Cr–1% V–0.7% C tool steel, austenitized at 1280°C, oil quenched, tempered for 2.5 hours, air cooled. *(Adapted from C. H. White and R. W. K. Honeycombe,* J.I.S.I., *vol. 197, p. 21, 1961.)*

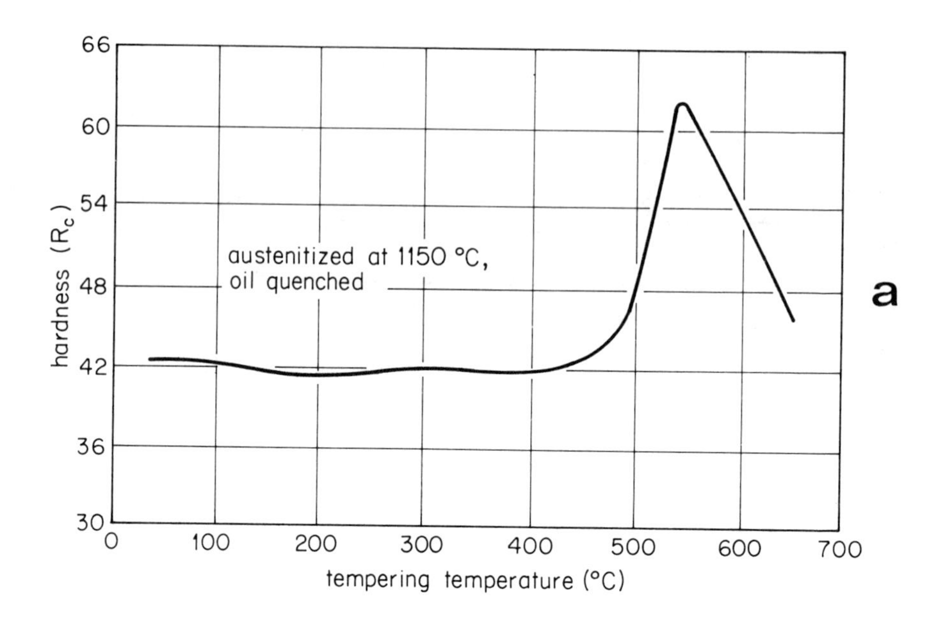

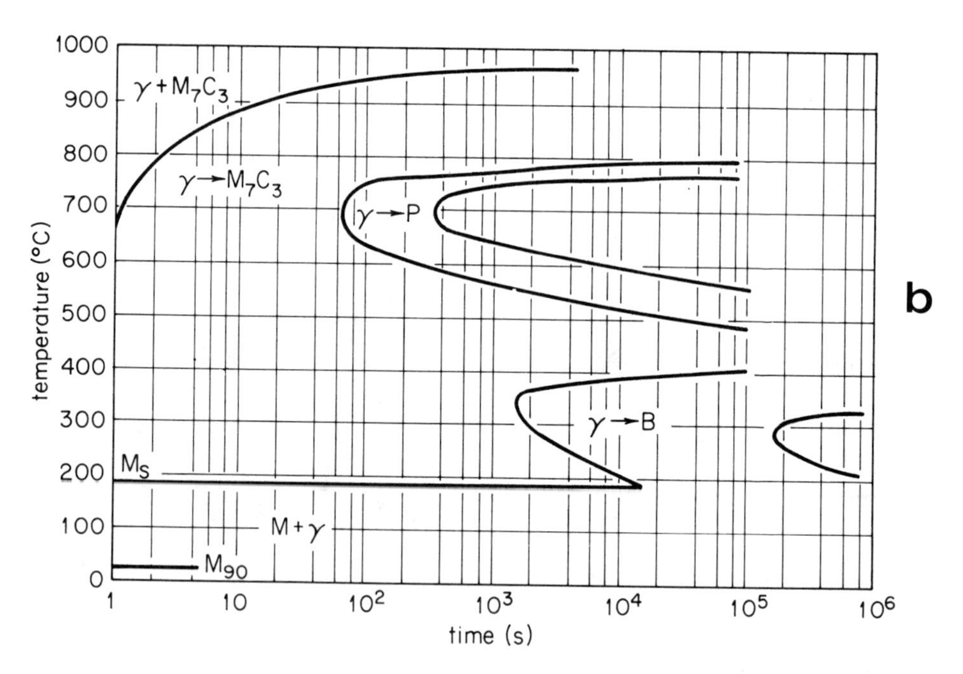

Figure 6-24 (a) Tempering curve for a 2.1% C–13.2% Cr–0.8% V tool steel, austenitized at 1150°C, oil quenched, tempered for 30 min. *(Adapted from J. P. Gill,* Trans. ASST, *vol. 15, p. 387, 1929.)* (b) Isothermal TTT diagram for a 2.1% C–11.5% Cr steel, austenitized at 970°C for 15 min. *(Adapted from A. Schrader and A. Rose, "De Ferri Metallographia," vol. II, W. Saunders Co, Philadelphia, 1966.)*

shows a tempering curve for a nominal 12% Cr-1.5% C tool steel for a high austenitizing temperature such that the as-quenched structure consists of a high percentage of retained austenite. Also shown is the isothermal TTT diagram for a similar steel. Tempering for 30 min should cause no significant transformation in the bainite range, and the tempering curve for the high austenitizing temperature remains about constant until around 500°C. The TTT diagram indicates that in the range 500 to 700°C the retained austenite should transform mainly to primary carbides. Hot hardness data, obtained by measuring the hardness at temperature at the end of the tempering time, show no increase in hardness due to this precipitation (Fig. 6-25). However, upon cooling to 25°C from the tempering temperature, the hardness shows a marked increase (Fig. 6-24*a*). The carbides that form upon tempering are richer in carbon and chromium than the retained austenite from which they precipitate. Thus the austenite adjacent to each carbide particle has its carbon and chromium content lowered; this region is then different chemically from the retained austenite far removed from the vicinity of the carbides, and due to this it now has higher martensite M_s and M_f temperatures. Upon cooling to 25°C, this austenite forms fresh martensite, raising the hardness at 25°C. Thus, in such cases the retained austenite contributes to secondary hardening.

The formation of the fresh martensite requires that the steel be reheated again to temper this fresh martensite. This will not only temper the martensite, but will also allow more carbides to precipitate if any retained austenite remains; this causes more fresh martensite to form upon cooling to 20°C. Thus it may be necessary to carry out multiple tempering operations to convert the retained austenite to martensite and to ensure tempering of all the martensite.

6-6 DIMENSIONAL STABILITY

Hardened and tempered steels usually have not attained equilibrium at 25°C, and they continue to undergo phase changes and relief of residual stresses with time. For many

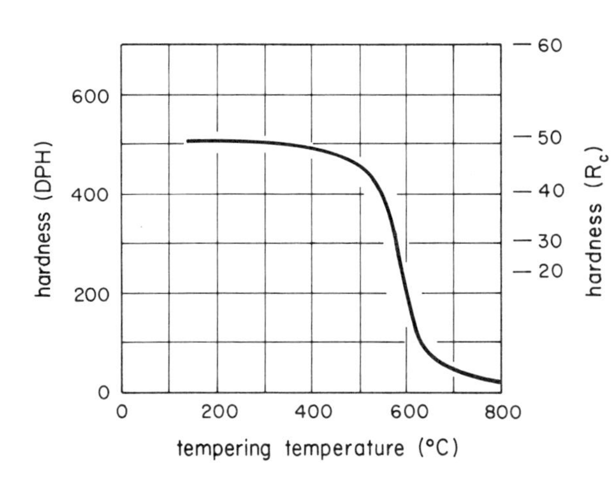

Figure 6-25 Hardness measured at temperature (hot hardness) as a function of temperature for a 1.5% C–12% Cr–3% Co tool steel, austenitized at 1150°C, oil quenched, reheated to tempering temperature for 20 min, then hardness measured. *(Adapted from O. E. Harder and H. A. Grove,* Trans. AIME, *vol. 105, p. 88, 1933.)*

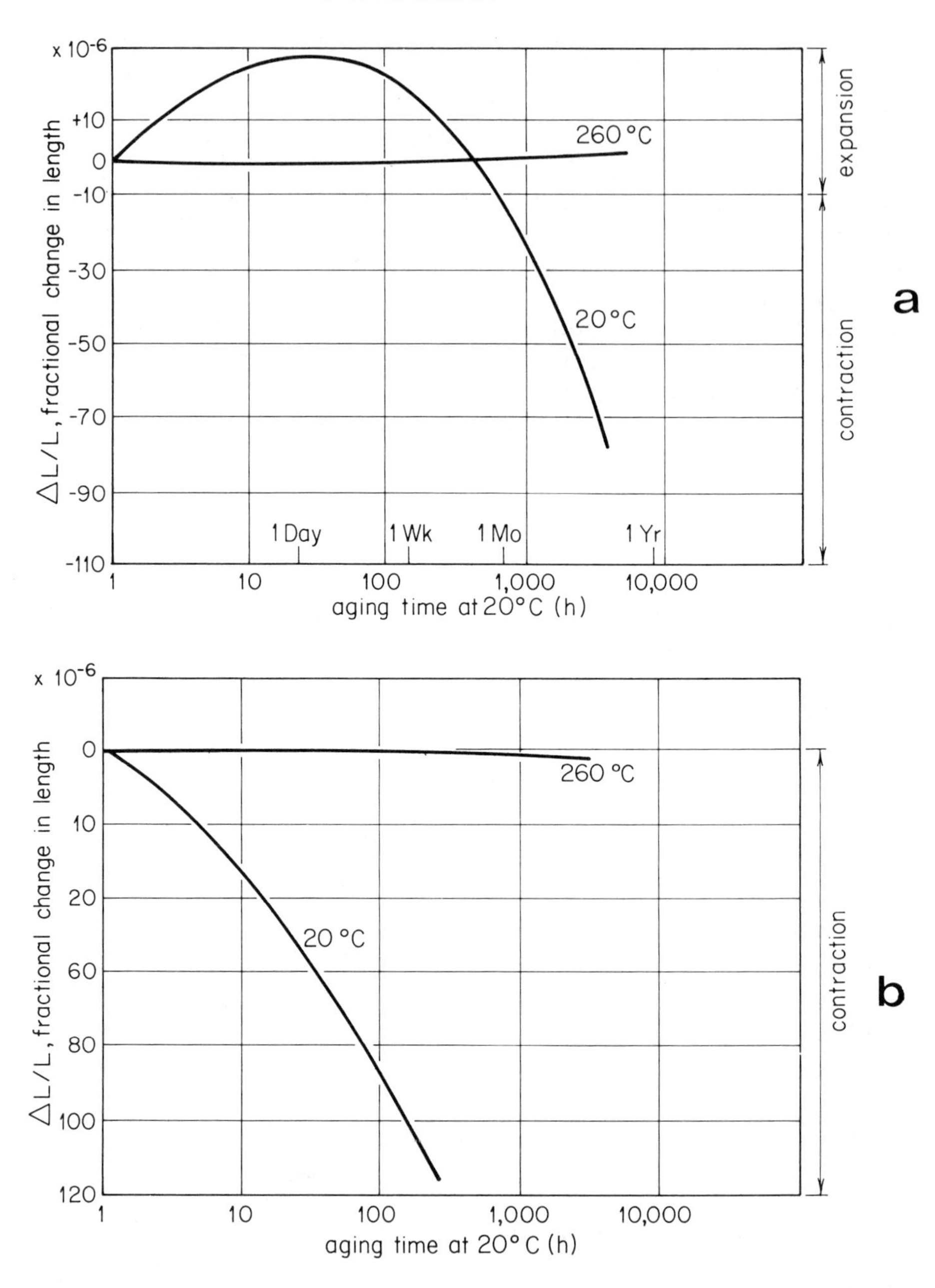

Figure 6-26 Change in length for a 1.0% C–1.5% Cr–0.2% V steel, initially austenitized at 845°C and quenched in oil to 20°C, followed by tempering at the temperature shown. In (a) the steel contained 4% retained austenite in the as-quenched condition. In (b) the as-quenched steel had been cooled from 20 to −195°C prior to tempering so that no retained austenite was present. *(Adapted from B. L. Averbach, M. Cohen, and S. G. Fletcher,* Trans. ASM, *vol. 40, p. 278, 1948.)*

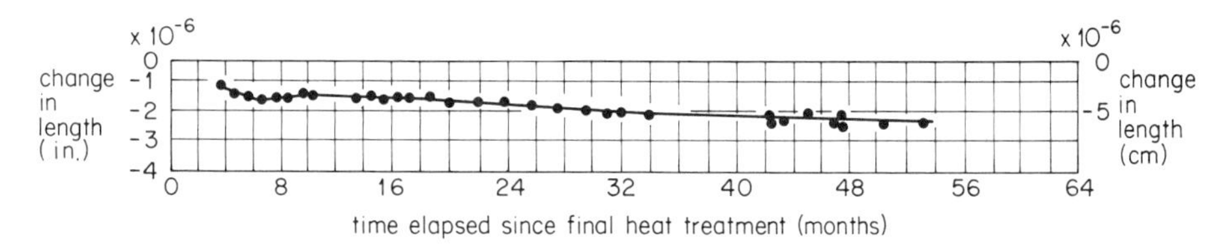

Figure 6-27 Change in length (contraction) with aging time at 20°C for a 1% C–1.5% Cr–0.3% Mo steel. *(Adapted from M. R. Meyerson and W. A. Pennington,* Trans. ASM, *vol. 57, p. 3, 1964.)*

applications these changes are sufficiently slow (and/or do not significantly affect properties) that the effect is of no consequence. In some applications, however, it must be carefully taken into account. To illustrate the problem, the use of tool steels as gage blocks will be considered.

First, let us look at some alterations at 25°C with time in the dimensions of hardened and tempered steels. The results of measurements on a 1% C-1.5% Cr-0.2% V tool steel are shown in Fig. 6-26. If the steel is quenched to 20°C, the structure consists of martensite, undissolved carbides, and about 4% retained austenite; the hardness is 64 R_c. Upon aging at 20°C, the steel undergoes an expansion, then a contraction. Tempering for 1 hour around 250°C makes the steel considerably more stable dimensionally. If the as-quenched steel is cooled to −195°C prior to tempering, to convert the retained austenite to martensite, then at 25°C the steel undergoes only an expansion. Again, tempering around 250°C for 1 hour improves the dimensional stability.

The contraction (Fig. 6-26) of the martensite is due to the formation of ϵ carbide. The retained austenite isothermally decomposes to martensite, with an expansion. Thus the curve in Fig. 6-26*a* is due to two factors: expansion due to the formation of martensite from the retained austenite, and contraction due to decomposition of martensite, both present initially in the as-quenched condition.

To illustrate the significance of the dimensional stability, we consider the heat treatment of gage blocks. These blocks are of precisely known geometry and are used to calibrate measuring equipment. For the machining of precision parts, tolerances approaching 2×10^{-5} cm are required. Thus if a gage block is to be used to calibrate a measuring instrument, it is desired that the block should not change its dimension more than this within the time span for which the block will be used. For example if the block is to be used for 10 years, then the average change must be about 2×10^{-6} cm per year.

Gage blocks are manufactured of hardened steels so as to be abrasion-resistant, but care is also required to make them quite dimensionally stable. As an example, we consider aging data of a 52100 steel, which contains about 1% C-1.5% Cr-0.3% Mo. The data in Fig. 6-27 illustrate measurements on this steel. Note that the block contracted about 2×10^{-6} cm in about 4 years. The heat treatment for this steel consisted of the following:

1. Austenitized in neutral salt at 843°C for 15 min.
2. Quenched in accelerated quenching oil.

3. Refrigerated immediately at −96°C for 18 to 24 hours.
4. Tempered at 120°C for 1 hour.
5. Refrigerated at −96°C for 18 to 24 hours.
6. Tempered at 120°C for 9 hours.
7. Ground to size preparatory to lapping, sharp corners and edges removed.
8. Demagnetized to remove possible magnetostrictive after-effects.
9. Stress relieved at 120°C for 3 hours. Final hardness 65 R_c.
10. Lapped to 5.000000 ± 0.000002 cm.

The refrigeration treatment, and the double tempering, are designed to convert as much retained austenite as possible to martensite. Step 6 causes the ϵ carbide to precipitate, so that the martensite remains stable at 20°C. Step 9 decreases the residual stresses generated by the grinding and tempers the fresh martensite formed in step 6.

CHAPTER
SEVEN
STAINLESS STEELS

Stainless steels are a class of iron-base alloys which are noted for their relatively high corrrosion and oxidation resistance. They contain from 11 to 30% chromium, with, in some cases, other additions, notably nickel (Table 7-1). They can be classified into three general categories, based on stable or metastable phases which are prevalent in each class. In this chapter, we will look at the influence of the chemistry and temperature on the phases present and at the phase transformations that occur, and the relationship of both of these to properties.

It must be kept in mind that the adjective *stainless* is a rather general term, and relative. Additions of chromium to iron usually will improve the resistance to corrosion and oxidation of iron, through the formation of a relatively adherent, passive oxide film. The choice of high chromium-content alloys is based on their relatively better corrosion resistance, which is illustrated in Fig. 7-1. There are environments, however, in which such steels are not suitable, and care must be taken when choosing stainless steels. It is thus quite important to understand the influence of the chemical composition of these steels, and their heat treatments, on corrosion resistance.

7-1 FERRITIC STAINLESS STEELS

Ferritic stainless steels are those composed mainly of a body-centered cubic phase. Body-centered cubic chromium causes the ferrite region of iron to be broadened; an element causing this effect is referred to as a ferrite stabilizer. The iron-chromium phase diagram is shown in Fig. 7-2. Note that the low-temperature ferrite phase and the high-temperature delta phase are now continuous. The face-centered cubic austenite region is closed off to form a loop. To obtain a structure that contains no austenite,

Table 7-1 Chemical compositions of common stainless steels

Chemical compositions of wrought ferritic chromium stainless steels*
(Not hardenable by heat treatment)

Chemical composition limits, %

SAE number[a]	C max.	Mn max.	Si max.	P max.	S max.	Cr range	Ni range	Other elements	AISI type number[a]
51405[b]	0.08	1.00	1.00	0.040	0.030	11.50–14.50	–	Al, 0.10–0.30	405
51409	0.08	1.00	1.00	0.045	0.045	10.50–11.75	0.50 max.	Ti, 6 × C or 0.75 max.	–
51430	0.12	1.00	1.00	0.040	0.030	14.00–18.00	–	–	430
51430F	0.12	1.25	1.00	0.06	0.15 min.	14.00–18.00	–	Zr or Mo, 0.60 max.[c]	430F
51430F Se	0.12	1.25	1.00	0.06	0.06	14.00–18.00	–	Se, 0.15 min.	430F Se
51434	0.12	0.30–0.90	0.50	0.040	0.030	16.00 min.	–	Mo, 0.75–1.25	–
51436	0.12	0.30–0.90	0.50	0.040	0.030	16.00 min.	–	Mo, 0.75–1.25; Cb. 0.25–0.75	–
51442	0.20	1.00	1.00	0.04	0.035	18.00–23.00	–	–	–
51446	0.20	1.50	1.00	0.04	0.030	23.00–27.00	–	N, 0.25 max.	446
51502	0.10	1.00	1.00	0.04	0.030	4.00–6.00	–	Mo, 0.40–0.65	502

*Based on information in SAE Standard J405b in the 1969 "SAE Handbook."

[a] Suffix F–denotes a free-machining steel; Se–denotes a free-machining steel with selenium addition.

[b] Essentially non-hardenable by heat treatment.

[c] At producer's option; reported only when intentionally added.

Chemical compositions of wrought martensitic chromium stainless steels*
(Hardenable by thermal treatment)

SAE number[a]	Chemical composition limits, % C max.	Mn max.	Si max.	P max.	S max.	Cr range	Ni range	Other elements	AISI type number[a]
51403	0.15	1.00	0.50	0.040	0.030	11.50–13.00	–	–	403
51410	0.15	1.00	1.00	0.040	0.030	11.50–13.50	–	–	410
51414	0.15	1.00	1.00	0.040	0.030	11.50–13.50	1.25–2.50	–	414
51416	0.15	1.25	1.00	0.06	0.15 min.	12.00–14.00	–	Zr or Mo, 0.60 max.[b]	–
51416 Se	0.15	1.25	1.00	0.06	0.06	12.00–14.00	–	Se, 0.15 min.	416 Se
51420	Over 0.15	1.00	1.00	0.040	0.030	12.00–14.00	–	–	420
51420F	0.30–0.40	1.25	1.00	0.06	0.15 min	12.00–14.00	–	Zr or Mo, 0.60 max.[b]	–
51420F Se	0.30–0.40	1.25	1.00	0.06	0.06	12.00–14.00	–	Se, 0.15 min.	–
51531	0.20	1.00	1.00	0.040	0.030	15.00–17.00	1.25–2.30	–	431
51440A	0.60–0.75	1.00	1.00	0.040	0.030	16.00–18.00	–	Mo, 0.75 max.	440A
51440B	0.75–0.95	1.00	1.00	0.040	0.030	16.00–18.00	–	Mo, 0.75 max.	440B
51440C	0.95–1.20	1.00	1.00	0.040	0.030	16.00–18.00	–	Mo, 0.75 max.	440C
51440F	0.95–1.20	1.25	1.00	0.06	0.15 min.	16.00–18.00	–	Zr or Mo, 0.75 max.[b]	–
51440F Se	0.95–1.20	1.25	1.00	0.06	0.06	16.00–18.00	–	Se, 0.15 min.	–
51501	Over 0.10	1.00	1.00	0.040	0.030	4.00–6.00	–	Mo, 0.40–0.65	501

Source: "The Making, Shaping and Treating of Steel," 9th ed., United States Steel Corporation, Pittsburgh, 1971.

*Based on information in SAE Standard J405b in the 1969 "SAE Handbook."

[a]Suffixes A, B and C denote differing carbon ranges for the same grade; F–a free-machining steel; Se–a free-machining steel with selenium addition.

[b]At producer's option; reported only when intentionally added.

Table 7-1 *(continued)*

Chemical compositions of wrought chromium nickel austenitic stainless steels*
(Not hardenable by thermal treatment)

SAE number[a]	Chemical composition limits, % C max.	Mn max.	Si max.	P max.	S max.	Cr range	Ni range	Other elements	AISI type number[a]
30201	0.15	5.5–7.5	1.00	0.060	0.030	16.00–18.00	3.50–5.0	N, 0.25 max.	201
30202	0.15	7.5–10.00	1.00	0.060	0.030	17.00–19.00	4.00–6.00	N, 0.25 max.	202
30301	0.15	2.00	1.00	0.045	0.030	16.00–18.00	6.00–8.00	–	301
30302	0.15	2.00	1.00	0.045	0.030	17.00–19.00	8.00–10.00	–	302
30302B	0.15	2.00	2.00–3.00	0.045	0.030	17.00–19.00	8.00–10.00	–	302B
30303	0.15	2.00	1.00	0.20	0.15 min.	17.00–19.00	8.00–10.00	Zr or Mo, 0.60 max.[c]	303
30303 Se	0.15	2.00	1.00	0.20	0.06	17.00–19.00	8.00–10.00	Se, 0.15 min.	303 Se
30304	0.08	2.00	1.00	0.045	0.030	18.00–20.00	8.00–12.00	–	304
30304L	0.03	2.00	1.00	0.045	0.030	18.00–20.00	8.00–12.00	–	304L
30305	0.12	2.00	1.00	0.045	0.030	17.00–19.00	10.00–13.00	–	305
30308	0.08	2.00	1.00	0.045	0.030	19.00–21.00	10.00–12.00	–	308
30309	0.20	2.00	1.00	0.045	0.030	22.00–24.00	12.00–15.00	–	309
30309S	0.08	2.00	1.00	0.045	0.030	22.00–24.00	12.00–15.00	–	309S
30310	0.25	2.00	1.50	0.045	0.030	24.00–26.00	19.00–22.00	–	310
30310S	0.08	2.00	1.50	0.045	0.030	24.00–26.00	19.00–22.00	–	310S

SAE number[a]	C max.	Mn. max.	Si max.	P max.	S max.	Cr range	Ni range	Other elements	AISI type number[a]
30314	0.25	2.00	1.50–3.00	0.045	0.030	23.00–26.00	19.00–22.00	–	314
30316	0.08	2.00	1.00	0.045	0.030	16.00–18.00	10.00–14.00	Mo, 2.00–3.00	316
30316L[d]	0.03	2.00	1.00	0.045	0.030	16.00–18.00	10.00–14.00	Mo, 2.00–3.00	316L
30317	0.08	2.00	1.00	0.045	0.030	18.00–20.00	11.00–15.00	Mo, 3.00–4.00	317
30321[e]	0.08	2.00	1.00	0.045	0.030	17.00–19.00	9.00–12.00	Ti, 5 X C min.	321
30330	0.15	2.00	1.50[b]	0.045	0.04	14.00–17.00	33.0–37.0	–	–
30347	0.08	2.00	1.00	0.045	0.030	17.00–19.00	9.00–13.00	Cb-Ta, 10 X C min.	347
30348	0.08	2.00	1.00	0.045	0.030	17.00–19.00	9.00–13.00	Cb-Ta, 10 X C min.; Ta, 0.10 max.	348
USS TEN-ELON	0.12	14.50–16.00	1.00	0.040	0.030	17.00–18.00		N, 0.35–0.50	
USS 18-18-2	0.07	2.00	1.80–2.20	0.040	0.020	17.00–19.00	17.50–18.50	–	

*Based on information in SAE Standard J405b in the 1969 "SAE Handbook."

[a]The suffixes with grade numbers denote: B–2.00–3.00 silicon range; Se–a free-machining steel with selenium addition; L–extra low-carbon grade; S–lower carbon grade.

[b]To minimize carbon or nitrogen pick-up 0.75–1.50 Si is recommended for high-temperature application involving carbon or nitrogen atmosphere.

[c]At producer's option; reported only when intentionally added.

[d]10.0–15.0 Ni permitted for tubular products.

[e]9.0–13.0 Ni permitted for tubular products.

the minimum chromium content must be 12.7%. The upper limit on the chromium content is dictated by the possibility of the formation of the σ phase. Its formation may not be rapid, but in general the ferritic stainless steels are limited in chemical composition to 27% Cr (see Table 7-1).

Carbon is a strong austenite stabilizer, so its influence on the formation of ferrite in these steels must be carefully considered. Figure 7-3 shows isopleths at increasing carbon contents. In Fig. 7-2 it is seen that a 15% Cr-Fe alloy is ferritic from about 500°C to the melting point. However, the addition of 0.05% C (Fig. 7-3) makes this composition two phase ($\alpha + \gamma$) from about 850 to 1350°C. Also, below about 850°C, the equilibrium structure is ferrite and $(Fe, Cr)_{23}C_6$. Increasing the carbon content will increase the amount of austenite present at any temperature. The austenite in these steels has a high hardenability, as evidenced by the isothermal TTT diagram in Fig. 7-4. Thus, unless cooled very slowly from the range of 1000°C, the austenite present will transform to martensite. The martensite is relatively low in carbon content and will have a hardness for 0.1% C of about 47 R_c. Tempering will eventually convert the martensite to ferrite and $(Cr, Fe)_{23}C_7$ carbides.

We now examine in more detail the heat treatment of 430 stainless steel. From Table 7-1 the carbon content is 0.12% maximum, and the chromium content is between 14 and 18%. We will use 17% Cr in our analysis. Let us first examine what treatments will give neither austenite nor martensite in the final structure. If the steel is heated below 900°C, Fig. 7-3 shows that the structure is ferrite and $(Fe, Cr)_{23}C_6$. This structure will be retained on cooling either rapidly or slowly to 20°C. If, however, the chromium content is at the lower limit, 14%, then at 900°C some austenite will be present. This can be converted on cooling to ferrite and carbides only by slowly cooling; the rate can be approximated from Fig. 7-4. For example, cooling between 15 to 30°C/hour to about 600°C, then cooling faster (e.g., air cooling) will give, at 20°C, ferrite and carbide.

If the steel is heated in the range 700 to 800°C, the phase diagram shows that ferrite and carbides will be present, and even for the lower chromium content no austenite will be present. Thus the steel can be air cooled or even water quenched from this range. It must be kept in mind that we are using the 0.10% C isopleth as a guide for a steel that contains 0.12% C. The carbon content has a strong influence, so the temperatures indicated here may have to be adjusted to make them conservative.

Two recommended heat treatments for 430 stainless steel to maintain a ferritic structure are given in Table 7-2. The temperatures and cooling rates in Table 7-2 are predicted quite well from the phase diagram.

These heat treatments are important to consider because of the heating processes involved in fabricating the steel. The two-phase structure of either austenite or martensite with the ferrite is avoided because of the reduced corrosion resistance. Note that increased carbon content requires generally higher chromium content to maintain as high a ferrite content.

Figure 7-5 illustrates microstructures for 430 stainless steel. The hot-rolled structure is developed by rolling in the range where austenite and ferrite were present; on cooling, part of the austenite transformed, by the eutectoid reaction, to ferrite and carbide (pearlite) and the remainder transformed to martensite. The other figure shows

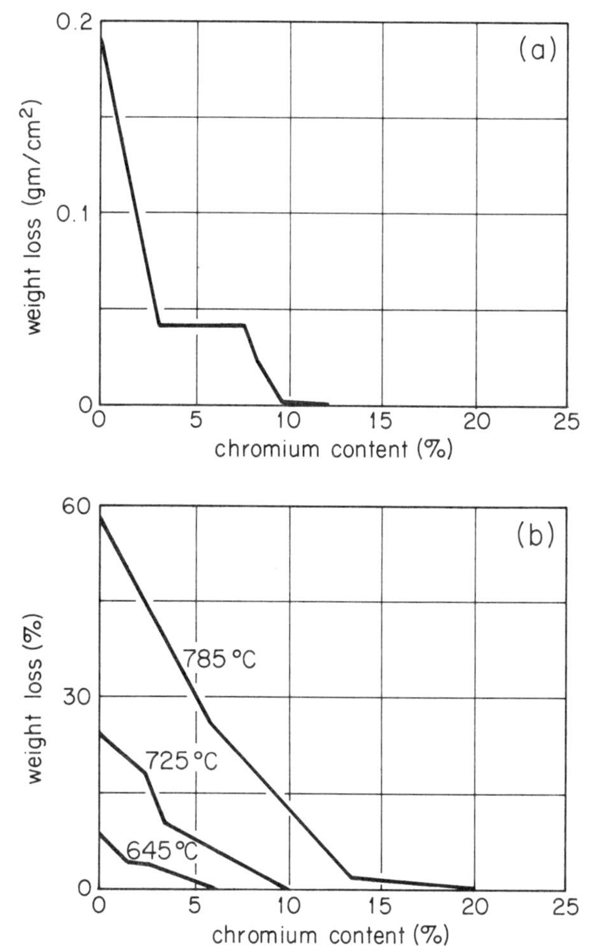

Figure 7-1 The effect of chromium content on iron-0.1% C steels. (a) Exposed at 20°C for 10 years to industrial atmosphere. (b) Exposed to air for 900 to 1000 hours at three different temperatures. *(Adapted from R. Franks,* Trans. ASM, *vol. 35, p. 616, 1945.)*

the microstructure developed by annealing at 843°C and cooling in air. At this temperature, the structure is ferrite and carbide, which does not change on cooling to 20°C.

In the ferrite and carbide condition, these steels are not hardenable by heat treatment, only by work hardening. The ferrite grain size can be controlled by proper annealing following cold working, and in applications requiring small grain size this problem must be considered.

Embrittlement

The iron-carbon-chromium phase diagrams we have been using do not show the formation of sigma phase, although this phase will form if held for sufficient time in the proper temperature range. In fact, the ferritic stainless steels containing more than about 12% Cr show embrittlement when held near 475°C; this effect has been referred to as 475°C (885°F) embrittlement, as this is the temperature at which the

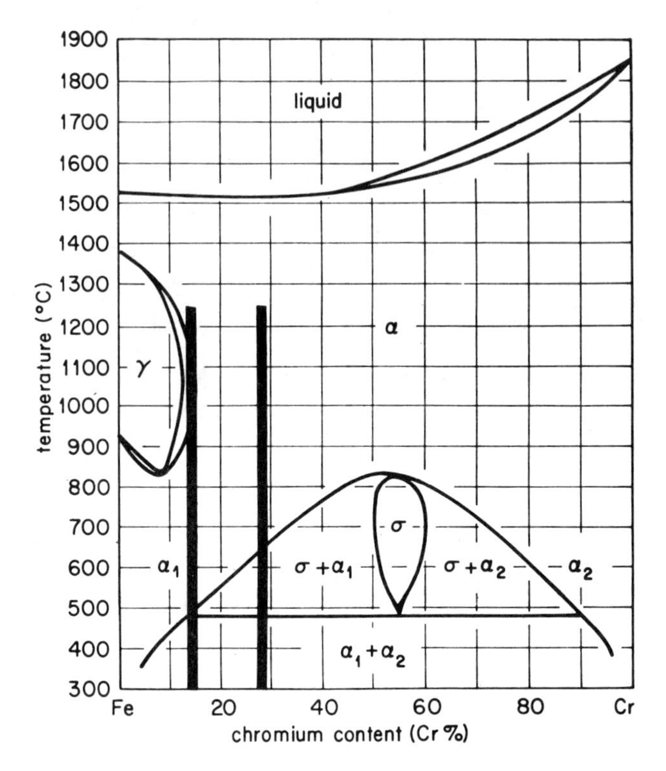

Figure 7-2 The iron-chromium phase diagram. The range of chromium content in commercial ferritic stainless steels falls between the bars. *(Adapted from "Metals Handbook," vol. 8, American Society for Metals, Metals Park, Ohio, 1973.)*

effect is most severe. The effect is more pronounced for higher chromium content. The embrittlement can be removed by reannealing above 650°C. Some data illustrating this embrittlement are shown in Fig. 7-6. For this steel, no precipitates were observed with optical microscopy to form upon aging, even though embrittlement occurred. Examination of the iron-chromium phase diagram (Fig. 7-2) shows that the embrittlement is probably associated with the formation of the chromium-rich α_2 phase. This precipitate forms slowly and is very fine and difficult to detect.

Heating above 475°C can, however, result in the precipitation of sigma, especially in the higher chromium alloys. Sigma is a very brittle phase, and the use of ferritic stainless steels under conditions of its formation must be considered carefully. The effect of temperature and chromium content on the rate of formation is shown in Fig. 7-7. The temperature above which no sigma forms, for a given chromium content, correlates with that given by the phase diagram (Fig. 7-2). The effect of sigma formation on the mechanical properties is illustrated in Fig. 7-8.

Another problem with the ferritic stainless steels is that their impact properties are sensitive to the ferrite grain size. This is a problem in body-centered cubic alloys, and in the ferritic stainless steels the grain size can be controlled only by cold working

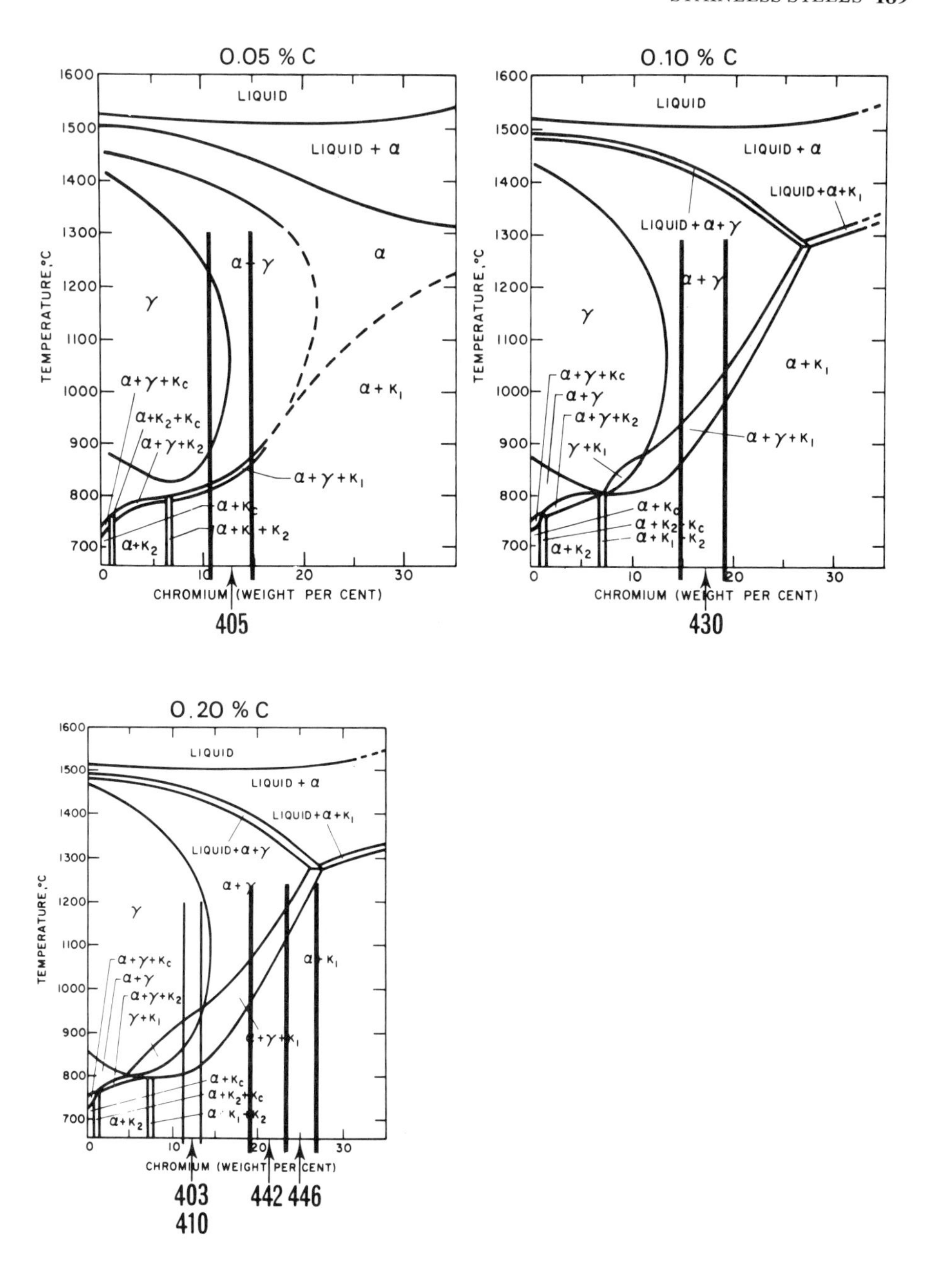

Figure 7-3 Isopleths from the iron-chromium-carbon system at 0.05% C, 0.10% C, and 0.20% C. The range of chromium contents for four commercial ferritic (405, 430, 446, 442) and martensitic (403, 410) stainless steels of similar carbon contents are shown. *(Adapted from H. E. McGannon (ed.), "The Making, Shaping and Treating of Steel," 9th ed., United States Steel Corporation, Pittsburgh, 1971.)*

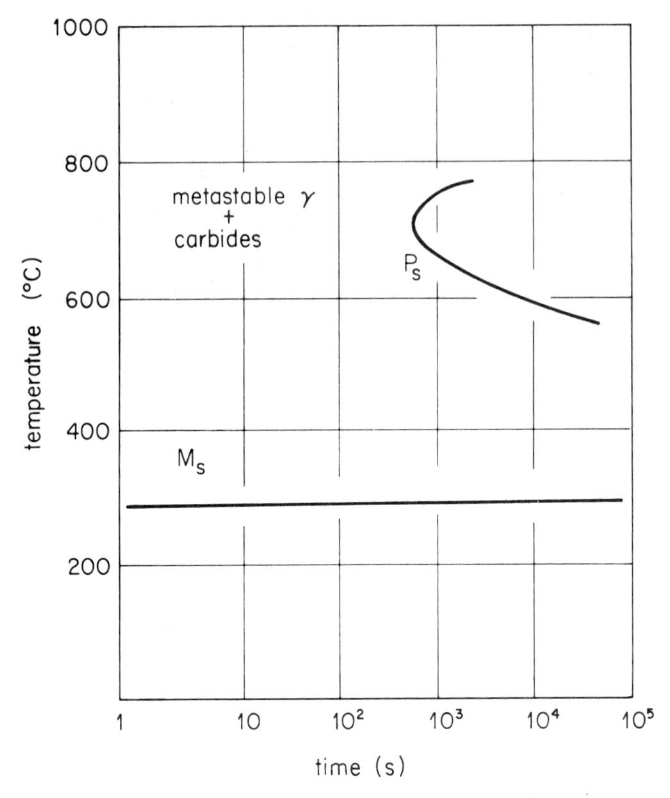

Figure 7-4 Isothermal TTT diagram for a 0.1% C–12% Cr steel. *(Adapted from K. J. Irvine, D. J. Crowe, and F. B. Pickering,* J.I.S.I., *vol. 195, p. 386, 1960.)*

and annealing. Thus prolonged heating at high temperature will cause grain growth, and hence lowering of the impact resistance. This can be a problem particularly in welding, where the cast structure or the heat-affected zone may contain large grains which cannot be altered if the fabricated part cannot be cold worked and annealed. The weld material, and perhaps the base plate, must be chosen to minimize this

Table 7-2 Two recommended heat treatments for 430 stainless steel for prevention of austenite or martensite formation

1. Heat to 705 to 790°C,
   air cool or water quench
2. Heat to 815 to 900°C,
   cool from 14 to 28°C/hour to 595°C,
   then air cool

Source: "Metals Handbook," 8th ed., vol. 2, p. 244, American Society for Metals, Metals Park, Ohio, 1964.

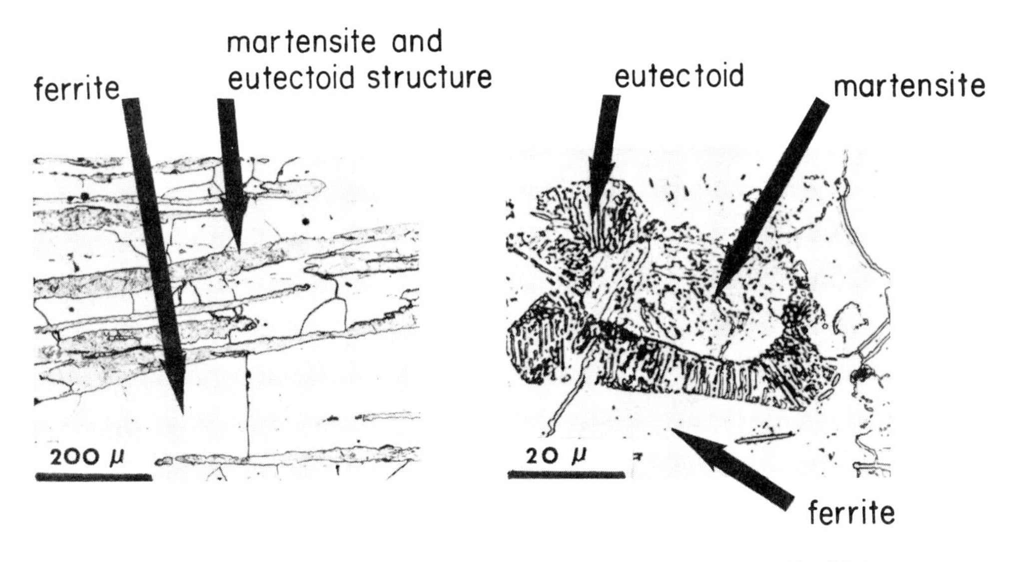

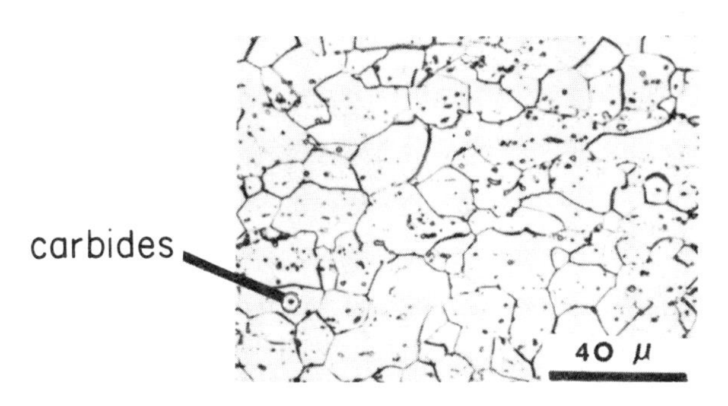

Figure 7-5 Microstructures of 430 stainless steel for two different heat treatments. *(From "Metals Handbook," vol. 7, 8th ed., American Society for Metals, Metals Park, Ohio, 1972.)*

problem. This may require a second phase (e.g., carbides) present at the high temperature to retard grain growth.

7-2 MARTENSITIC STAINLESS STEELS

The martensitic stainless steels must have the proper combination of chromium and carbon (and other elements) so that austenite is present at high temperature, and

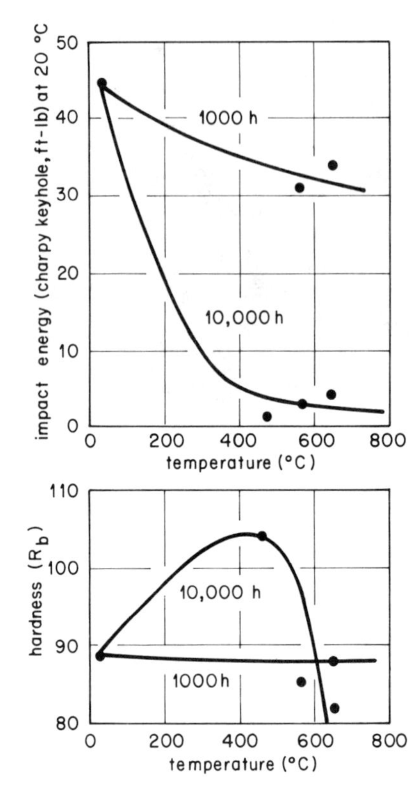

Figure 7-6 Impact energy and hardness (both measured at 20°C) as a function of aging temperature for two aging times for a 430 stainless steel. *(Adapted from A. B. Wilder and J. O. Light,* Trans. ASM, *vol. 41, p. 141, 1949.)*

martensite present upon cooling to 20°C. These steels are stronger than the ferritic and austenitic class, and their utilization relies on the high strength along with corrosion resistance.

To illustrate their heat treatment, consider a 0.14% C–12% Cr steel. This composition is essentially that for a 403 martensitic stainless steel (Table 7-1). From Fig. 7-3, it is seen that this steel will have to be heated between 950 and 1250°C to be all austenite. If the steel being considered here is solution-austenitized for about 1 hour at 1050°C, then air cooled, the structure will be all martensite with a hardness of about 46 R_c. The high hardenability allows the use of air hardening without formation of pearlite or bainite (Fig. 7-4). Note that the hardness of the martensite is predicted well from Fig. 1-31 based only on the carbon content. The M_f temperature is above 20°C, in spite of the high chromium content of the austenite, because of the low carbon content.

The tempering behavior of this steel is shown in Fig. 7-9. Cooling slowly may allow some autotempering, causing some fine $(Fe, Cr)_3C$ precipitates to be present in the fresh martensite. These, however, do not significantly influence the hardness. Tempering below about 350°C causes more of these carbides to form and coarsen, causing only a very slight drop in hardness (for tempering times up to 100 hours). The chromium in the martensite is effective in preventing rapid growth of these carbides.

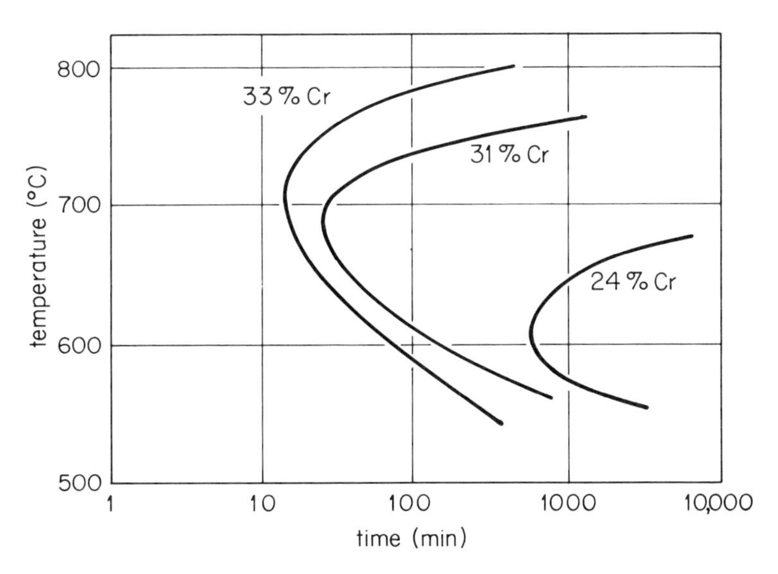

Figure 7-7 The effect of temperature and chromium content on the start of sigma formation in three iron-chromium-chromium alloys each containing about 0.05% C. *(Adapted from F. J. Shortsleeve and M. E. Nicholson,* Trans. ASM, *vol. 43, p. 142, 1951.)*

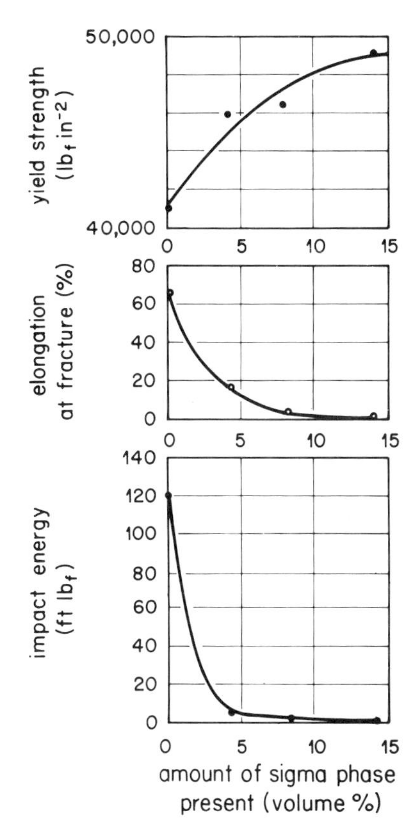

Figure 7-8 The effect of the amount of sigma on the mechanical properties (at 20°C). The steel contained about 0.15% C–25% Cr–21% Ni. *(Adapted from J. I. Morely and H. W. Kirkby,* J.I.S.I., *vol. 172, p. 129, 1952.)*

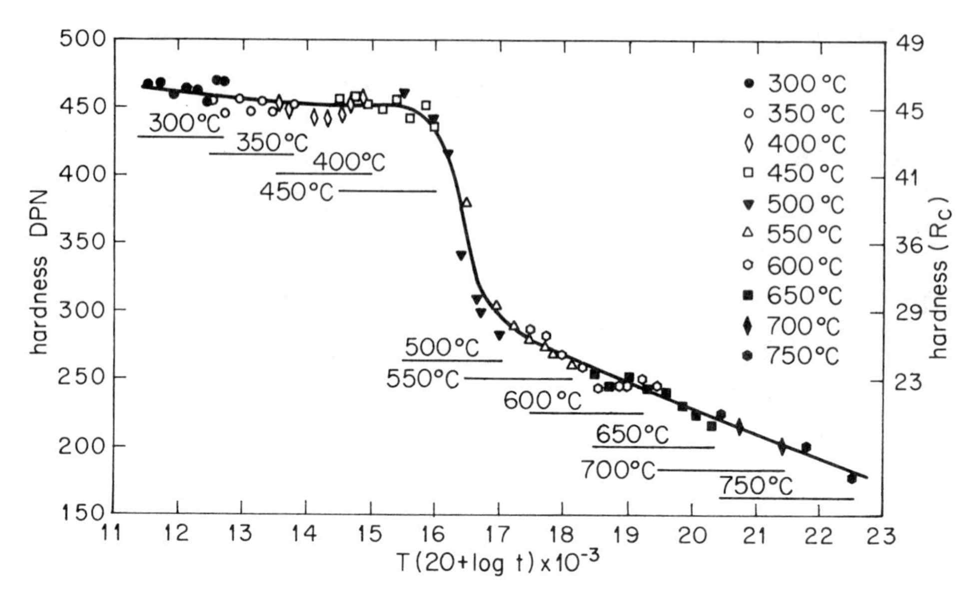

Figure 7-9 Master tempering curve for a 0.1% C–12% Cr steel. *(Adapted from K. J. Irvine, D. J. Crowe, and F. B. Pickering,* J.I.S.I., *vol. 195, p. 386, 1960.)*

Tempering from 300 to 450°C causes secondary hardening, although the increase is slight; the most important effect is retardation of softening. In this range the (Fe, $Cr)_3C$ carbides dissolve and are replaced by fine (Cr, $Fe)_7C_3$ carbides. This is not the equilibrium carbide, which Fig. 7-3 shows to be (Fe, $Cr)_{23}C_6$. A Cr_2C-type carbide is also formed in this range, which makes a major contribution to the secondary hardening.

At 500°C and above, the equilibrium carbide (Cr, $Fe)_{23}C_6$ forms, and coarsens. The coarse carbides, together with the simultaneous formation of ferrite, cause marked softening (Fig. 7-9). In this range the non-equilibrium carbides will eventually disappear.

The impact properties may be of concern along with the strength (or hardness); the effect of tempering temperature on these for this steel is shown in Fig. 7-10. The impact energy was measured at 20°C, and it is especially low for a tempering temperature of around 500°C. This minimum is associated with the formation of rather coarse carbide precipitates along the grain boundaries. It is not the presence of the carbides, per se, which is a problem, but the lowering of the chromium content of the ferrite adjacent to the carbides. This depleted region is consequently weaker and is a site for easier crack initiation. The addition of elements such as molybdenum to cause solid-solution strengthening of the ferrite improves the impact properties in this temperature range.

The martensitic stainless steels must be softened for machining. This can be accomplished by cooling very slowly from the austenite region (although the high hardenability makes the time involved long), or the hardened steel can be tempered.

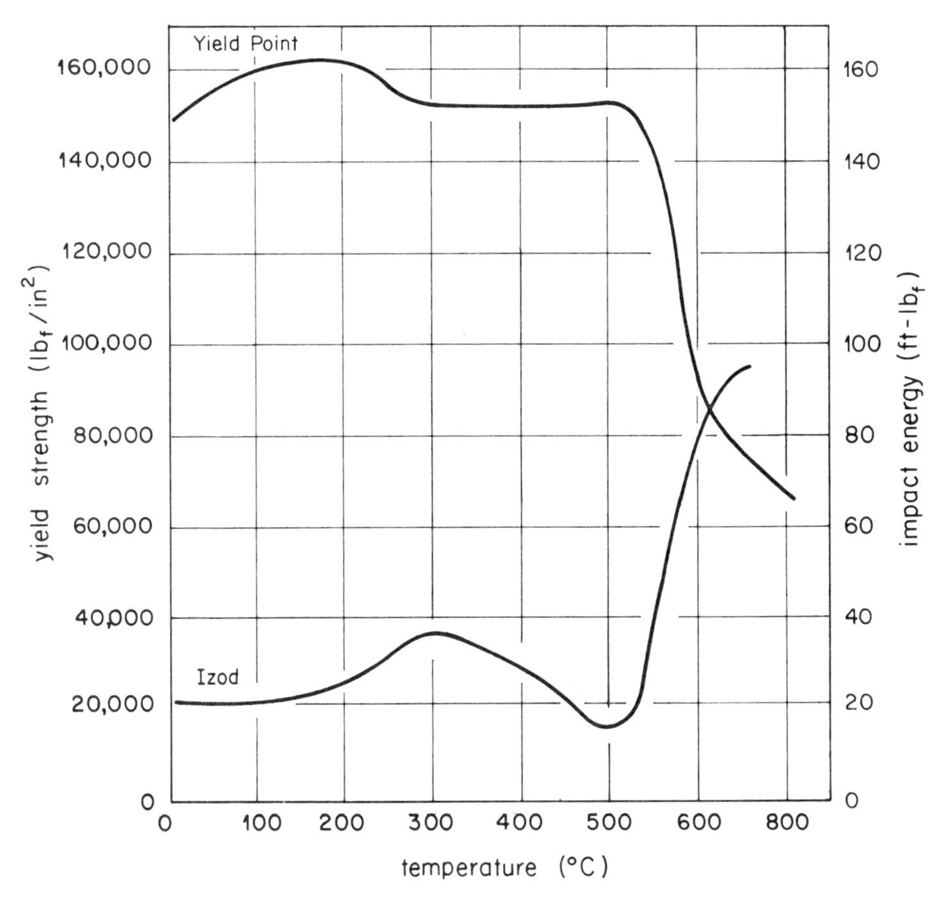

Figure 7-10 Effect of a tempering temperature on the impact value and yield strength (at 20°C) for a 0.10% C–11.7% Cr steel. *(Adapted from N. L. Mochel,* Trans. ASST, *vol. 10, p. 353, 1926.)*

The corrosion resistance of the martensitic stainless steels is sensitive to the tempering temperature. Figure 7-11 illustrates the behavior of 410 stainless steel, which was quenched from about 1000°C to 20°C to form all martensite. Tempering in the range of about 400 to 900°C lowers the corrosion resistance. Above about 950°C, the steel is back in the austenite region. Thus it is seen that the martensitic steels should be used in the all-martensitic condition to retain the best corrosion resistance, although a low tempering temperature (e.g., below about 300°C) retains good corrosion resistance. Note that the low temperature range of loss in corrosion resistance corresponds approximately to that of embrittlement (Fig. 7-10).

Table 7-3 gives typical heat treatments for 403 stainless steel (which are similar to those for 410 stainless steel). They correlate well with the phase diagram, the TTT diagram, and the tempering data. Figure 7-12 illustrates the recommended heat treatments and hardness attained for martensitic stainless steels.

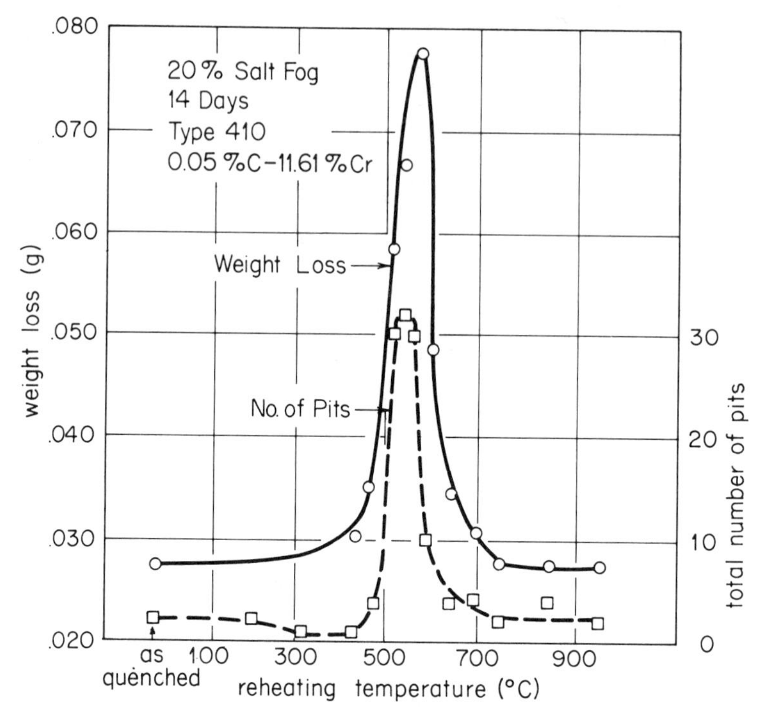

Figure 7-11 The effect of tempering temperature on the corrosion resistance of 410 stainless steel. *(Adapted from F. K. Bloom,* Corrosion, *vol. 9, p. 56, 1953.)*

Table 7-3 Typical heat treatments for 403 stainless steel

Hardening 930 to 1010°C for 1 hour, air or oil quench	*Tempering* 205 to 370°C or 565 to 610°C; avoid the range of 370 to 565°C
Full annealing 830 to 885°C for 1 hour, furnace cool to 595°C at 14 to 28°C/hour, air cool to 20°C	*Isothermal annealing* 830 to 885°C for 1 hour, hold 6 hours at 705°C, air cool to 20°C

Source: Adapted from "Metals Handbook," 8th ed., vol. 2, p. 245-247, American Society for Metals, Metals Park, Ohio, 1964.

7-3 AUSTENITIC STAINLESS STEELS

The austenitic stainless steels are those whose structure is normally the face-centered cubic austenite. This is achieved by the addition of strong austenite stabilizers which either make austenite the stable phase, or retard the formation of ferrite so that the metastable austenite is the normal phase present. To promote the stability of the

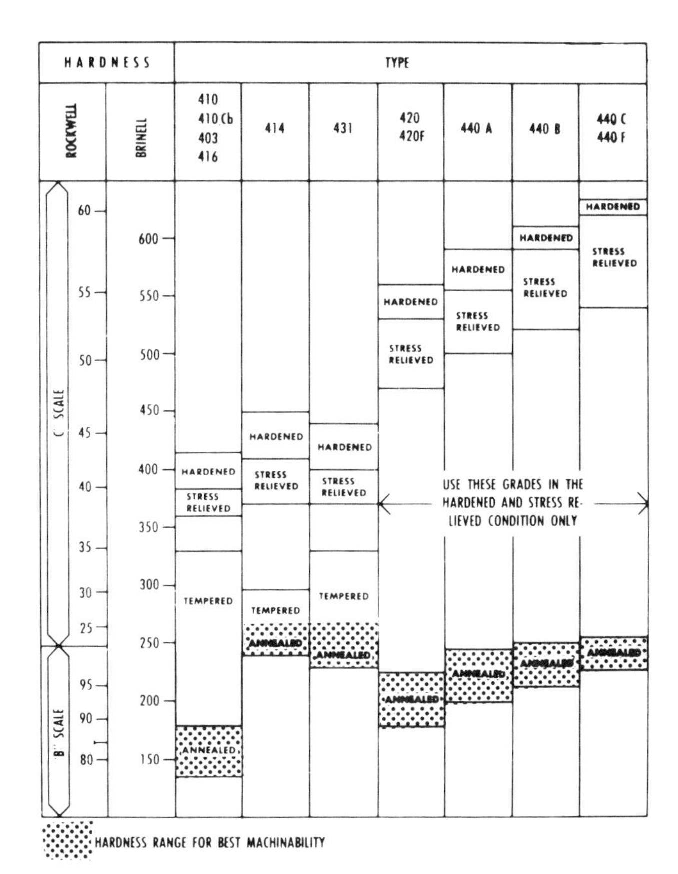

Figure 7-12 Recommended heat-treating practices for martensitic stainless steel. *(From "Heat Treating Armco Stainless Steels," Armco Steel Corporation, Middletown, Ohio, 1966.)*

austenite, nickel is the most common addition to the basic iron-chromium composition. Thus these steels are frequently thought of as 18-8 stainless steels, referring to 18% Cr–8% Ni common to many of them. Carbon is another important element.

To introduce the physical metallurgy of this class of stainless steels, we will center attention first on the ternary alloy of iron, chromium, and nickel. Figure 7-13 presents isotherms from the ternary system at three temperatures. Taking an 18% Cr–8% Ni composition for discussion, it is seen that this composition is just in the austenite region (small circle) at 1100°C. Cooling to 800°C establishes ferrite and austenite as the stable phases. At 650°C, austenite, a small amount of ferrite, and the brittle sigma phase are stable. More sigma is present at 550°C, along with ferrite and austenite. The triangular composition-temperature diagram is shown schematically in Fig. 7-14 to illustrate the origin of the shape of the isotherm.

The formation of the sigma is sufficiently slow that we will not consider its formation now but treat this in a later section. Isopleths then reveal only ferrite and austenite. Isopleths for increasing chromium content are shown in Fig. 7-15. It is seen

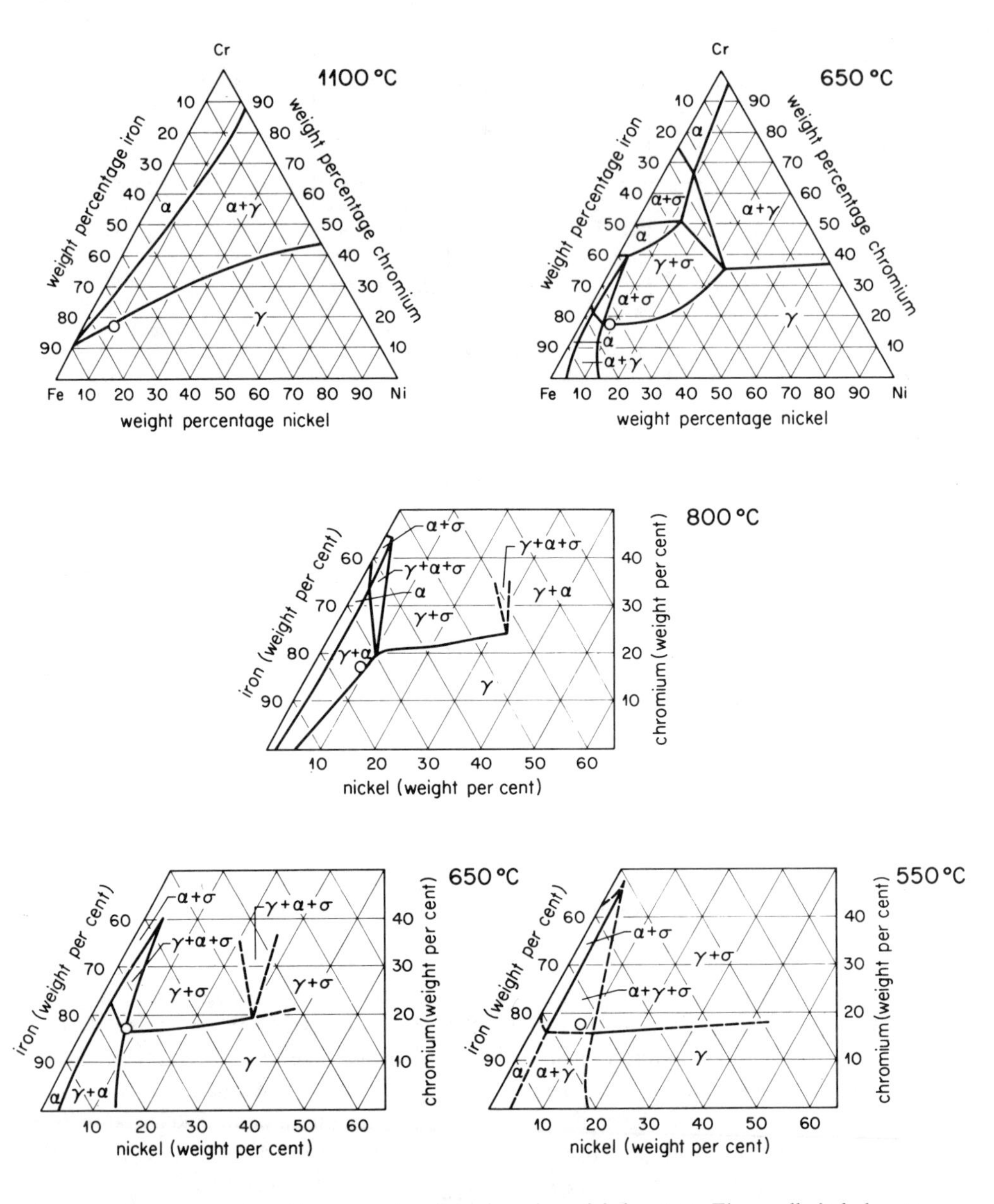

Figure 7-13 Isothermal sections from the iron-chromium-nickel system. The small circle locates the composition 18% Cr–8% Ni. *(Upper diagrams: from "Metals Handbook," vol. 8, 8th ed., American Society for Metals, Metals Park, Ohio, 1973, Lower diagrams: from W. P. Rees, B. D. Burns, and A. J. Cook,* J.I.S.I., *vol. 162, p. 325, 1949.)*

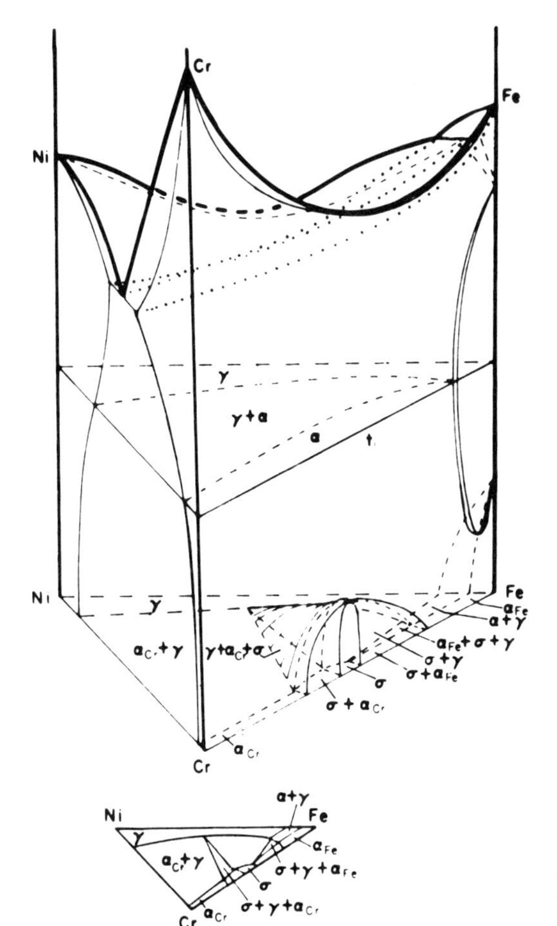

Figure 7-14 The iron-chromium-nickel phase diagram, shown schematically. *(From W. Hume-Rothery, "The Structures of Alloys of Iron," Pergamon Press, London, 1966.)*

at 18% Cr that less than about 7% Ni will place the alloy in either the ferrite region or the austenite and ferrite region, depending on the temperature. Above about 7% Ni, the alloy is austenitic at high temperatures (above about 800°C) but ferritic at low temperatures. However, the formation of the ferrite is suppressed by sufficiently rapid cooling from the austenite region. This shows the basis for a minimum of about 8% Ni for most of the austenitic stainless steels (see Table 7-1). Table 7-1 also shows that the higher chromium steels have higher nickel contents. This is required to compensate for the ferrite stabilizing effect of the increasing chromium content. For example, 309 stainless steel contains from 22 to 24% Cr. Figure 7-15 shows that at 24% Cr the nickel content must be above about 12% to attain an all-austenite structure. Thus, 309 stainless steel contains 12 to 15% Ni.

The effect of carbon is critical. (As will be discussed, the precipitation of chromium-rich carbides is to be avoided.) Because of the effect of nickel on the influence of carbon, the iron-chromium-carbon phase diagram cannot be used to evaluate the phases present, but instead the quaternary iron-chromium-nickel-carbon diagram

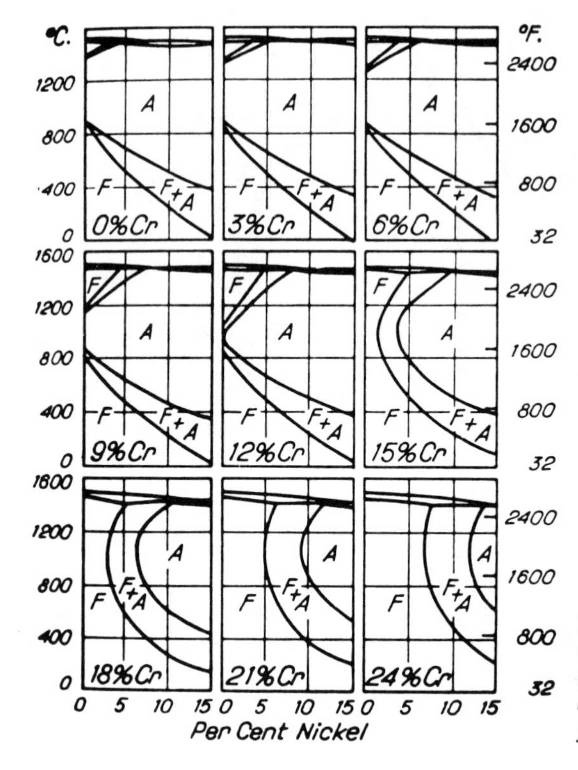

Figure 7-15 Isopleths from the iron-chromium-nickel phase diagram. *(From "Metals Handbook," 1939 edition, p. 421, American Society for Metals, Metals Park, Ohio, 1939.)*

should be examined. The most convenient way to do this is to examine the phase relationships for varying temperature and carbon content, with the chromium and nickel content fixed. Figure 7-16 shows two such isopleths, one for 18% Cr–4% Ni and the other for 18% Cr–8% Ni. For very low (e.g., 0.02%) carbon contents, the stable phase at low temperature is ferrite and carbide. Increasing the carbon content to the range 0.03% or greater will allow austenite also to be stable at low temperature. The rate of formation of the ferrite is low, so that its presence is easily avoided. The carbide formation is less easily controlled. However, for a steel containing 0.08% C, austenitizing for 1 hour at 1000°C and then water quenching (if the part is not too large) will suppress carbide formation, and the structure at 20°C will be austenite.

Examination of the pertinent phase diagrams shows that the composition can be adjusted to attain only austenite at high temperatures, but at sufficiently low temperatures ferrite and carbides may exist at equilibrium with the austenite. We now examine the significance of the formation of these phases.

Formation of Ferrite

Rapidly cooling an austenitic stainless steel from the austenite region will suppress the formation of ferrite. However, the phase diagram (Fig. 7-13) shows ferrite to be stable at sufficiently low temperatures, so slow cooling from the austenite region or prolonged heating in the lower temperature range will allow ferrite to form. It is also

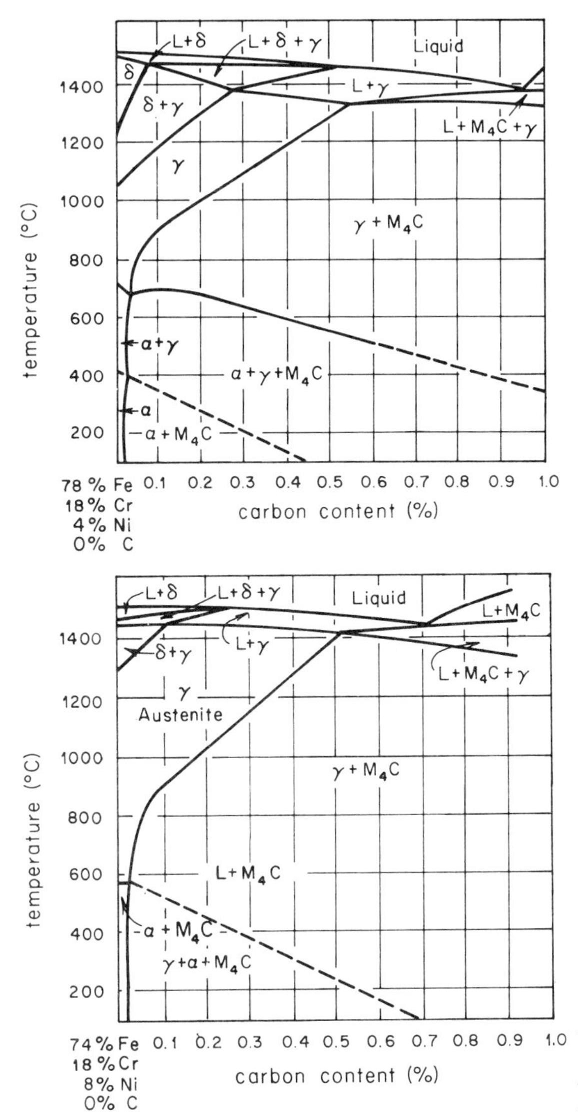

Figure 7-16 Isopleth for 18% Cr–4% Ni and for 18% Cr–8% Ni. *(Adapted from V. N. Krivobok, in E. E. Thum (ed.), "The Book of Stainless Steels," American Society for Metals, Metals Park, Ohio, 1935.)*

possible to have the body-centered cubic phase present in as-cast structures by the formation of the high-temperature ferrite, called δ ferrite, during the solidification process. This is an important factor in the welding of austenitic stainless steels.

The austenitic stainless steels will transform to ferrite martensitically if cooled to a sufficiently low temperature. This is illustrated in Fig. 7-17 for three steels. Although the austenite transforms by a martensite reaction, the phase is body-centered cubic and hence also is referred to as ferrite. This creates some confusion as to what terminology to utilize; however, here we will refer to the phase as martensite. Unlike the formation of martensite in higher carbon steels, in which the martensite will

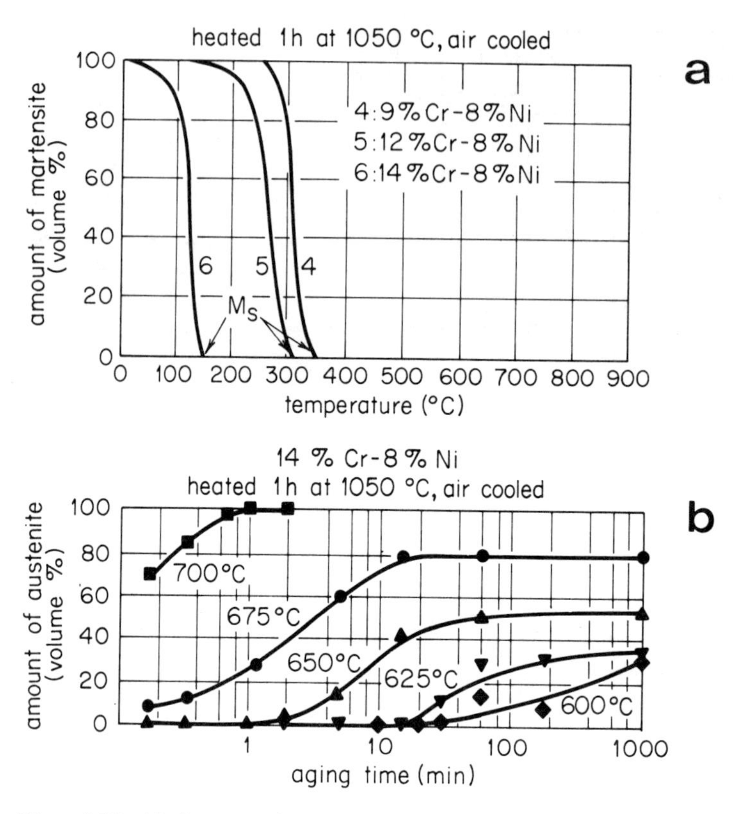

Figure 7-17 (a) Amount of martensitic ferrite formed as a function of temperature for three austenitic stainless steels. (b) Amount of austenite present as a function of aging time at several temperatures. The austenite is forming from the martensitic ferrite, which formed upon the initial cooling to 20°C. *(Adapted from K. Bungardt, W. Spyra, and G. Lennartz, in "Steel-Strengthening Mechanisms," Climax Molybdenum Co., Ann Arbor, Michigan, 1969.)*

transform to carbides and ferrite on reheating, the martensite in these stainless steels will revert to austenite. This is shown also in Fig. 7-17. Note, however, that the two phases α and γ exist in equilibrium in the lower temperatures; this is consistent with the phase diagram (see Fig. 7-13).

An approximate formula* which predicts the M_s temperature is

$$M_s\,(^\circ\mathrm{C}) = 42(14.6 - \%\,\mathrm{Cr}) + 61(8.9 - \%\,\mathrm{Ni}) + 33(1.33 - \%\,\mathrm{Mn}) \\ + 28(0.47 - \%\,\mathrm{Si}) + 1677[0.068 - (\%\,\mathrm{C} + \%\,\mathrm{N})] - 18$$

This relation is valid for the composition limits of 10 to 18% Cr, 6 to 12% Ni, 0.6 to 5% Mn, 0.3 to 2.6% Si, 0.004 to 0.12% C, and 0.01 to 0.06% N. For example, the M_s temperature is approximately –50°C for a 301 stainless steel, and –250°C for a 304 steel.

*From G. H. Eichelman and F. C. Hull, *Trans. ASM,* vol. 45, p. 77, 1953.

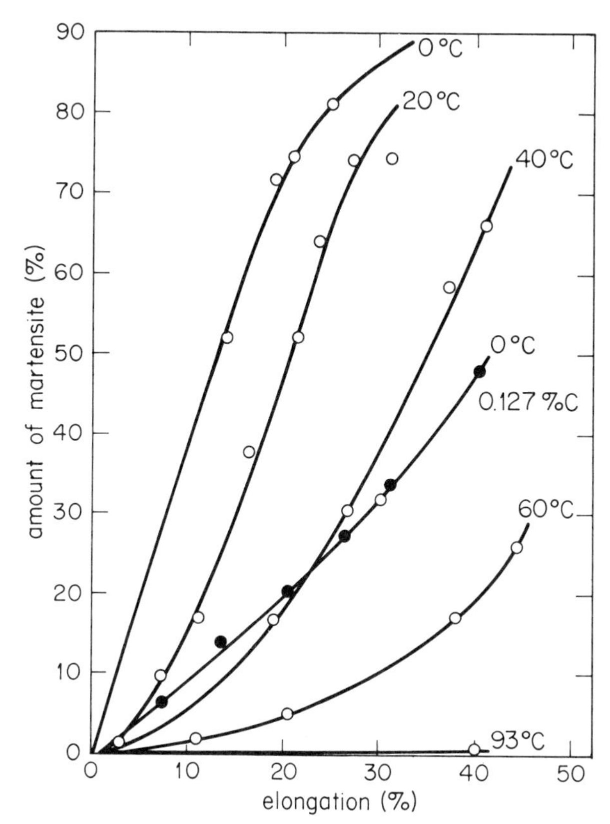

Figure 7-18 The amount of martensite produced in a 0.016% C–8.6% Ni–18.5% Cr stainless steel by tensile plastic deformation at various temperatures. The M_s for this steel is approximately —10°C. Also shown is the curve for a 0.127% C–8.1% Ni–19.8% Cr stainless steel; the M_s for this steel is approximately -227°C. *(Adapted from H. C. Fiedler, B. L. Averbach, and M. Cohen,* Trans. ASM, *vol. 47, p. 267, 1955.)*

Plastic deformation introduces shearing strains which initiate the martensite transformation and hence raise the temperature at which it may occur. However, there is a temperature limit above which martensite does not form upon plastic deformation. This is called the M_d temperature, and it is a function of the composition and the amount of strain. Figure 7-18 gives data for strain-induced martensite formation in an 18% Cr–8% Ni steel with 0.016% C. The M_d temperature for approximately 40% elongation is about 90°C and only a small amount of strain-induced martensite is formed. As the temperature is decreased, the critical strain for martensite formation decreases, and increased amounts of martensite may be formed. The large effect of an increase in the carbon content is evident by comparing the effect of strain at 0°C for the two compositions. The decreased amount of martensite results from the effect of carbon on the M_d temperature.

The initiation of the martensite transformation is clearly shown by examination of the stress-strain curves as a function of temperature for types 301 and 304 stainless steels shown in Fig. 7-19. The type 304 is higher in nickel (Table 7-1) which lowers the M_d below −78°C such that normal stress-strain behavior is observed at 20° and −78°C. However, at −196°C, an increase in flow stress is observed near a true strain of 0.2 due to the onset of strain-induced martensite formation. An increase in flow

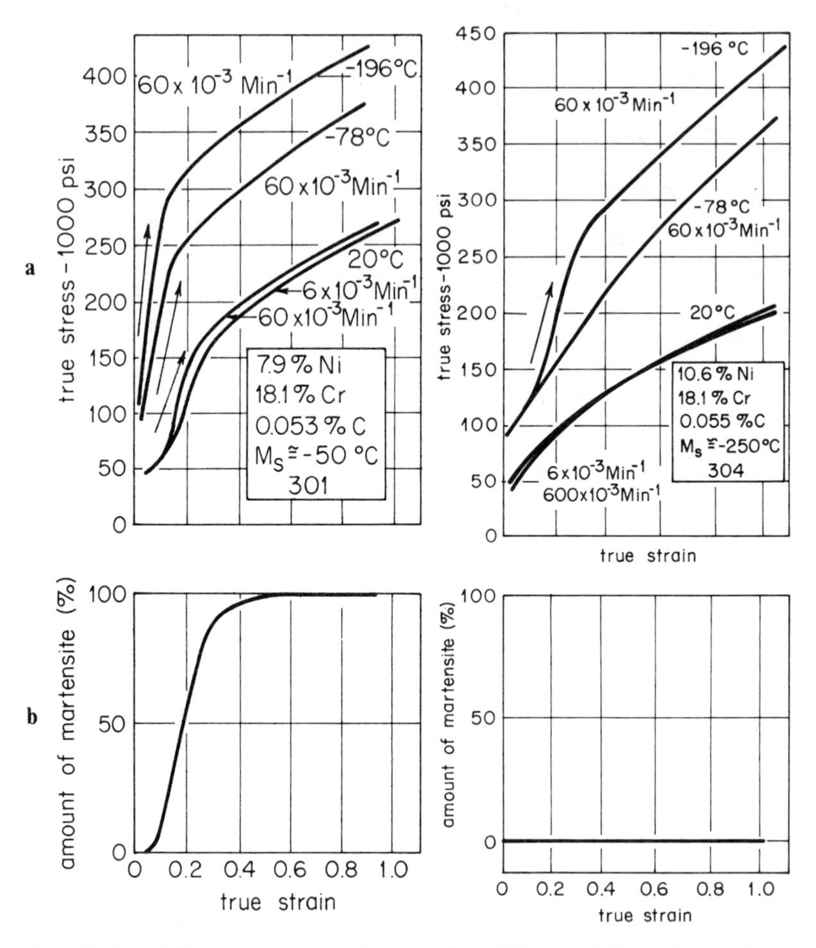

Figure 7-19 (a) True stress-true strain curves for a 301 and a 304 stainless steel tested at different temperatures and strain rates. The arrows show the region on the curves at which plastically induced ferrite forms. (b) The amount of ferrite formed as a function of strain for testing at 20°C and a strain of 60 × 10^{-3} min^{-1}. *(Adapted from G. W. Powell, E. R. Marshall, and W. A. Backofen,* Trans. ASM, *vol. 50, p. 478, 1958.)*

stress greater than that expected for normal work hardening occurs as martensite formation progresses to 80% at a true strain of 0.4. Beyond this strain the amount of martensite forming decreases, and the slope of the stress-strain curve decreases to a value characteristic of the work hardening of the martensitic structure.

By a slight change in analysis to that of the type 301, martensite formation may be strain-induced at 20°C to the extent that 80% transformation occurs at a true strain of approximately 0.3. At −196°C, 100% martensite is formed at a true strain of about 0.1. Note that M_s is above −78°C in the type 301 but is below −196°C in type 304.

The influence of the formation of strain-induced martensite on the strength is

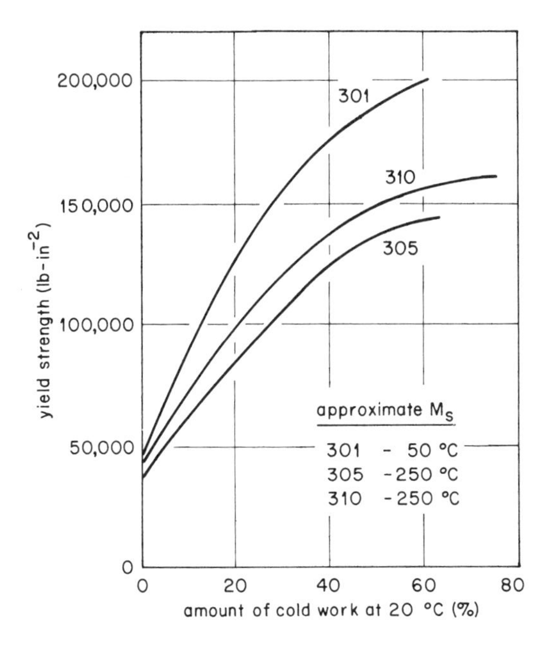

Figure 7-20 Yield strenth at 20°C as a function of reduction in thickness by rolling (% cold work) at 20°C for three austenitic stainless steels. The 301 steel forms martensite upon deformation, and hence shows greater strengthening from plastic deformation. *(Adapted from K. G. Brickner, in "Selection of Stainless Steels," American Society for Metals, Metals Park, Ohio, 1968.)*

clearly seen by examining the strength of the austenitic stainless steels as a function of cold working. Figure 7-20 shows data for three steels. The 301 stainless steel has an M_s of −50°C, and deformation at 20°C, which is below the M_d for this steel, causes the deformation of martensite. This steel thus shows considerably more strengthening than 305 and 310 stainless steels, which have a very low M_s and an M_d temperature below 20°C. The increased work hardening due to martensite formation is important in considering the plastically deformed austenitic steels for use where high strength is required. The strengthening effect is further illustrated by the data in Fig. 7-21. The yield strength can be increased by a factor of about 6 by increasing the amount of martensite present to 90%.

Formation of Sigma

As in the ferritic stainless steels, sigma can form in the austenitic steels at low temperatures (Fig. 7-13). In general, the precipitation rate is slow, so that aging is required to generate measurable amounts. Figure 7-22 shows the influence of the formation of sigma on the impact energy at 20°C. The embrittling effect is obvious. As the amount of sigma increases, the impact properties deteriorate, as shown in Fig. 7-23. Also, in certain cases, the presence of sigma will decrease the corrosion resistance.

Figure 7-24 shows an isothermal TTT diagram for the beginning of the precipitation of sigma in the 309 stainless steel. For this composition (23% Cr–13% Ni–0.13% C) Fig. 7-13 shows sigma stable at 800°C and below. The TTT diagram (Fig. 7-24) indicates about 880°C as the upper temperature of stability of sigma. The maximum rate of precipitation occurs around 1380°F (750°C) for this alloy. Also shown is the sigma precipitation start curve for samples which were obtained by filing the annealed

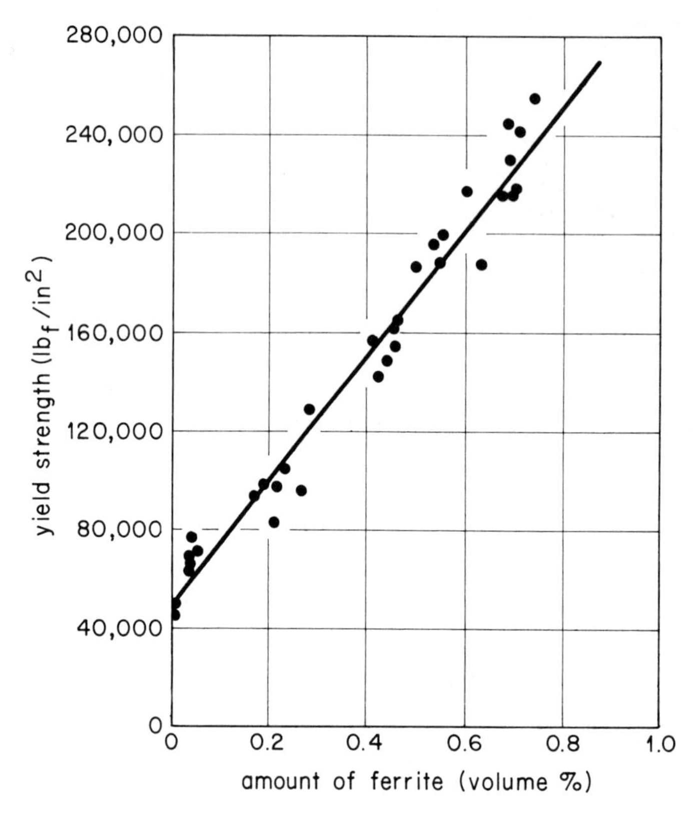

Figure 7-21 The yield strength (measured at 20°C) of a 304 stainless steel as a function of the amount of martensitically formed ferrite present. The steel contained about 18% Cr, 8.5% Ni, and 0.07% C. The M_s temperature was below −269°C, and the M_d temperature was about 20°C. The amount of martensite present was varied by controlling the deformation temperature and subsequent aging. Thus the data points represent a variety of amounts of plastic deformation and aging treatments. *(Adapted from P. L. Mangonon and G. Thomas,* Met. Trans., *vol. 1, p. 1587, 1970.)*

material prior to aging. This highly cold-worked structure precipitates the sigma much faster than the annealed material; this can be an important consideration in design applications of these austenitic stainless steels.

The precipitation reactions in these austenitic stainless steels are frequently quite complex, with the formation of carbides (discussed in the next section) and sigma as well as sulfides, nitrides, and other intermetallic phases similar to sigma. Figure 7-25 illustrates this behavior for a 316 stainless steel. At the solution temperature of 1200°C, TiC, TiN, and $Ti_4C_2S_2$ are present in a very small amount (about 0.4 wt. % total). Aging this alloy causes the precipitation first of the carbide $M_{23}C_6$, followed by σ, then another intermetallic phase χ. For long-time, high-temperature applications these reactions must be carefully considered.

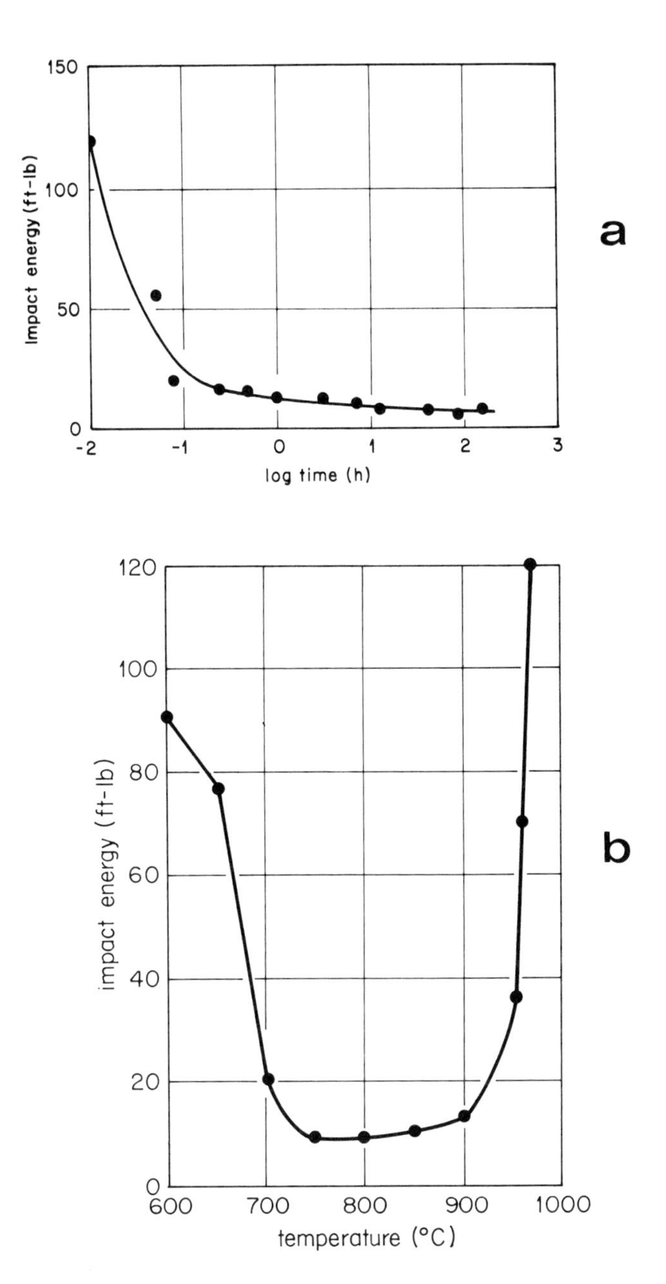

Figure 7-22 (a) Impact energy measured at 20°C as a function of aging time at 850°C. The steel contained 18% Cr–8% Ni–3% Mo, and was water quenched from 1100°C prior to the aging treatment. (b) Impact energy measured at 20°C as a function of reheating temperature, for 7 hours at each temperature. The steel contained 18% Cr–8% Ni–3% Mo, and was water quenched from 1100°C prior to the aging treatment. *(Adapted from L. Smith and K. W. J. Bowen,* J.I.S.I., *vol. 158, p. 295, 1948.)*

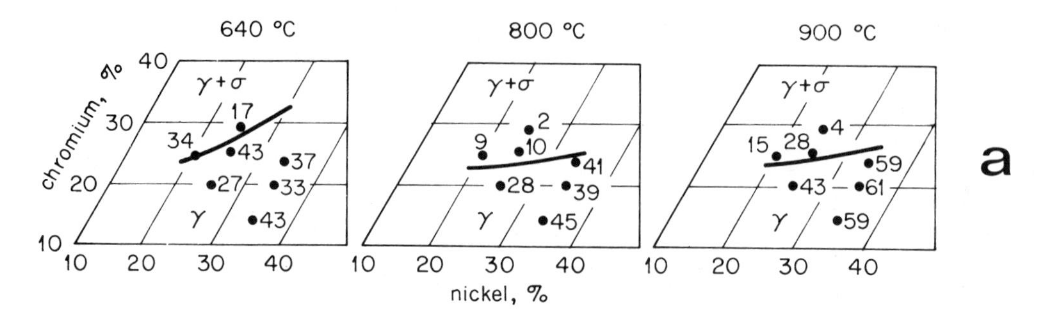

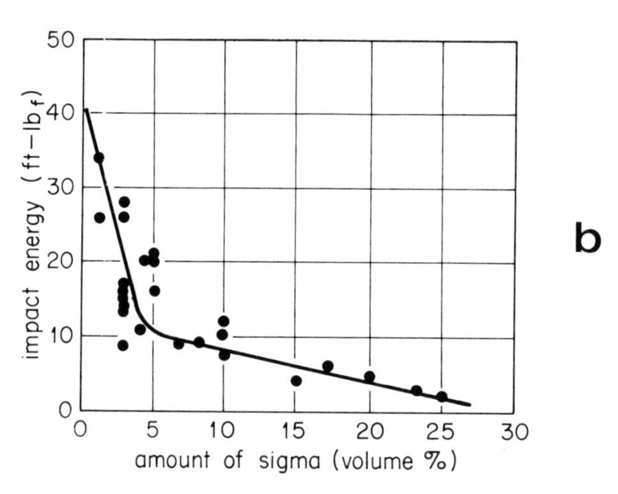

Figure 7-23 Relationship between the impact energy at 20°C and sigma formation in austenitic stainless steels. The nickel content was between 20 and 35% and the chromium content was between 14 and 29%. (a) Isothermal sections of the iron-chromium-nickel ternary phase diagram illustrating the influence of the presence of sigma on the 20°C impact energy. The steels were solution-annealed for one hour at about 1060°C, water quenched, then aged at the indicated temperatures for 3000 hours. (b) Relationship between the amount of sigma present and the 20°C impact energy. *(Adapted from A. M. Talbot and D. E. Furman,* Trans. ASM, *vol. 45, p. 429, 1953.)*

The formation of sigma will be accelerated by the presence of ferrite, as the ferrite will decompose to the sigma. This again is an important consideration in welded structures where either α ferrite or δ ferrite may be present prior to aging.

Carbide Precipitation

The formation of carbides in austenitic stainless steel is dictated by the phase relations, as shown in Fig. 7-16. Although the carbide precipitation can be suppressed by sufficiently rapid cooling from the austenite region, the utilization of this steel for high-temperature applications, or the occurrence of slower cooling during fabrication, makes consideration of the effect of the carbides on properties of importance.

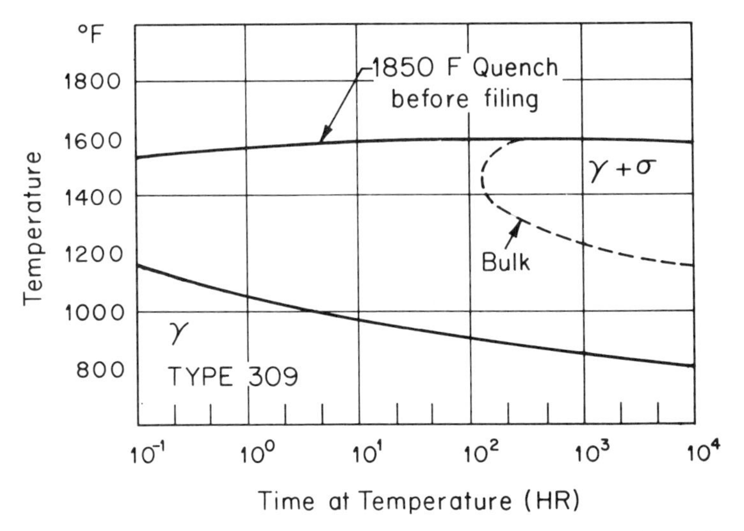

Figure 7-24 Isothermal TTT diagram for the start of the precipitation of sigma in a 309 austenitic stainless steel. The steel contained 23.21% Cr, 13.40% Ni, and 0.13% C. The curve displaced to shorter times was obtained on filings (thus highly plastically deformed) made from the annealed material prior to aging. *(From G. F. Tisinai, J. K. Stanley, and C. H. Samans,* Trans. AIME, *vol. 206, p. 600, 1956.)*

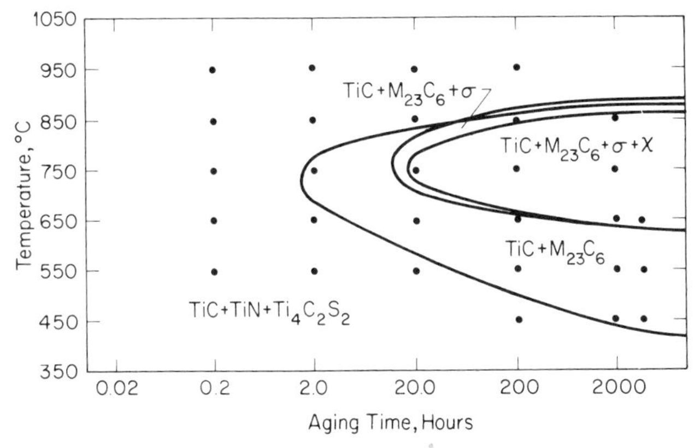

Figure 7-25 Isothermal TTT diagram for the start of precipitation of phases in aging a 316 titanium modified stainless steel. The steel contained about 0.06% C, 17% Cr, 14% Ni, 2.5% Mo, 0.3% Ti, and 1.4% Mn. It was solution-annealed for about 1 hour at 1200°C, then water quenched. The solution-annealed condition contained about 0.4 wt. % total of TiC, TiN, and $Ti_4C_2S_2$. *(From A. S. Grot and J. E. Spruiell,* Met. Trans., *vol. 6a, p. 2023, 1975.)*

Consider the process of precipitation of a carbide at an austenite grain boundary (Fig. 7-26). Assume that the austenite matrix contains about 18% Cr, 8% Ni, and 0.08% C. The carbide is much richer in carbon and in chromium; the exact composition depends on the chemistry of the steel, but we will assume reasonable estimate

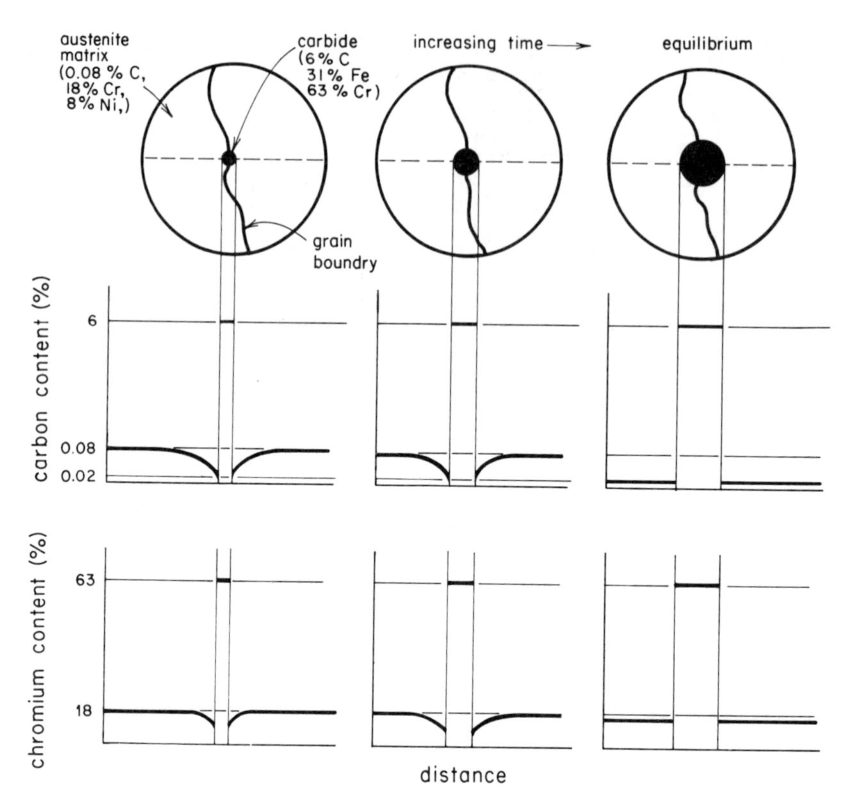

Figure 7-26 Schematic representation of the growth of a carbide, and the development of the chomium and carbon profiles along the dashed line. Note the chromium-depleted region *adjacent* to the carbide.

values of 6% C, 31% Fe, and 63% Cr. When the precipitate first forms, its higher carbon and chromium content comes from the surrounding, adjacent austenite, thus lowering the carbon and chromium content of the austenite. Hence, as the carbide grows, a carbon and chromium gradient is developed which allows these two elements to diffuse to the austenite-carbide interface where the transformation occurs. As the carbide gets larger, its rate of growth slows as the equilibrium amount of carbides is approached. The gradient becomes less, and eventually the carbide is in equilibrium with the austenite, each having uniformly their equilibrium compositions. These events are depicted in Fig. 7-26.

Note that in the earlier stages of precipitation, the chromium content in the austenite adjacent to the carbide is low. This region is thus susceptible to corrosion relative to the austenite far removed from the carbide. Further, the carbides will nucleate at the austenite grain boundaries, and this may form a continuous film of low chromium-content austenite around each grain (Fig. 7-27). An austenitic stainless steel in this condition is referred to as *sensitized*, because when placed in a corrosive environment this depleted region will be attacked, and corrosion rates are high, especially in the vicinity of the grain boundaries. This effect is illustrated in Fig. 7-28.

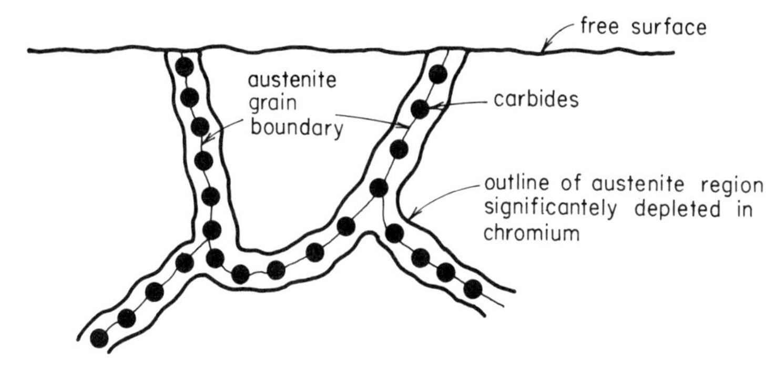

Figure 7.27 Schematic illustration of the chromium-depleted region around the austenite grain boundaries due to high-chromium carbide precipitation.

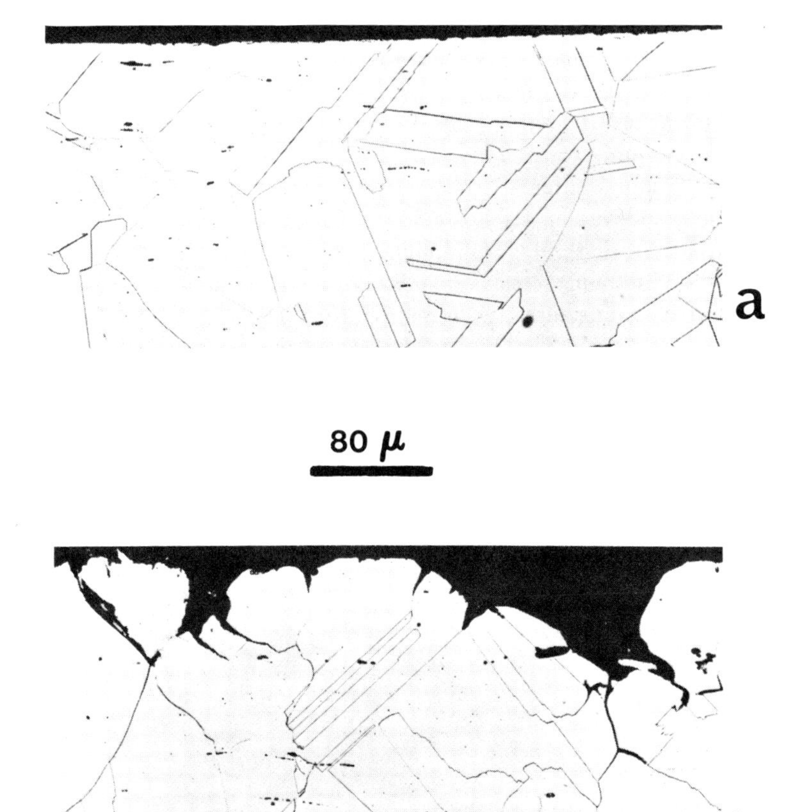

Figure 7-28 Microstructures of a 304 stainless steel after exposure to boiling 65% nitric acid. The surface in contact with the acid is normal to the plane, and at the top of each picture. (a) Steel, water quenched from 1065°C. (b) Steel, water quenched from 1056°C, then held 3 hours at 677°C and air cooled. *(Adapted from J. F. Mason,* Met. Engr. Quart., *vol. 8, no. 2, p. 67, 1968.)*

The grain boundary attack is referred to as intergranular penetration; the corrosion rate is obtained from microstructures such as Fig. 7-28*b* by measuring the average penetration normal to the surface in contact with the corrosive medium.

The intergranual penetration depends on the presence of the low chromium content of the austenite adjacent to the grain boundaries. As shown in Fig. 7-27, if the aging is sufficiently long, the gradient decreases. That this lowers the corrosion rate is illustrated in Fig. 7-29, where it is seen that the corrosion rate passes through a maximum with time at 650°C.

An isothermal TTT diagram can be obtained for the beginning of measured (within the resolution of optical microscopy) intergranular penetration, and Fig. 7-30 shows curves for three steels.

The problem of avoiding sensitization, or deleting it, must be considered. One possibility is to remove the carbon so that no carbide precipitates. The influence of the carbon content is felt strongly, as shown in Fig. 7-31. Depending somewhat on the nickel and chromium content, the carbon content should be below about 0.03% to make the corrosion rate negligible; this value, of course, depends on the corrosive environment. Low carbon grades of austenitic stainless steels can be purchased with the purpose of minimizing sensitization; Table 7-1 shows both 304L and 316L to contain 0.03% carbon maximum.

Attaining very low carbon contents is not always the most economical method of avoiding sensitization, and instead steels are used which have elements added to form more stable carbides than the high-chromium carbides. Thus, even if carbide precipitation occurs, the chromium remains in solution in the austenite so that the steel retains its high corrosion resistance. This is referred to as *stabilization*. Figure 7-32 shows corrosion data for an 18.1% Cr–8.9% Ni–0.08% C stainless steel and an 18.1%

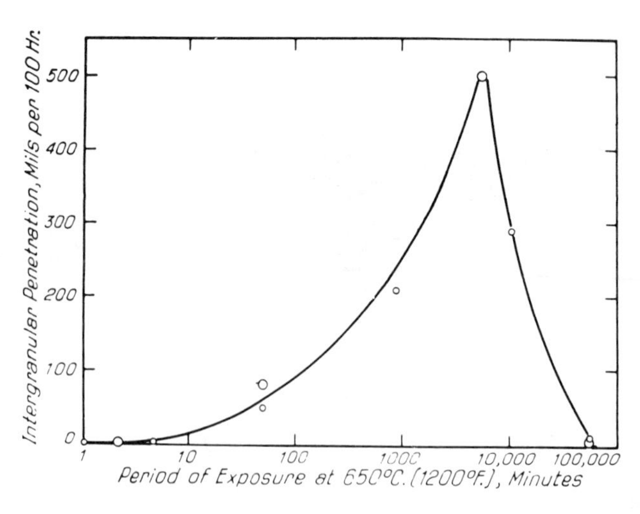

Figure 7-29 The intergranular penetration as a function of time at 650°C for a 16.5% Cr–11.3% Ni–0.08% C austenitic steel. *(From E. C. Bain, R. H. Aborn, and J. J. B. Rutherford,* Trans. ASST, *vol. 21, p. 481, 1933.)*

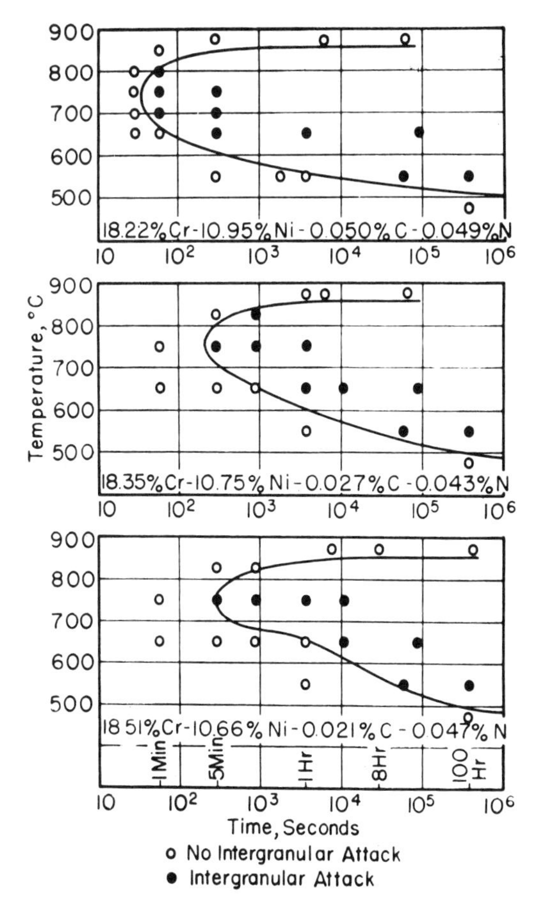

Figure 7-30 Isothermal TTT diagrams in terms of the time for first detectable intergranular penetration. Tests were made in boiling acidified copper sulphate solutions. *(From W. O. Binder, C. M. Brown, and R. Franks,* Trans. ASM, *vol. 41, p. 1301, 1949.)*

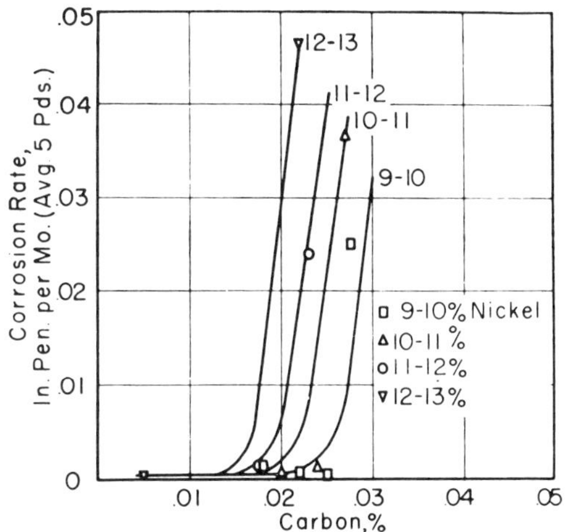

Figure 7-31 Influence of carbon content on the average corrosion rate of 18% Cr austenitic stainless steels of various nickel content. The steels were solution annealed at 1075°C, air cooled, then aged for 100 hours at 550°C, prior to the corrosion test. *(Adapted from W. O. Binder, C. M. Brown, and R. Franks,* Trans. ASM, *vol. 41, p. 1301, 1949.)*

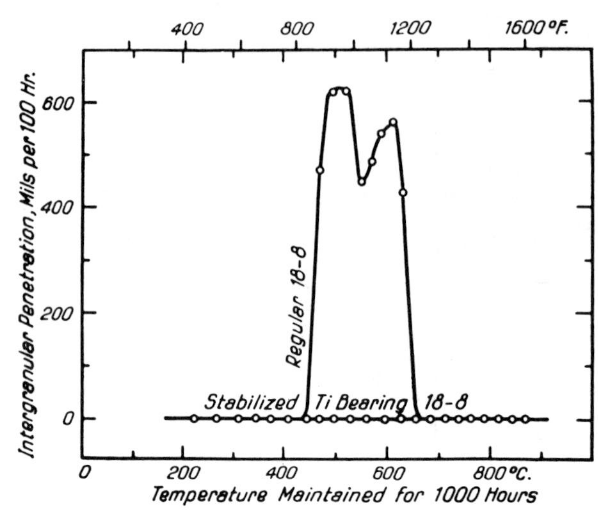

Figure 7-32 The influence of titanium addition (0.38%) on the corrosion behavior of 18% Cr–8% Ni–0.07% C stainless steels. *(From E. C. Bain, R. H. Aborn, and J. J. B. Rutherford,* Trans. ASST, *vol. 21, p. 481, 1933.)*

Cr-9.1% Ni-0.38% Ti-0.07% C steel. The corrosive medium was a copper sulphate-sulphuric acid solution. The corrosion rate was low for both steels for 1000 hours' aging below about 450°C, and above 700°C. However, between 450 and 700°C the titanium-containing steel showed negligible corrosion rate, due to the formation of TiC instead of the high-chromium carbide. Type 321 (titanium additions) and 347 (coloumbium and/or tantalum additions) are standard stainless steels with stabilizing element additions (see Table 7-1).

The previous two methods prevent the precipitation of chromium-rich carbides. However, if these cabides do precipitate, they can be redissolved in the austenite region, given sufficient time. This may take several hours at temperatures in the range of 1200°C. Further, it may not be feasible to anneal large parts or complicated and large parts joined by welding. In such cases, the low-carbon stainless steels or the stabilized stainless steels must be used.

Plastic deformation prior to aging in the sensitizing range will allow the carbides to precipitate along slip bands and hence be quite homogeneously distributed and not concentrated at the grain boundaries. Thus, even though a depleted region forms, it does not have the continuity that the grain boundary has and hence will have better corrosion resistance. This effect is illustrated in Fig. 7-33. Such deformation is usually not very practical except in special cases requiring very special geometries, such as plates that can be rolled or tubes that can be extruded.

The importance of avoiding the precipitation of the high-chromium carbides lies in the effect the depleted region has on reducing the corrosion resistance. Since the grain boundaries of the austenite will be the prime nucleating sites, the depleted zone is essentially continuous along the boundary, and it is this region that is removed, leaving the grains unsupported. The precipitation can occur to a measurable degree in some steels in quite short times. Figure 7-33*a* compares the effect of aging time. Even as short as 1.8 min will allow sufficient precipitation so that intergranular pene-

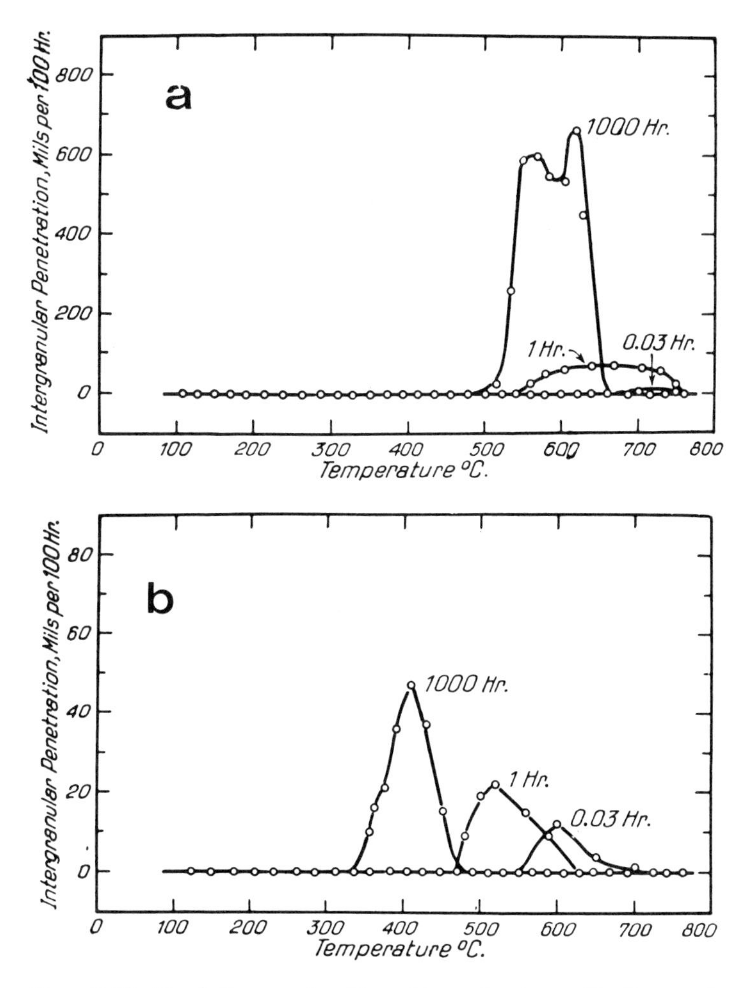

Figure 7-33 The influence of cold working prior to sensitization on the corrosion resistance. The steel contained 18.1% Cr, 8.9% Ni, and 0.08% C. The data in (a) are for the alloy not cold worked, and in (b) it has been cold worked. *(From E. C. Bain, R. H. Aborn, and J. J. B. Rutherford,* Trans. ASST, *vol. 21, p. 481, 1933.)*

tration is observable in some cases. This means that attention must be paid to cooling rates from the solution-annealing temperature. In welding, even with the rapid cooling rate of the metal, precipitation may be significant, forcing resolution-annealing of the part, or the choice of a low carbon or a stabilized steel.

The precipitation of the carbides, either the chromium-rich carbides or the more stable carbides in the stabilized grades, will affect the mechanical properties. There is a slight hardening and slight improvement of the creep properties, and slight embrittling. However, these factors are usually not as important as the corrosion resistance consideration.

7-4 COMPARISON OF MECHANICAL PROPERTIES OF STAINLESS STEELS

In this section a comparison is given of the properties of the three classes of stainless steels. The comparison must be made with caution as the properties depend on the heat treatment and the amount of plastic deformation. However, it is useful to compare the temperature dependence of the yield strength, impact strength, and creep properties of one specific stainless steel from each category.

The yield strength as a function of temperature is shown in Fig. 7-34 for an annealed ferritic steel (430), an annealed austenitic steel (304), and two martensitic steels (414 and 416). For this short-time mechanical test, the austenitic stainless steel is inferior to the others below about 800°C. Above this range, the superior creep properties of the austenitic stainless steel make it more suitable for high-temperature, long-time applications. Figure 7-35 compares some creep data for ferritic and austenitic steels.

The impact behavior of austenitic steels is compared to those of a quenched and tempered martensitic steel in Fig. 7-36. The tempered martensitic steel, and the annealed steel, consisting of a microstructure of carbides and ferrite, have an impact curve typical of body-centered cubic metals and alloys: a relatively high impact energy at high temperature, but a drastic decrease over a rather small temperature interval at low temperatures. The face-centered cubic austenitic steels have a high impact energy with only a relatively low decrease with temperature, retaining a high value to absolute zero; this is typical for this crystal structure. For this reason, some low-temperature applications require the use of the austenitic stainless steel, even though corrosion resistance is not a factor in this application.

7-5 SUMMARY

The classification of stainless steels in terms of their structure, and the influence of the chemistry on this, is summarized in Fig. 7-37. In this chapter we have seen how this classification is related to the phase diagram and to the kinetics of the phase changes.

Within a given class of stainless steel, the chemistry is altered for specific purposes, and this is illustrated for each class in Figs. 7-38, 7-39, and 7-40. Note that the addition of sulphur improves machinability, and these grades are chosen where intricate parts must be machined (consistent with other restraints). In general, the addition of chromium improves oxidation and corrosion resistance, but the formation of sigma must be reckoned with. Additions of nickel, chromium, and molybdenum to the basic 18% Cr–8% Ni austenitic stainless steels improves the high-temperature properties and the corrosion resistance.

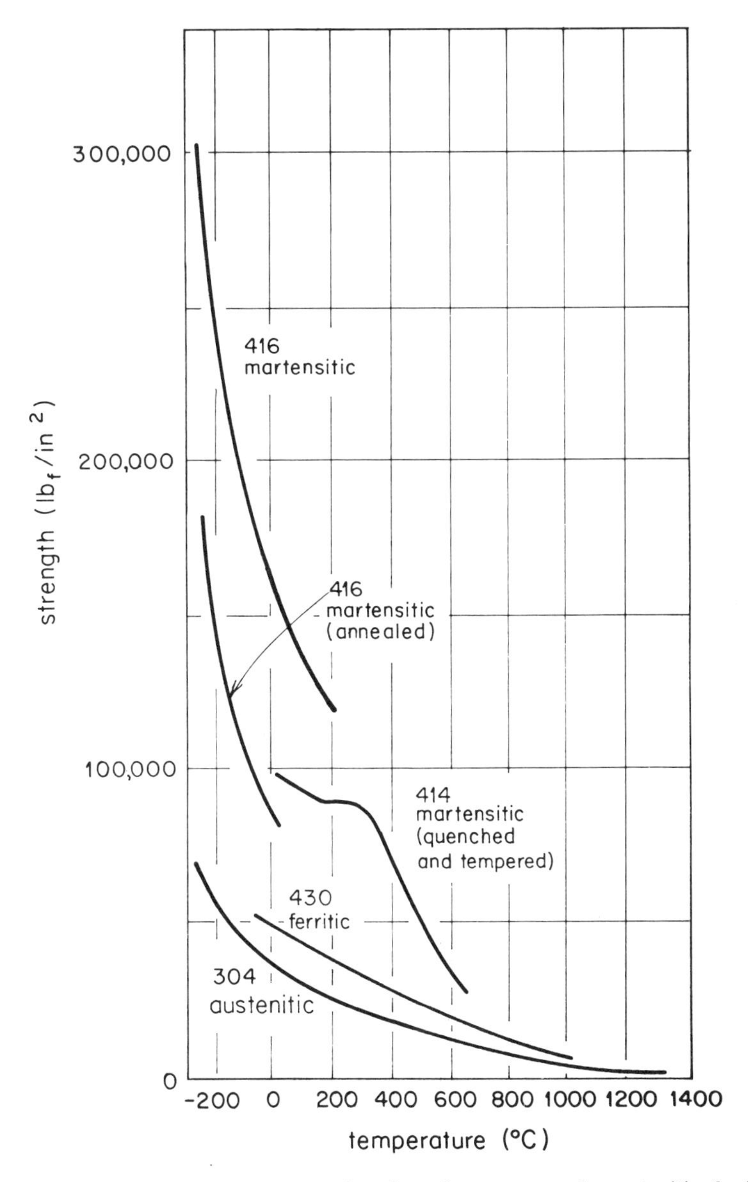

Figure 7-34 Yield strength as a function of temperature for austenitic, ferritic, and martensitic stainless steels. *(Compiled from various sources.)*

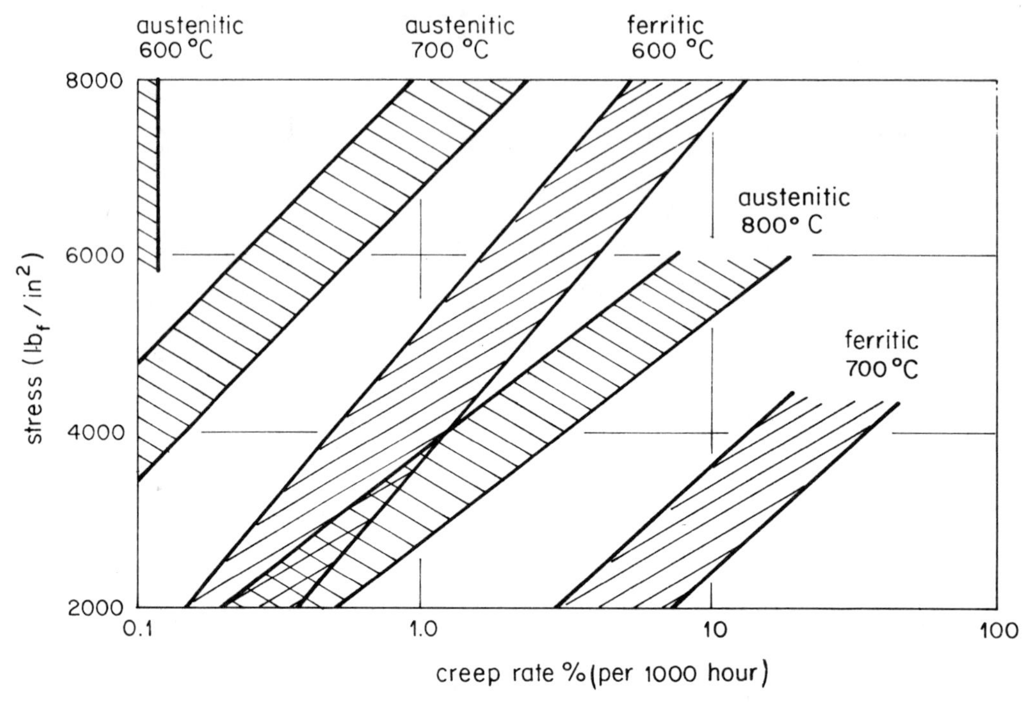

Figure 7-35 Comparison of the effect of stress on creep rate of austenitic and ferritic stainless steels. *(Adapted from C. R. Austin and H. D. Nichol,* J.I.S.I., *vol. 37, p. 117p, 1938.)*

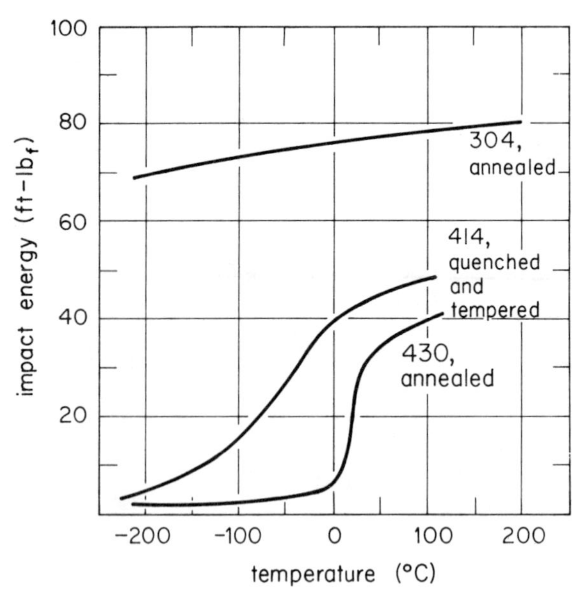

Figure 7-36 Impact energy versus temperature for austenitic (304), martensitic (414, quenched and tempered), and ferritic (430) stainless steels. *(From various sources.)*

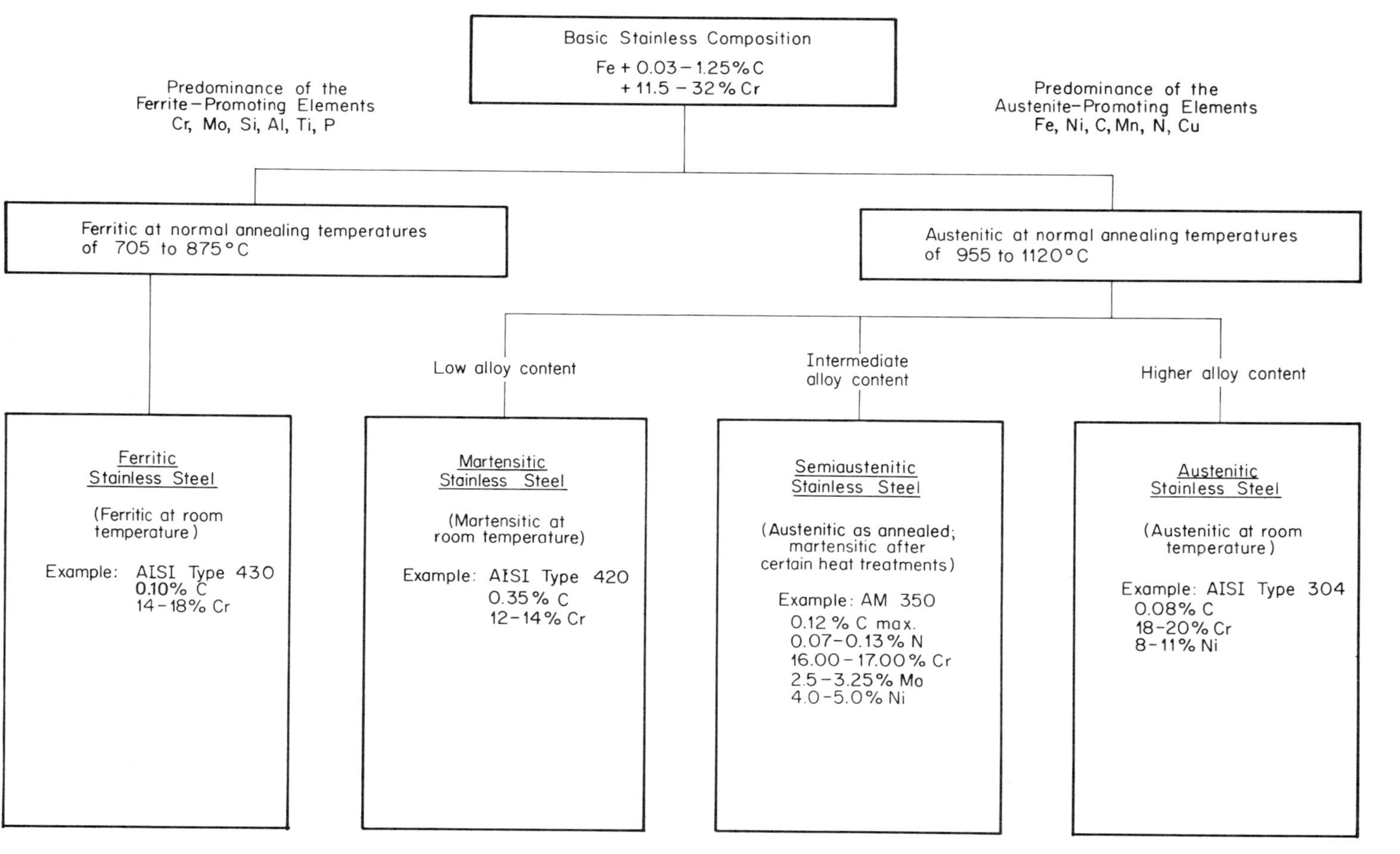

Figure 7-37 Classification of the stainless steels, based on chemical composition. *(Adapted from D. C. Lidwigson and A. M. Hall, "DMIC Report 111," April 20, 1959, Battelle Columbus Laboratories, Columbus, Ohio.)*

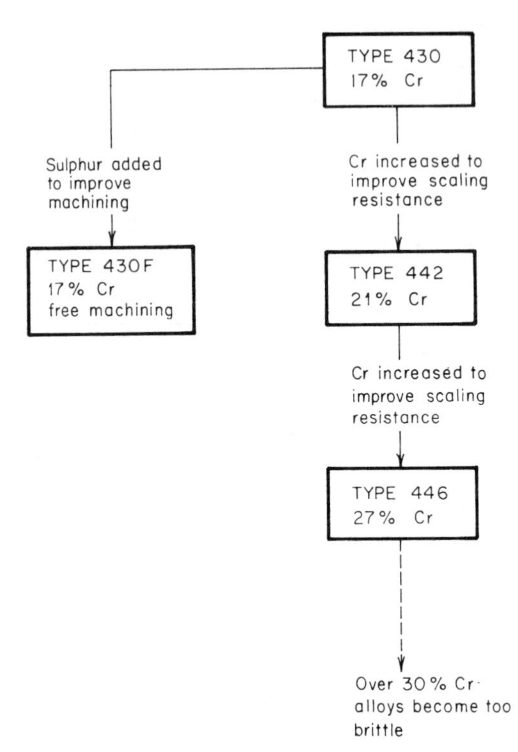

Figure 7-38 Diagram illustrating the development of ferritic stainless steels. *(Source unknown.)*

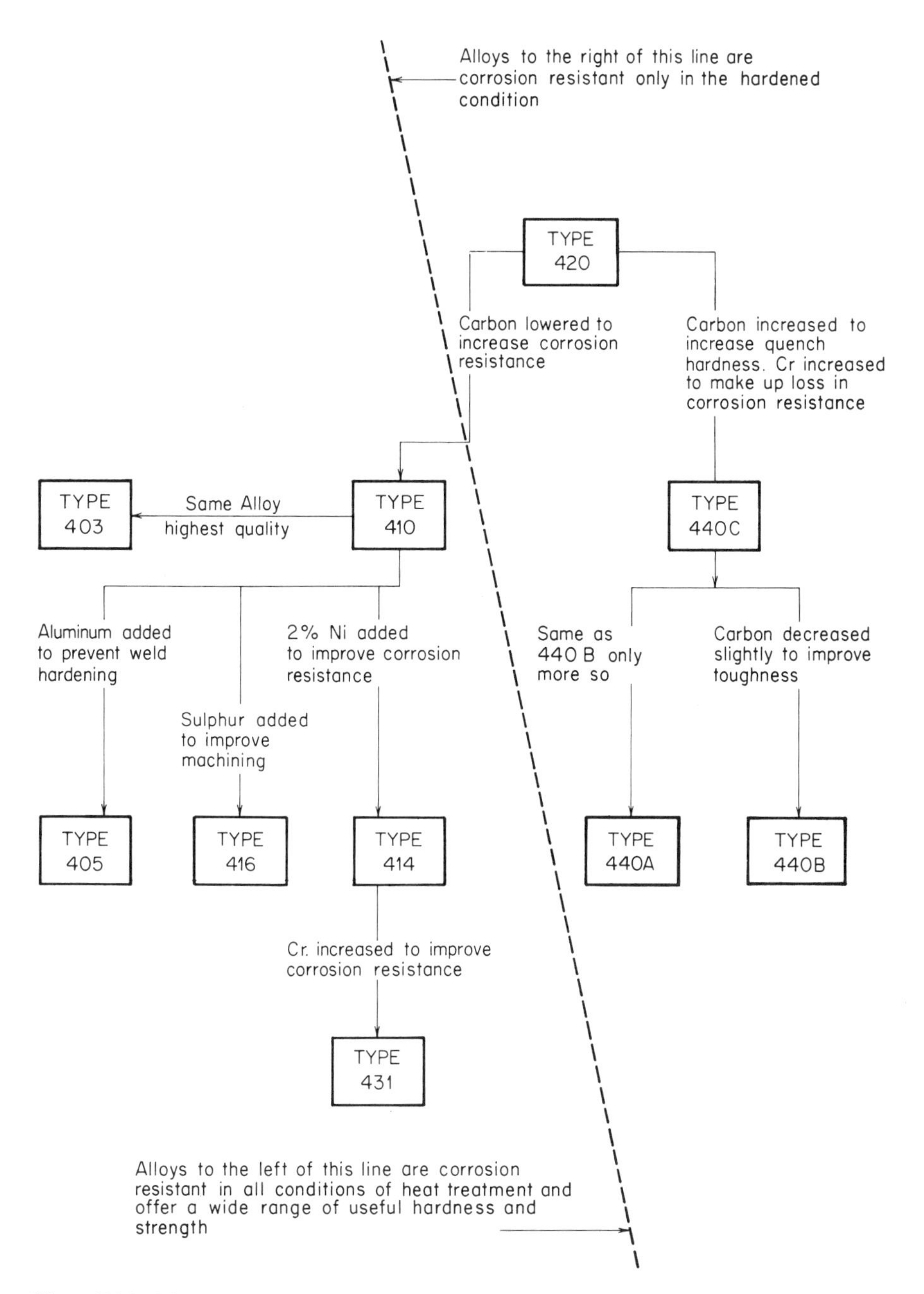

Figure 7-39 Diagram illustrating the development of martensitic stainless steels. *(Source unknown.)*

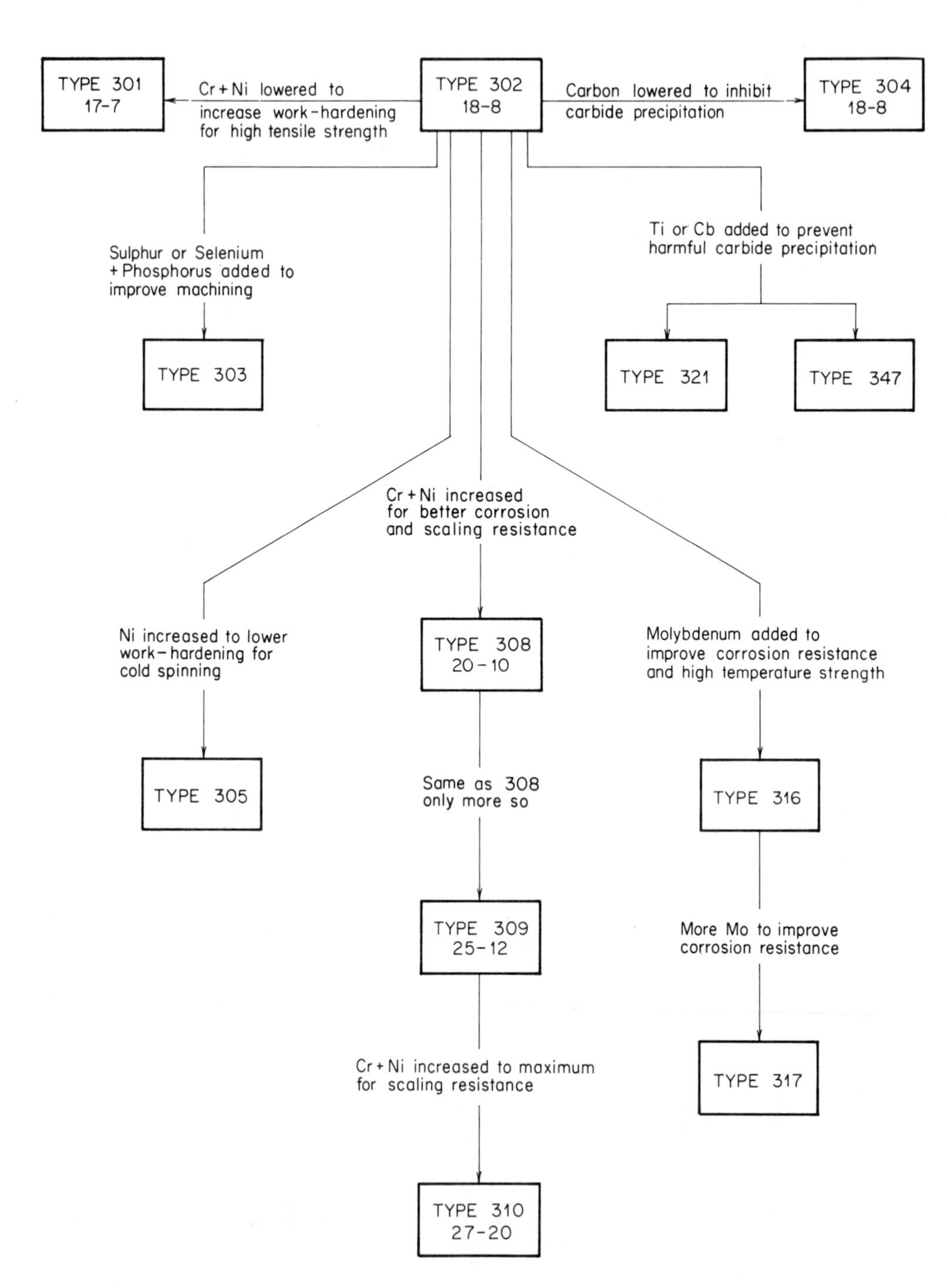

Figure 7-40 Diagram illustrating the development of austenitic stainless steels. *(Source unknown.)*

CHAPTER

EIGHT

STRUCTURAL STEELS

In this chapter, we will discuss the physical metallurgy of the steels that are used to fabricate rather large shapes that are frequently not amenable to heat treatment by forming martensite and then tempering the steel to the desired properties. The type of shapes involved are I beams, large pipes, and pressure vessels. In the last 15 years considerable progress has been made in improving the properties of these steels by specialized alloying additions and by combining heat treatment with fabrication practice. The general goal is the highest strength consistent with sufficient impact properties and fabricability by welding.

The structures to be examined first are those consisting of primary ferrite and pearlite. The strength of such a structure can be increased by solid-solution strengthening of the ferrite by additions of certain elements, and by heat treatment to reduce the primary ferrite grain size. The reduction of the size of these grains also lowers the transition temperature. Further studies showed that certain elements could be added to allow increased strengthening by precipitation hardening. Then, hot fabrication processes to give a fine austenite grain size, and hence a fine primary ferrite grain size, were developed to obtain a very fine primary ferrite grain size. These effects are all combined to various degrees in the commercial steels.

In general, the types of steels being discussed here are referred to as high-strength structural steels and sometimes as high-strength, low-alloy structural steels. The term "microalloyed steel" is also sometimes used. It describes the strong effect some elements have on the properties when added to simple unalloyed structural steel in amounts one or two powers of ten less than they are added in the conventional sense.

8-1 EFFECT OF PRIMARY FERRITE GRAIN SIZE

To examine the effect of ferrite grain size, the steels must be heat treated so that the amount of primary ferrite and pearlite is kept constant, and the pearlite spacing is fixed. This can be achieved by varying the austenite grain size from which the primary ferrite-pearlite structure forms; the steels are then cooled to develop a structure of the same amount of primary ferrite and pearlite, but with varying ferrite grain size (d). The variation of yield strength with grain size will be expected to vary linearly with $d^{-1/2}$ (see Chap. 3). Data for a manganese-containing steel are shown in Fig. 8-1. (Note that the grain-size range is much smaller than that encountered for the austenite grain size in normal heat treating, which runs from about 0.05 to 0.003 cm.)

Reducing the primary ferrite grain size also lowers the transition temperature, as shown in Fig. 8-1. Thus a reduction in the ferrite grain size is favorable to both improved strength and impact properties.

8-2 EFFECT OF CARBON AND MANGANESE CONTENT

In these primary ferrite-pearlite structures, increasing the carbon content increases linearly the amount of pearlite, and hence the tensile strength increases linearly with the carbon content. However, the yield strength is insensitive to the amount of pearlite present. In these steels, the pearlite content will not exceed about 25%. This is too small an amount to act as an impediment to significant initial plastic deformation. However, as deformation continues, the pearlite influences the work hardening, and hence the tensile strength.

Although improved strength is achieved by increasing the carbon content, and

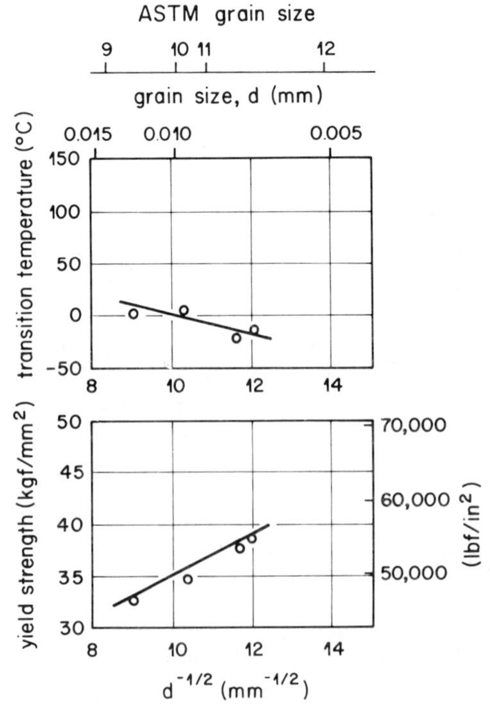

Figure 8-1 Effect of ferrite grain size on yield strength and impact transition temperature for a 0.14% C–0.30% Si–1.30% Mn (Al killed) steel. *(From I. Kozasu and T. Osuka, in J. M. Gray (ed.), "Processing and Properties of Low Carbon Steel," The Metallurgical Society, New York, 1973.)*

hence the amount of pearlite, the transition temperature also increases. Further, the carbon content is limited to about 0.25% in order to make the steel weldable.

The effect of manganese is more complicated. Some solid-solution hardening occurs due to the presence of manganese, which increases the strength but also the transition temperature. Manganese lowers the carbon content of the eutectoid composition so that more pearlite forms. It also lowers the eutectoid temperature, so that the transformation to primary ferrite begins at lower temperatures with increasing manganese content. This in turn increases the ferrite nucleation rate, giving a finer primary ferrite grain size.

Figure 8-2 summarizes the effect of grain size and of carbon and manganese content.

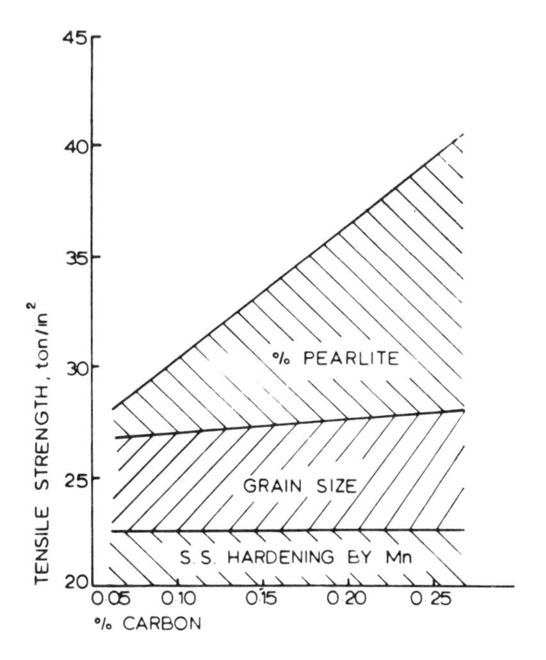

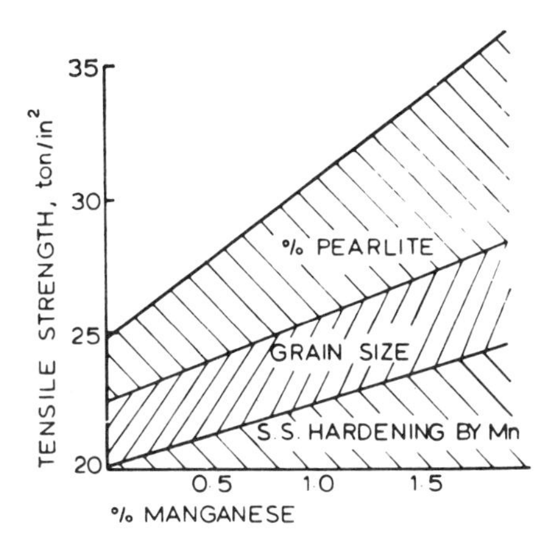

Figure 8-2 Factors contributing to the strength of carbon-manganese steel, heat treated to develop a primary ferrite-pearlite structure. *(From K. J. Irvine, in "Strong, Tough Structural Steels," Iron and Steel Institute, London, 1967.)*

The meaning of these diagrams can be explained by example. A 0.15% C steel will contain about 80% primary ferrite and 20% pearlite, and it will have a tensile strength of about 38 kgf/mm^2. This can be increased by about 7 kgf/mm^2 by the addition of manganese, the effect being an increase of about 3 kgf/mm^2 for each 1% increase in manganese content. Thus if the 0.15% C steel here has about 1.5% Mn, then the tensile strength will be increased from about 38 to about 42 kgf/mm^2. This value is a minimum based on a primary ferrite grain size of about 0.003 cm. The grain size can be controlled to be as small as 0.008 cm, increasing the tensile strength to about 46 kgf/mm^2.

Table 8-1 summarizes the empirical relations for the yield strength, tensile strength, and impact transition temperature for these steels. Also included here is the effect of silicon.

8-3 EFFECT OF CONTROLLED PRECIPITATION

The control of the austenite grain size, discussed in Chap. 2, was attributed to the presence of oxide precipitates which impede the austenite grain growth. It has been found that controlled additions allow the formation of grain-growth inhibiting precipitates, such as carbides and nitrides. As an example, we will examine the effect of niobium. If niobium is added to the 0.15% C–1.5% Mn steel and austenitized at 950°C, the austenite grain size is reduced, which in turn reduces the subsequent primary ferrite grain size. The yield-strength dependence on grain size is shown in Fig. 8-3. It is seen that the curve obtained is an extension of that for the plain carbon steel. The effect of the niobium addition is the formation of Nb (CN) at the austenitizing temperature, which retards the growth of the austenite grains much more than without the niobium.

If the niobium-containing steel is austenitized at a higher temperature (e.g., 1250°C) some of the carbonitrides go into solution in the austenite, and there is less precipitate present to inhibit grain growth. Thus, even with air cooling, ferrite grain sizes cannot be obtained as small as in the plain carbon steel (see Fig. 8-3). However, the yield strength is considerably higher than for the plain carbon steel. This increase has been found to be due to precipitation hardening of the ferrite during

Table 8-1 Empirical relations for strength properties and variables for steels of a primary ferrite-pearlite structure

Lower yield point (kgf/mm^2)	=	1.57 [6.74 + 2.11(% Mn) + 5.44(% Si) + 0.255($d^{-1/2}$)]	(d in microns)
Tensile strength (kgf/mm^2)	=	1.57 [19.1 + 1.78(% Mn) + 5.35(% Si) + 0.523(% pearlite) + 0.100($d^{-1/2}$)]	(d in microns)
Impact transition temperature (°C)	=	[63 + 44.1(% Si) + 2.2(% pearlite) − 2.3($d^{-1/2}$)]	(d in microns)

Source: K. J. Irvine, in "Steel Strengthening Mechanisms," Climax Molybdenum Co., Ann Arbor, Michigan, 1969.

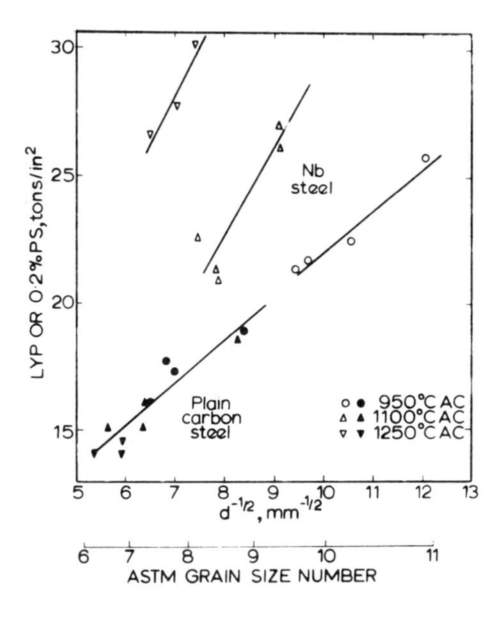

Figure 8-3 Effect of austenitizing temperature on the strength-grain size relationship of niobium-containing steel. *(From K.J. Irvine,* J.I.S.I., *vol. 207, p. 837, 1969.)*

cooling. At the high austenitizing temperature most of the niobium is in solution in the austenite. Upon cooling, the austenite transforms to pearlite and primary ferrite, and the ferrite inherits the niobium content of the austenite from which it forms. However, the niobium carbonitride precipitates from the ferrite as cooling continues, causing precipitation hardening. Again, although this increases the strength, it also increases the impact transition temperature.

Figure 8-4 illustrates the type of heat treatment designed to take advantage of the precipitation hardening. The TTT diagram is for a 0.10% C steel containing manganese, aluminum, vanadium, and niobium grain refiners which also contribute to precipitation hardening. Note that the steel is first austenitized at 1100°C for 5 min to dissolve most of the carbides and nitrides. Then it is cooled to 910°C, and held for 1 min. For this short time, few of the carbonitrides reprecipitate, and the lower temperature (910°C) is used as the quenching temperature for which the TTT diagram applies. Note that in the region where primary ferrite-pearlite forms on cooling, faster cooling

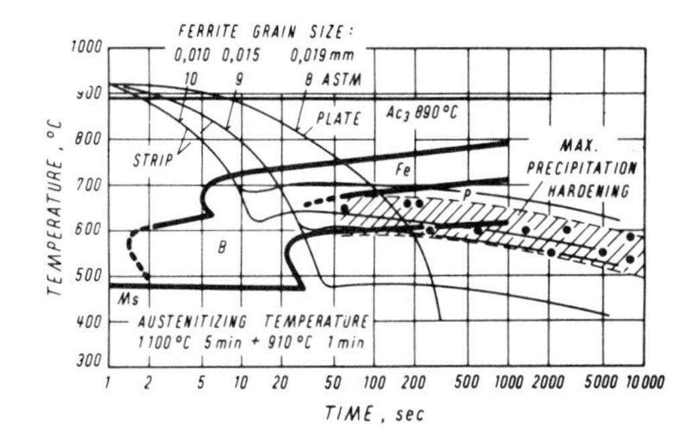

Figure 8-4 Continuous cooling TTT diagram for a 0.10% C–0.33% Si–1.23% Mn–0.04% Al–0.06% V–0.035% Nb steel. The steel was first austenitized at 1100°C for 5 min., then cooled to 910°C and held for 1 min. *(From W. Vor dem Esche, K. Daup and H. Wladika, in J. M. Gray (ed.), "Processing and Properties of Low Carbon Steel," The Metallurgical Society, New York, 1973.)*

gives a smaller ferrite grain size. However, to take advantage of the precipitation-hardening effect in the ferrite, the cooling is interrupted around 650°C and the steel then cooled more slowly. The maximum precipitation-hardening effect temperature-time region is superimposed on the TTT diagram in Fig. 8-4. For commercial production of strip material, this precipitation is carried out in the final step where the steel is coiled from the last hot-rolling mill. The coiling speed is adjusted to give the temperature-time curve necessary to allow precipitation hardening.

8-4 EFFECT OF DEFORMATION PROCESSING

It is clear that a further reduction in the ferrite grain size would be desirable, but it seems to be limited to around 0.007 mm for the niobium-containing steel austenitized at 950°C. Using higher austenitizing temperatures to take advantage of the precipitation-hardening effect increases the austenite grain size, and hence the subsequent ferrite grain size, so that the latter effect cannot be exploited fully. The question is whether the austenite grain size can be reduced further and yet use a high (e.g., 1250°C) austenitizing temperature.

In principle, a method of controlling the austenite grain size does exist. Following plastic deformation, the deformed structure will eventually recrystallize, and proper choice of the heating temperature and time should allow development of a desired austenite grain size. This effect is illustrated schematically in Fig. 8-5. The recrystallized grains are austenitic, and the cooling rate is controlled so that these small austenite grains transform to primary ferrite before the austenite grains have an opportunity to grow very large. The fine austenite grains give even finer primary ferrite grains. (This rolling process goes by various names, such as microduplex processing and controlled rolling.) The effect of the rolling and cooling processes is illustrated schematically in Fig. 8-6.

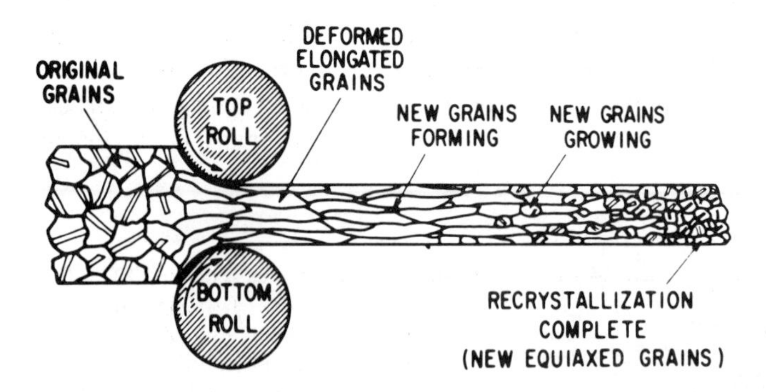

Figure 8-5 Schematic illustration of recrystallization following hot deformation. *(From P. A. Grange, in W. A. Backofen, J. J. Burke, L. F. Coffin, N. T. Reed, and V. Weiss (eds.), "Fundamentals of Deformation Processing," Syracuse University Press, Syracuse, New York, 1964; as adapted from J. M. Camp and C. B. Francis. "The Making, Shaping and Treating of Steel," 5th ed., United States Steel Corporation, Pittsburgh, 1940.)*

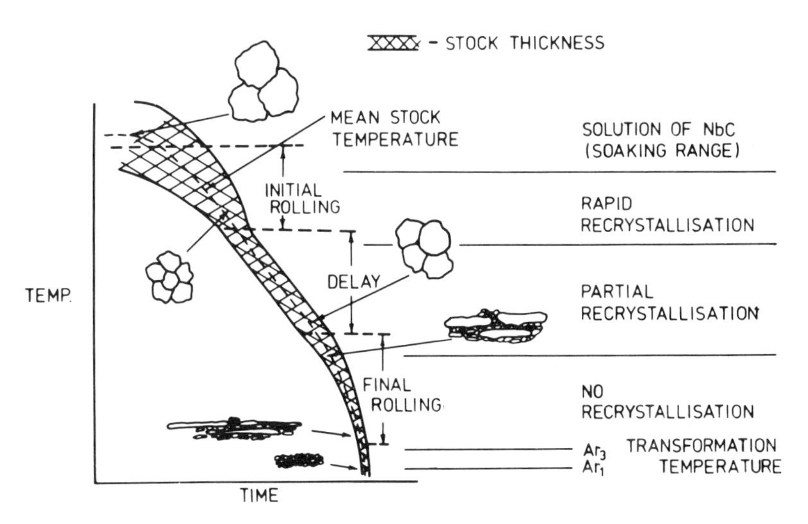

Figure 8-6 Schematic illustration of the controlled-rolling process for a carbon-manganese-niobium steel. *(From J. K. Baird and R. R. Preston, in "Processing and Properties of Low Carbon Steel," The Metallurgical Society, New York, 1973.)*

In the final rolling operation, the steel is first austenitized in the high-temperature range (e.g., 850 to 1250°C) to dissolve most of the niobium carbide. The plate is then cooled to a low final rolling temperature so that the deformed austenite recrystallizes to a fine austenite grain size. Thus, what is desired is that recrystallization is just completed when the primary ferrite begins to form. This can be accomplished by controlling the cooling rate of the plate during and/or following deformation. A typical process is shown in Fig. 8-7.

The effect of the rolling practice is summarized in Fig. 8-8 for the niobium-containing steels. Control of the finishing temperature allows the development of a very fine ferrite grain size and yet a high strength due to precipitation hardening. The effect of these on the impact properties must be considered, and Fig. 8-9 summarizes the results for these steels. The increase in strength increases the transition temperature, so that a steel with a yield strength of about 50 kgf/mm^2 has a transition temperature around 0°C.

8-5 BAINITIC STEELS

The strength can be increased above that of ferrite-pearlite structures by forming bainite. It is necessary to choose a steel with a CCT diagram such that neither primary ferrite nor martensite form. This requires elements that retard the ferrite formation, but not the bainite reaction. Steels with molybdenum additions around 0.5% are suitable, and these frequently have boron added (about 0.06%) to develop the required CCT diagram. Figure 8-10 shows a CCT diagram for this type of steel. Note that bainite will be produced for a rather wide range of cooling rates, in the range corresponding to

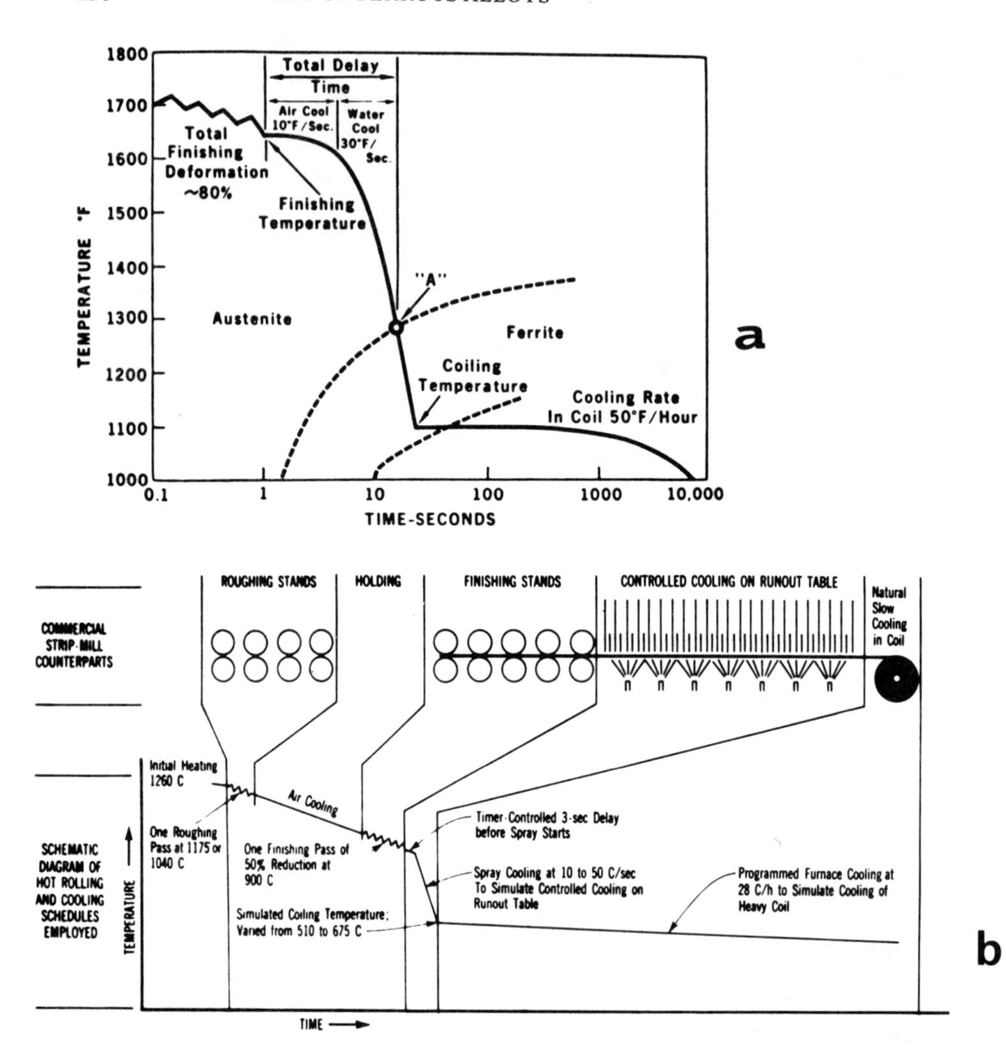

Figure 8-7 Illustration of the process for development of fine-grained ferrite-pearlite structure. (a) Approximate process for a carbon-manganese-niobium steel. (b) Schematic illustration of temperature-time path and rolling mill operation. *(a: adapted from M. Korchynsky and H. Stuart, in "Symposium: Low Alloy, High Strength Steels," The Metallurg Co., Dusseldorf, 1970; b: A. P. Coldren, R. L. Cryderman, and M. Semchyshen, in "Steel Strengthening Mechanisms," Climax Molybdenum Co., Ann Arbor, Michigan, 1969.)*

air cooling for sizes around 0.1 to 10 cm. In this range only about 10% or less martensite will form. The dashed line shows a typical controlled-cooling process used. Note that this gives about 93% bainite, and 7% martensite. However, the slow cooling in the final stage tempers the martensite during cooling, so that the structure is basically bainitic.

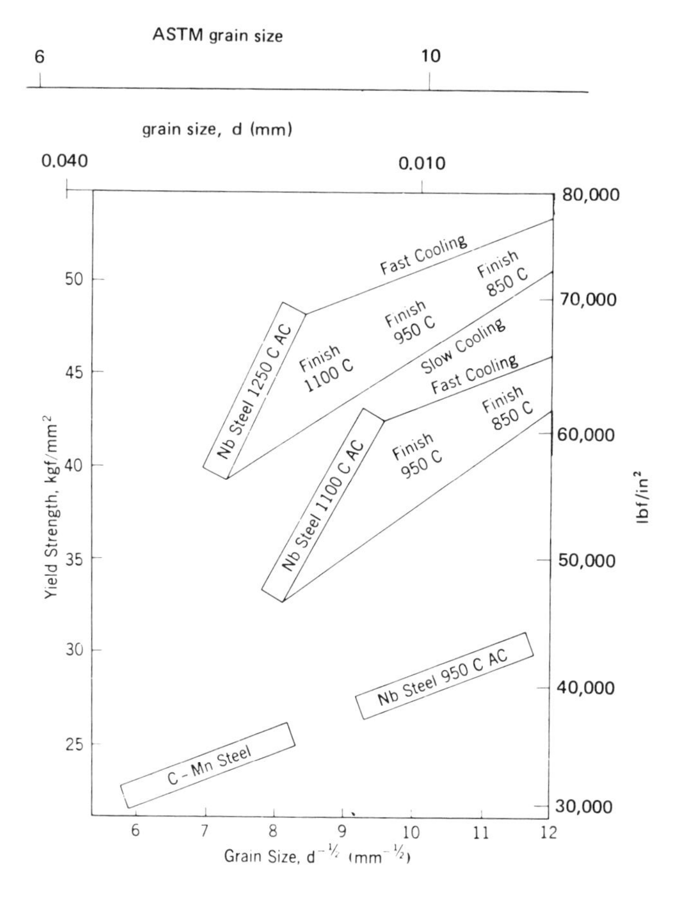

Figure 8-8 Effect of rolling variables on the yield strength of niobium-containing steels. *(From K. J. Irvine, in "Steel Strengthening Mechanisms," Climax Molybdenum Co., Ann Arbor, Michigan, 1969.)*

Using the bainite lath size as a measure of the ferrite grain-size effect, a comparison can be made to the properties of the fine-grained, ferrite-pearlite steels. This is shown in Fig. 8-11, where it is seen that the yield strength is extended above that of the niobium-containing steels. However, the effect of the bainite structure on the impact properties depends upon the type of bainite, upper or lower, formed. The reasons for this were discussed in Chap. 2. Adjacent upper-bainite laths have a close crystallographic relationship, so that a crack can easily propagate across the low-angle boundaries, until the high-angle, prior austenite boundary is reached. Thus the impact properties correlate best with the prior austenite grain size. In lower bainite, the carbides precipitated within the ferrite laths impede fracture, and impact properties improve. Figure 8-12 compares the impact properties of bainite structures to those of ferrite-pearlite and of martensite.

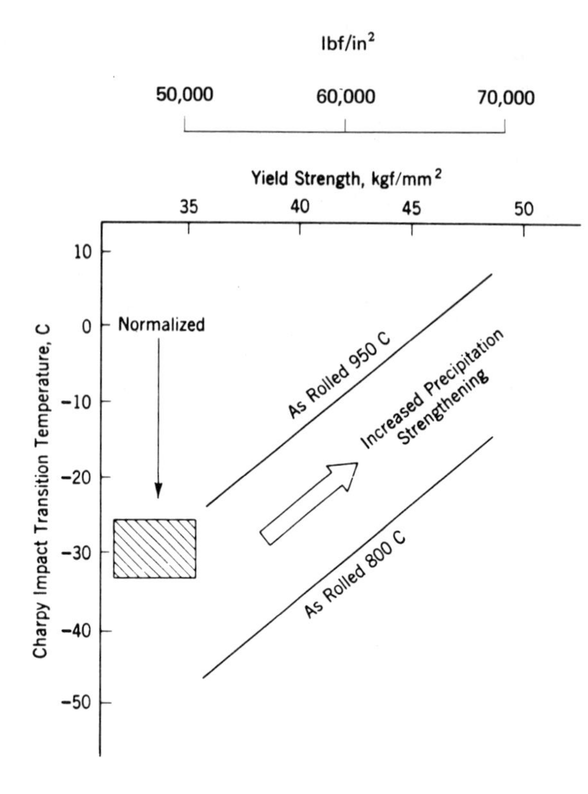

Figure 8-9 Relation between yield strength and transition temperature for a niobium-containing steel. *(From K. J. Irvine, in "Steel Strengthening Mechanisms," Climax Molybdenum Co., Ann Arbor, Michigan, 1969.)*

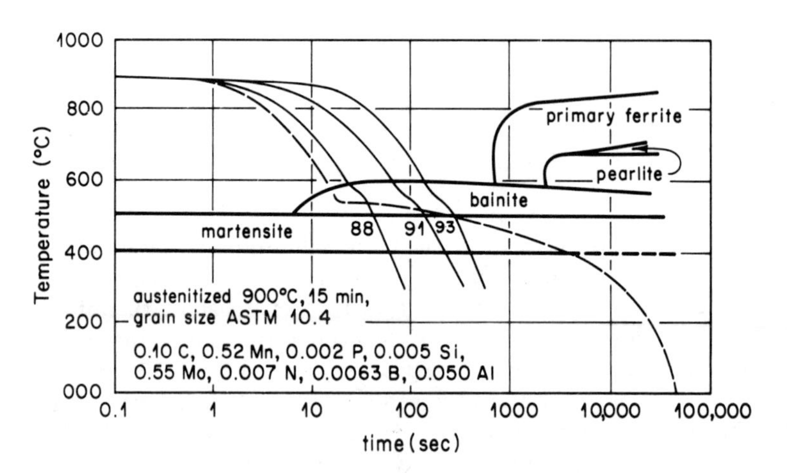

Figure 8-10 Continuous cooling transformation (CCT) diagram for a 0.10% C–0.55% Mo–0.063% B steel. The numbers in the fields give the amount (%) of bainite present. *(Adapted from Metal Progress Data Sheet,* Met. Progr., *vol. 96, February, 1969.)*

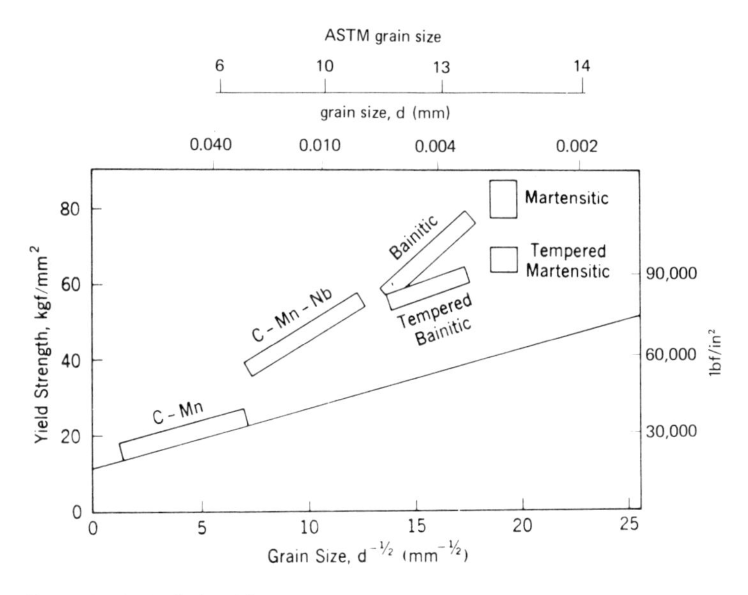

Figure 8-11 Relationship between grain size and yield strength for several microstructures. *(From K. J. Irvine, in "Steel Strengthening Mechanisms," Climax Molybdenum Co., Ann Arbor, Michigan, 1969.)*

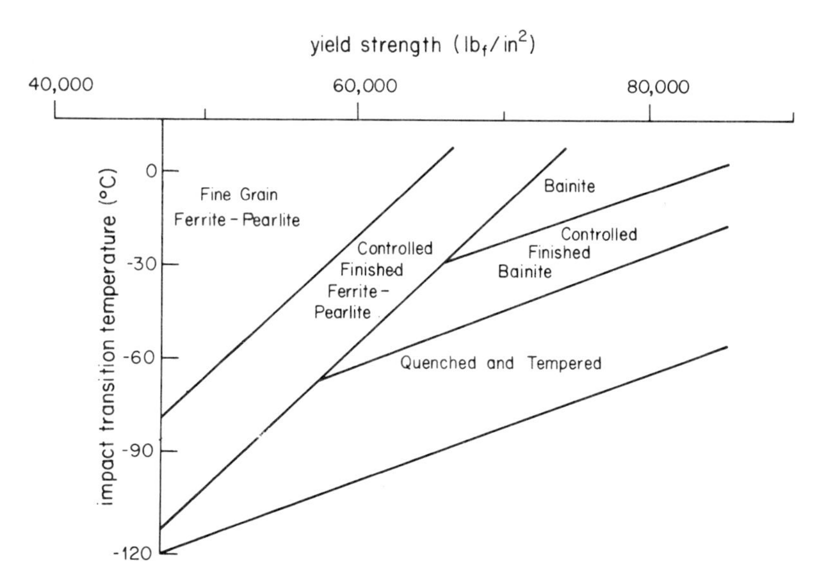

Figure 8-12 Relationship between yield strength and impact transition temperature for several microstructures. *(From K. J. Irvine, in "Strong, Tough Structural Steels," Iron and Steel Institute, London, 1967.)*

8-6 MARTENSITIC STEELS

Steels cooled to give a martensite structure are similar in structure to the bainite because of autotempering (Chap. 2). The carbides are fine, so that the strength is higher (Fig. 8-11). However, the martensite regions are separated by sufficiently high-angle boundaries that they impede fracture, and hence this structure shows markedly better impact properties than bainite (Fig. 8-12).

Thus the tempered martensite structure gives the best combination of strength and impact properties. However, in structural steel this structure may not be feasible to form by heat treatment (e.g., the part may be too complex to quench, or construction may not allow it such as in continuous pipelines). Then, in structural steels, most heat treatments involve the formation of ferrite-pearlite or bainite structures.

CHAPTER

NINE

CAST IRONS

Cast irons are a class of iron-carbon alloys which have sufficiently high carbon content (near the eutectic composition value) to attain a relatively low melting temperature. These alloys are used to cast objects which may be machined into final form and may be heat treated, but which are not fabricated by plastic deformation. Thus their properties are derived by control of the casting process and of subsequent heat treatments. Further, special elements (e.g., silicon, chromium, etc.), in addition to carbon, are added to allow development of properties. Cast irons are classified according to the type of microstructure developed, and in this chapter each class will be treated separately.

9-1 GRAY CAST IRON

The structure of cast irons is related to the kinetics of the solidification process, and particularly to whether graphite or whether iron carbide form during the cooling process. Recall that iron carbide is not a stable phase but, given sufficient time at sufficiently high temperature, will decompose to iron and graphite. (For most steels this is not a problem, except in certain high-temperature applications.) Figure 9-1 compares the equilibrium iron-carbon phase diagram to the metastable iron-iron carbide phase diagram. Note that the eutectic temperature and composition are about the same for both cases, but the hypereutectic liquidus is considerably different.

In examining the structures that develop upon cooling from the liquid, both diagrams must be considered. Solidification may involve the formation of graphite

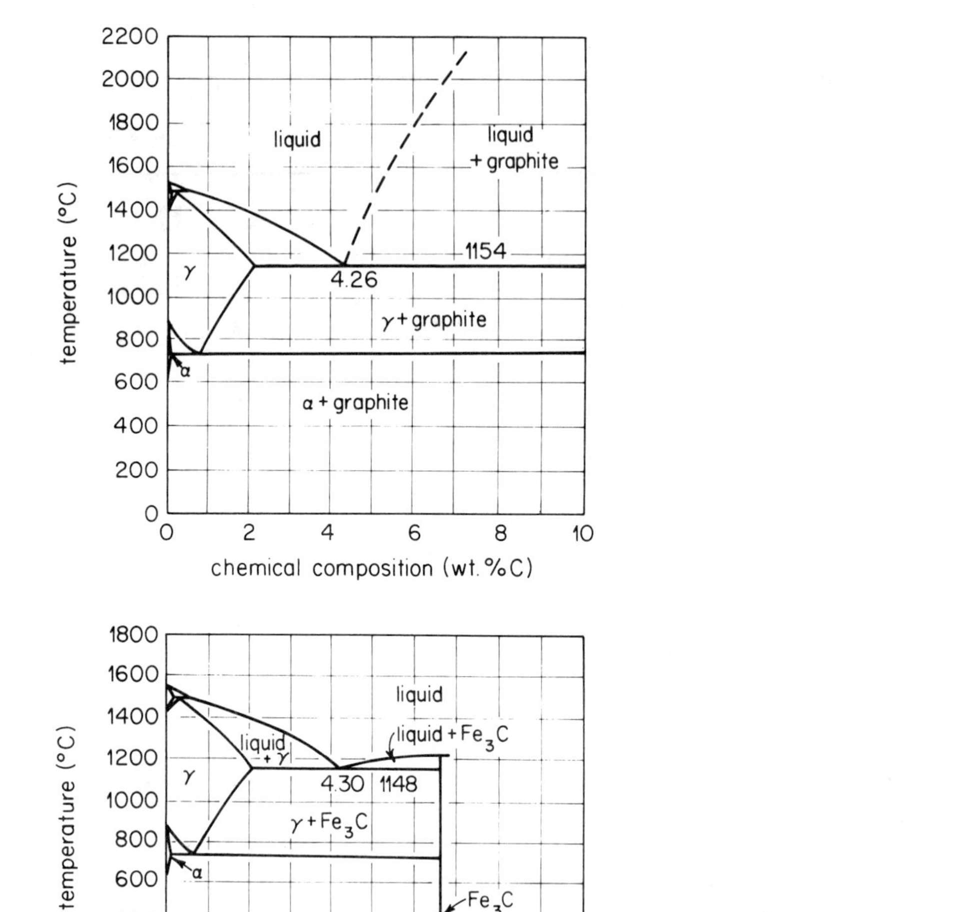

Figure 9-1 The iron-carbon and iron-iron carbide phase diagrams. *(Adapted from "Metals Handbook," vol. 8, Metallography, Structures and Phase Diagrams, American Society for Metals, Metals Park, Ohio, 1973.)*

as a hypereutectic phase, but upon further cooling into the lower temperature range the austenite may decompose into ferrite and carbide. The stability of the carbide is promoted by certain elements, and these can be used to develop structures in which iron carbide is prominent; other additions promote the decomposition of iron carbide, so that it is more difficult to obtain.

To illustrate the formation of the structure of gray cast iron, we will discuss first the solidification of an iron-5% carbon alloy. Upon cooling, the primary graphite crystals will precipitate from the liquid in the region *a-b* (Fig. 9-2), so that just above the eutectic temperature the structure appears similar to that shown. Upon further cooling to 1154°C, the liquid (now of eutectic composition) undergoes the eutectic

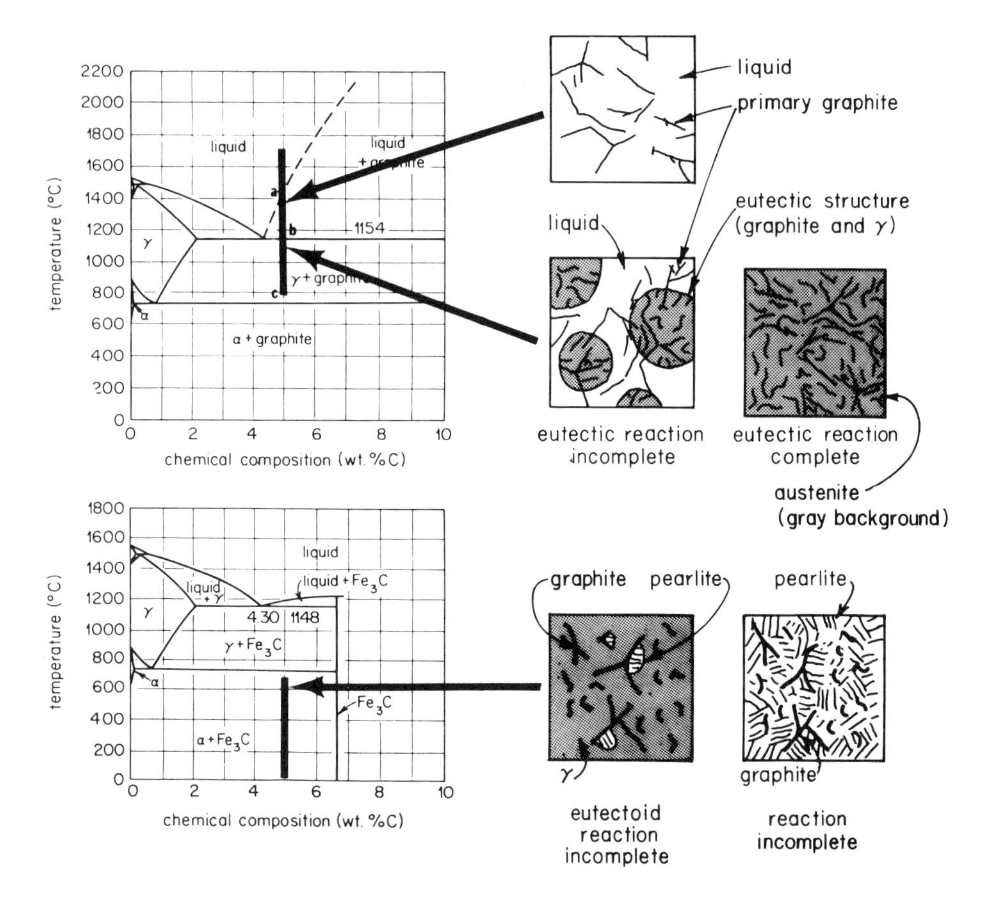

Figure 9-2 Schematic illustration of the formation of the microstructure of a hypereutectic cast iron. *(Adapted from H. Morrogh and W. J. Williams,* J.I.S.I., *vol. 155, p. 321, 1947.)*

reaction to form austenite and graphite simultaneously. This nucleates around the primary graphite crystals, giving the microstructure shown in Fig. 9-2. Thus, when solidification is complete, the structure consists of primary graphite and a graphite-austenite eutectic mixture (Fig. 9-2).

Upon cooling from point *b* to point *c* (Fig. 9-2) the austenite solubility decreases, and the excess carbon is deposited upon the already existing graphite crystals, increasing their size. Thus, at point *c*, the structure is still graphite and austenite, but the austenite has the composition of the eutectoid, 0.8% C. Upon cooling further the eutectoid reaction should occur, with the austenite decomposing into ferrite and graphite. However, in this lower temperature range even slow cooling is sufficiently rapid to suppress graphite formation, and instead the austenite decomposes to pearlite. The final structure is shown schematically in Fig. 9-2.

The actual microstructure obtained for such cooling as described above is shown in Fig. 9-3. Note that there are two different kinds of graphite "flakes" present, those

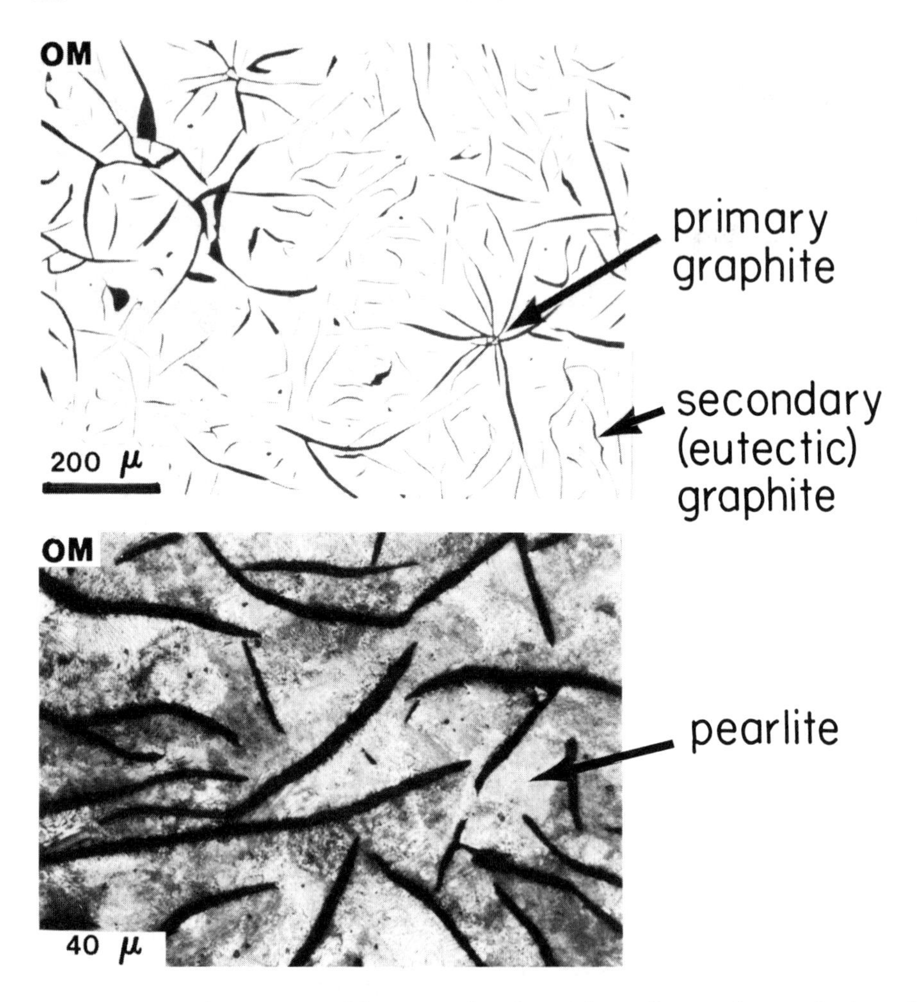

Figure 9-3 The microstructure of hypereutectic gray cast iron. The top photograph shows the unetched structure at low magnification. The bottom photograph shows the etched structure at higher magnification, resolving the pearlite. *(Adapted from A.R. Bailey and L.E. Samuels, "Foundry Metallography," Metallurgical Services, Betchworth, Surrey, England, 1971.)*

which formed as proeutectic graphite (sometimes called "kish") and those which formed as part of the eutectic reaction. These latter flakes are somewhat smaller. The remaining structure is pearlite. The origin of the name "gray cast iron" lies in the gray appearance imparted to the surface by the graphite flakes.

If the carbon content is less than the eutectic value, upon cooling the primary austenite forms first, and then the austenite-graphite eutectic. In most cases, subsequent cooling converts the austenite to pearlite. Thus, the structure is similiar to that described for hypereutectic irons. An example of the microstructure is shown in Fig.

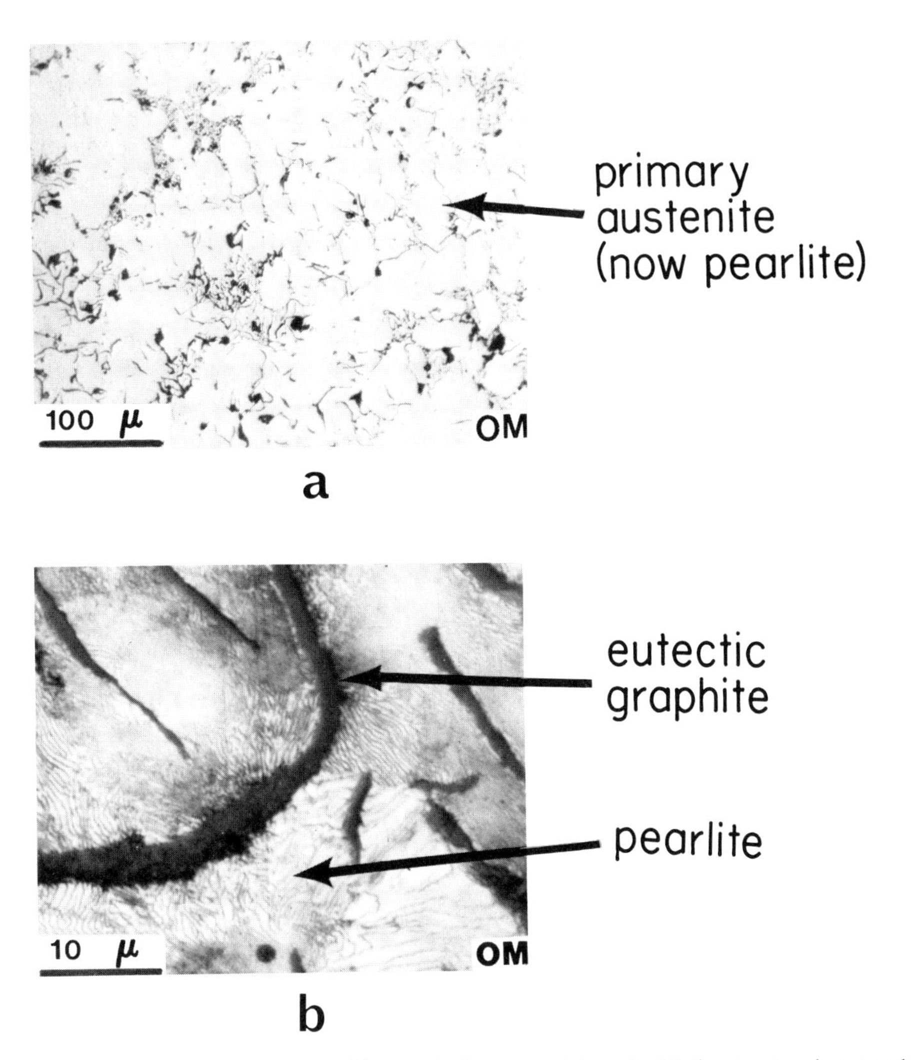

Figure 9-4 The microstructure of hypoeutectic gray cast iron. In (a) the structure is not etched. The dendrites of primary austenite are seen clearly. In (b), the structure has been etched to reveal the pearlite.

9-4. However, since the graphite orginates only from the eutectic reaction, the flakes are smaller than in hypereutectic irons.

The size and shape of the primary graphite flakes does depend upon the cooling rate as solidification occurs. If cooling is very slow, the graphite flakes begin to form with little undercooling. Under such conditions, the "driving force" for crystallization is relatively low, and hence the nucleation rate of the graphite flake is low. Thus relatively few flakes form, but these have time to grow rather large before the eutectic relation begins. If cooling is faster, greater undercooling is achieved before the graphite flakes precipitate. At this lower temperature, the driving force is larger, and the nucleation rate increases, giving more flakes than for slower cooling. The general shape of each flake will be finer also. A comparison can be made between the primary flakes in Fig. 9-3 and those formed upon more rapid cooling shown in Fig. 9-5. These graphite flakes impart important properties to the cast iron, such as excellent machinability and the ability to dampen vibrations and sound.

The formation of graphite in cast irons is quite sensitive to the chemical composition. Graphite stability is favored by the addition of silicon, nickel, and copper. Also, these elements affect the eutectic temperature and composition. Thus for iron-carbon-silicon irons the ternary phase diagram should be examined to predict the process of solidification. However, this can become quite complicated and instead it is useful to speak in terms of an "equivalent carbon content." One formulation of this gives

$$\text{Equivalent carbon content} = (\%\,\text{C}) + 0.33(\%\,\text{P}) + 0.30(\%\,\text{Si})$$

Thus, an iron containing 4.2% C and 2.1% Si has a carbon equivalent of 4.8% C, which means that this will be hypereutectic iron and will form primary graphite upon slow cooling.

The alloying additions in gray cast iron also can allow the formation of phases other than graphite, iron carbide, and ferrite. For example, sufficient phosphorous additions allow a ternary eutectic structure of austenite, iron carbide, and iron phosphide (Fe_3P) to form, in addition to primary graphite.

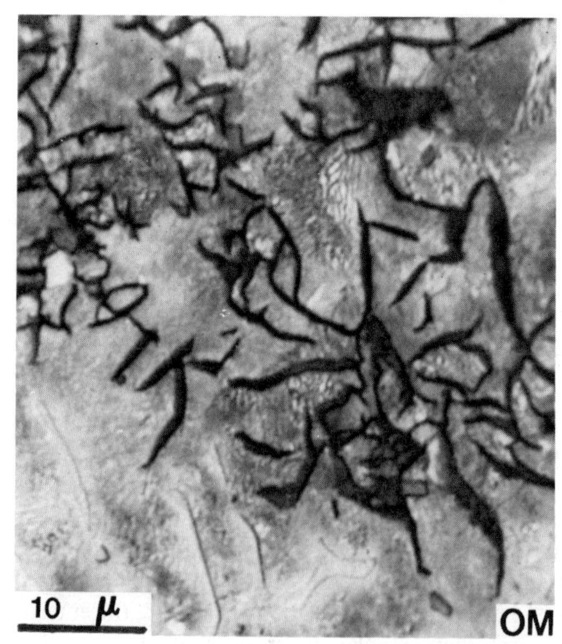

Figure 9-5 The microstructure of a hypereutectic gray cast iron which was cooled faster than the iron shown in Fig. 9-3, and hence has finer primary graphite flakes.

9-2 WHITE CAST IRON

If the cooling rate is sufficiently high, the formation of graphite during solidification is suppressed, and instead Fe_3C will form. If the composition of the alloy is hypoeutectic, then the primary phase is austenite, and the eutectic structure is austenite and iron carbide. This eutectic structure is sometimes called ledeburite. The austenite will decompose to pearlite at low temperatures. Figure 9-6 illustrates schematically the sequence of events, and Fig. 9-7 shows a microstructure of a hypoeutectic white cast iron. Note that the majority of the structure is iron carbide, the remainder being ferrite.

If the iron is hypereutectic, primary Fe_3C will be present instead of primary austenite. Figure 9-8 shows a microstructure of such a white cast iron. Note that the primary Fe_3C crystals are quite long. The preponderance of iron carbide (and the lack of graphite) gives the fracture surface of white cast irons a white, crystalline appearance, which is the origin of the name "white cast iron."

The formation of white cast iron is favored by rapid cooling during solidification, and by the addition of certain alloying elements, such as manganese and chromium. Because of the high proportion of iron carbide present, these cast irons are quite hard and brittle and are not machinable.

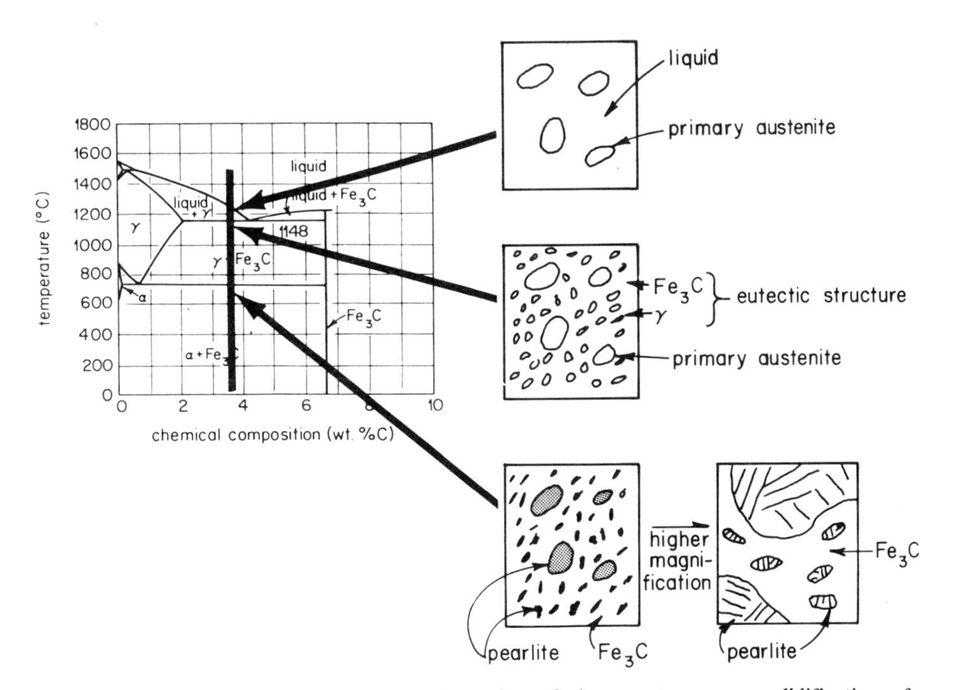

Figure 9-6 Schematic description of the formation of the structure upon solidification of a hypoeutectic white cast iron.

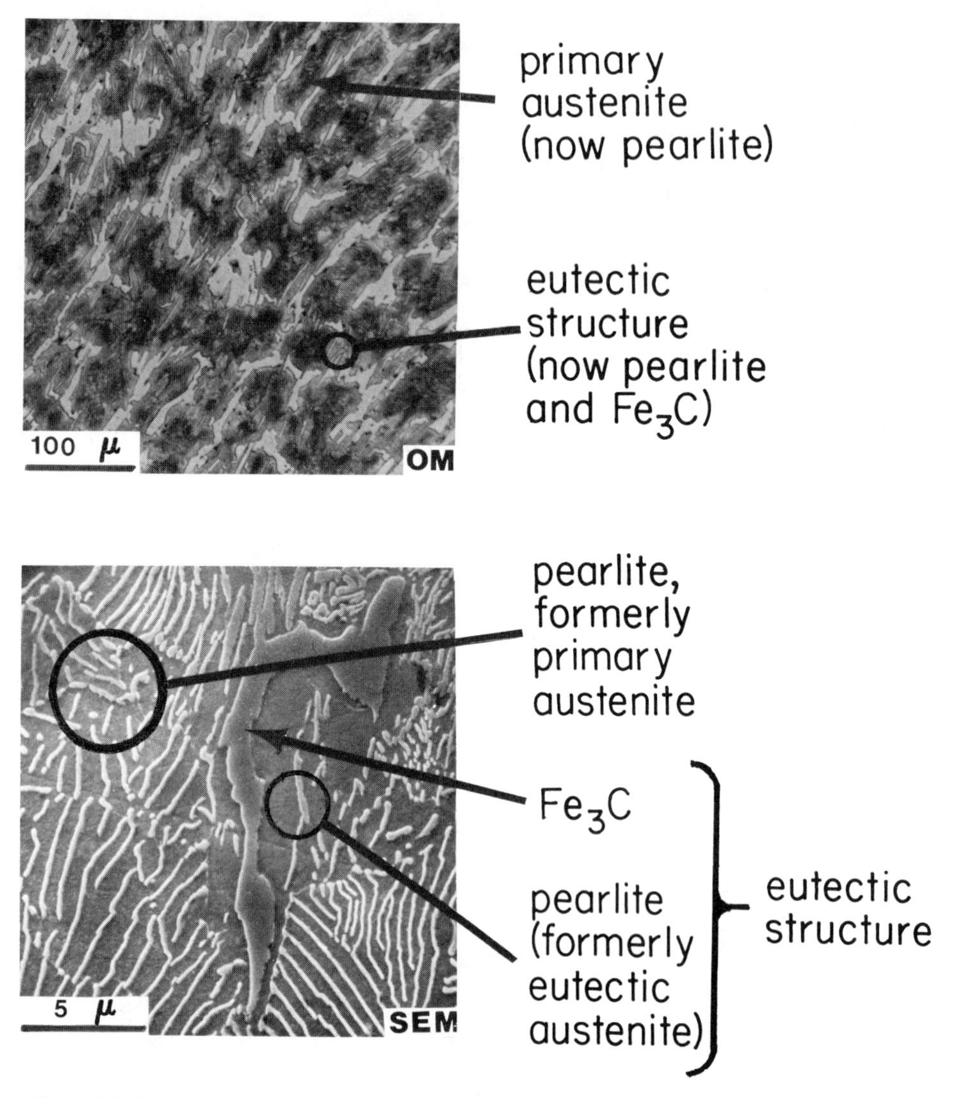

Figure 9-7 Microstructure of a hypoeutectic white cast iron. The dark areas at low magnification *were* austenite (primary austenite and eutectic austenite) at high temperature, but at 25°C they are now pearlite. This is apparent at higher magnification.

9-3 NODULAR (DUCTILE) CAST IRON

For some applications, the presence of graphite in the form of flakes is undesirable. By the addition of certain elements (e.g., magnesium, calcium, and cerium) to the liquid alloy, upon solidification the graphite will form as spherulites, or nodules. This

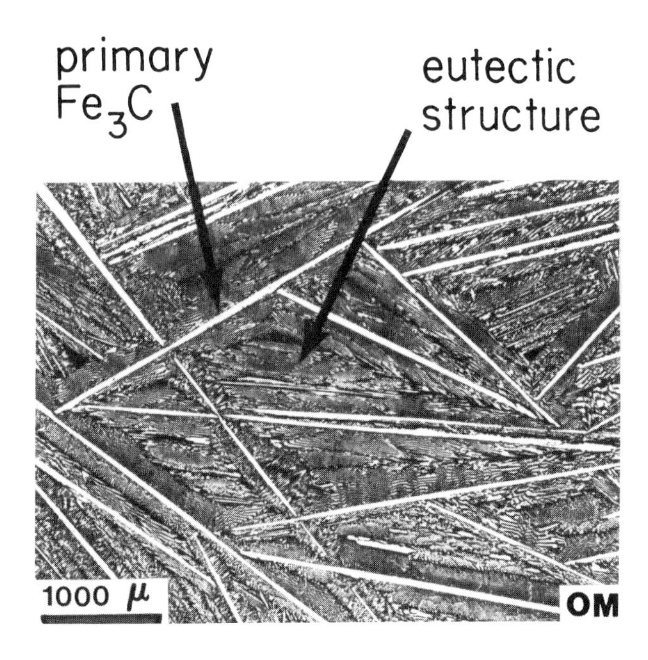

Figure 9-8 Microstructure of hypereutectic white cast iron. The long white rods are primary Fe_3C crystals. The eutectic structure consists of dark pearlite (formerly secondary austenite) and Fe_3C. *(Adapted from A. R. Bailey and L. E. Samuels, "Foundry Metallography," Metallurgical Services, Betchworth, Surrey, England, 1971.)*

configuration consists of crystals of graphite radiating from a central point, with the close-packed planes tangent to the spherulite. A typical microstructure is shown in Fig. 9-9. If the alloy is hypoeutectic, then the primary phase is austenite, and the eutectic reaction involves the formation of graphite as spherulites.

9-4 MALLEABLE CAST IRON

Another method to obtain a graphite structure, but avoid the graphite flakes, is to heat treat a white cast iron so that the carbides decompose to ferrite and graphite (graphitization). (The graphite formed is sometimes called temper carbon.) If done properly, the graphite will be in the form shown in Fig. 9-10. The heat treatment may be quite involved. For example, the iron may be heated slowly to 850 to 930°C (in the austenite plus Fe_3C range) over a period of 24 hours then held for up to 60 hours, during which time the Fe_3C converts to graphite. If the iron is cooled very slowly, graphite will precipitate on the existing graphite regions upon cooling through the "austenite plus graphite" range; then during the eutectoid reaction, the graphite formed will deposit on these graphite regions. This leaves a structures of graphite in a ferrite matrix, as shown in Fig. 9-10. If cooling is faster, the austenite may form pearlite, as shown in Fig. 9-11.

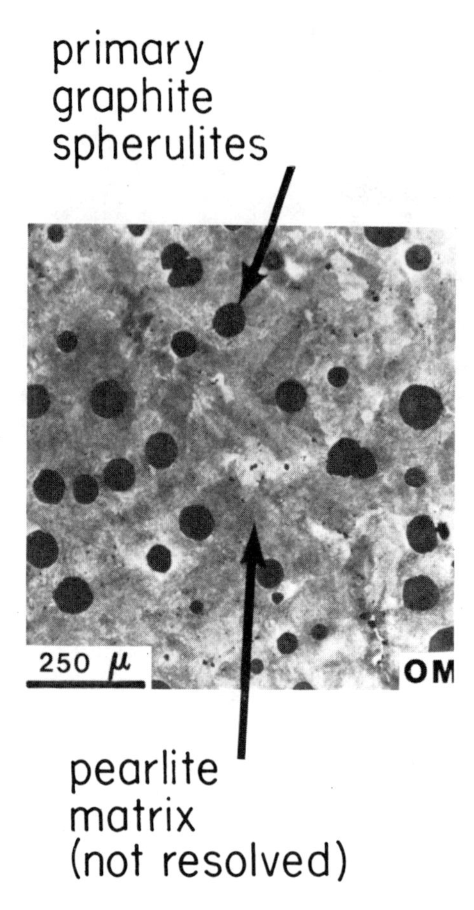

Figure 9-9 Microstructure typical of hyper-eutectic nodular (ductile) cast iron. *(Adapted and reprinted with permission from W. Hume-Rothery, "The Structure of Alloys of Iron," Pergamon Press, New York, 1966.)*

9-5 PROPERTIES

Because of the ability to control the structure through chemical composition, casting procedure, and heat treatment, the properties of cast iron are quite varied. There are several factors which allow cast irons to have wide uses. One is castability. Another is that cast irons in many applications have superior wear and abrasion resistance. Yet another unique property is the ability of gray cast irons to dampen mechanical (and acoustic) vibrations. In this section, these properties will be briefly examined.

The mechanical properties of cast iron cover a wide range of values since the structure can be varied from graphite flakes in ferrite to iron carbide in martensite. Table 9-1 lists typical tensile properties and ranges of hardness values for various structures. The white cast irons are, of course, the hardest, with the structures containing large amounts of ferrite being the softest. The ductility of white cast iron is extremely low (essentially zero) due to the preponderance of the brittle iron carbide. Also, the gray cast irons with primary graphite flakes are quite brittle in tension,

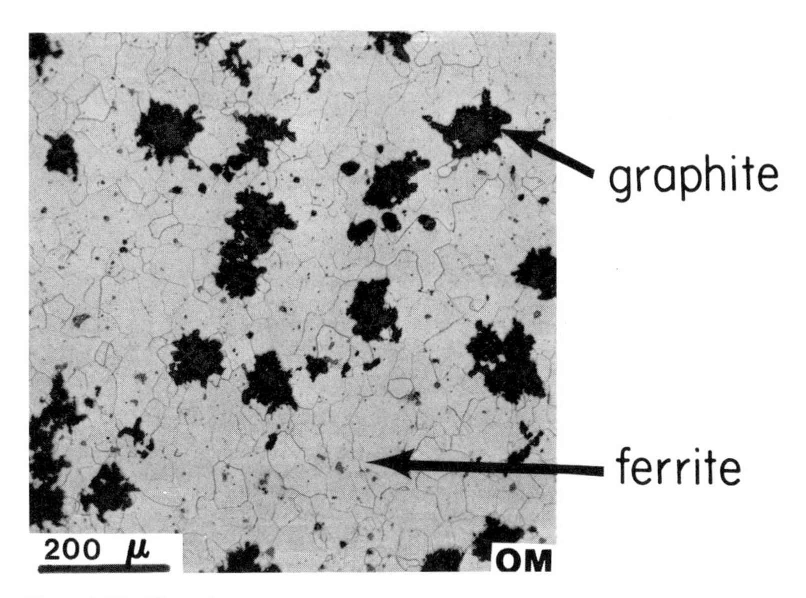

Figure 9-10 The microstructure of malleable cast iron in which graphite is in a ferrite matrix.

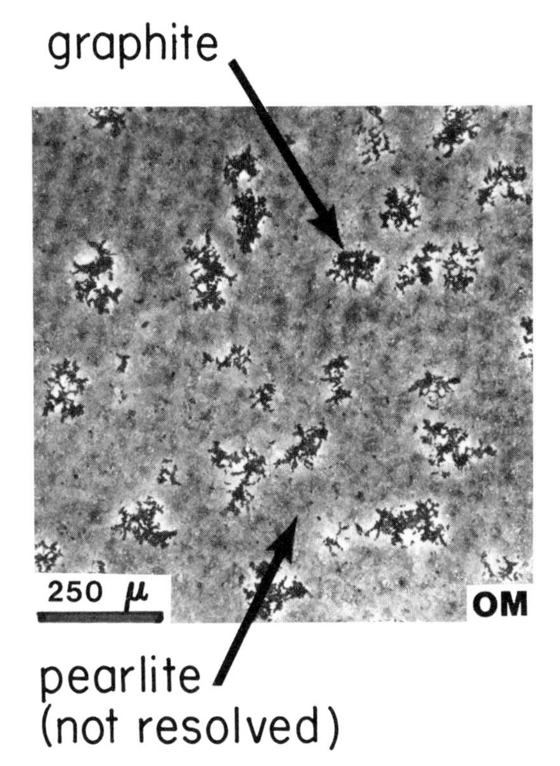

Figure 9-11 The microstructure of malleable cast iron in which graphite is in a matrix of pearlite. *(Adapted and reprinted with permission from W. Hume-Rothery, "The Structure of Alloys of Iron," Pergamon Press, New York, 1966.)*

Table 9-1 Typical tensile mechanical properties of cast irons, and hardness ranges

Type	Microstructure	Yield strength, psi	Tensile strength, psi	Elongation in 2 in., %	Hardness, DPH
Gray cast iron	Ferrite + graphite	–	25,000	0.5	
	Pearlite + graphite	–	45,000	0.5	
Ductile cast iron	Ferrite + graphite	40,000	60,000	18	160–200
	Ferrite + pearlite + graphite	45,000	65,000	12	180–220
	Pearlite + ferrite + graphite	55,000	80,000	6	200–270
Malleable cast iron		35,000	35,000	18	110–140
White cast iron	Primary iron carbide + pearlite	–	60,000	0	350–600

because of the weakness of the graphite flakes and the relatively large quantity of them present.

The damping capacity (Fig. 9-12) of a material is a measure of the ability to damp vibrations. The presence of graphite flakes or nodules in the microstructure of cast irons gives them a relatively high damping capacity, which can be seen from

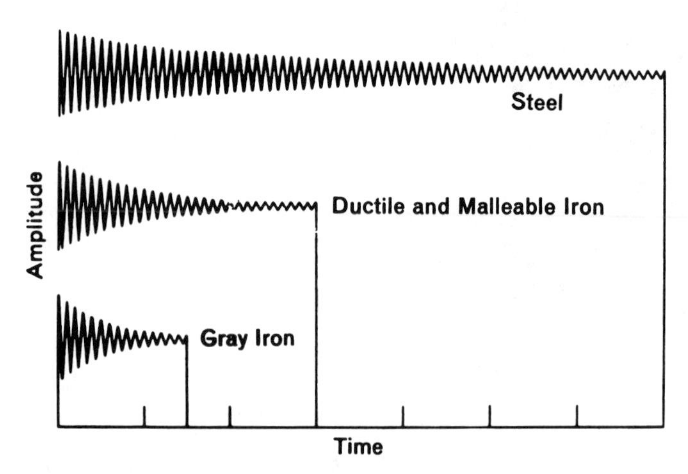

Figure 9-12 Schematic illustration of the relative ability of ferrous alloys to dampen vibration. *(From C. F. Walton (ed.), "Iron Castings Handbook," p. 155, Iron Founder's Society, Inc., Cleveland, Ohio, 1971.)*

the comparison of gray cast iron and ductile and malleable cast iron to steel in Fig 9-12. This ability is important as it minimizes the tendency for moving parts to attain resonance, and it allows the material to operate with lower noise generation. Table 9-2 compares the damping capacity of cast irons and a steel. The hypereutectic cast irons, with a higher carbon equivalent, have a higher quantity of primary graphite flakes present.

Another property that cast irons frequently possess is wear resistance. Wear is a complicated process, and the response of the material chosen depends upon a number of factors, such as the presence of lubricants, the material of the mating surface, and the presence of abrasive particles. Thus, a detailed comparison of the wear resistance of cast irons and other materials cannot easily be made. However, two examples will be cited to illustrate the superiority of cast irons. Figure 9-13 shows the volume of wear debris generated when mating several materials against hardened 52100 steel. The low wear rate of the gray and ductile cast irons is associated with the lubricating qualities of the graphite, which may spread a thin film over the surface so that contact with the mating surface and the ferrite matrix is limited. In many applications involving abrasive wear, it is important to have a surface that is hard and that will maintain its hardness. White cast iron, heat treated to give primary carbides in a martensite matrix, gives good service in such cases. Table 9-3 compares the relative wear rates under abrasive wear conditions of several materials.

Table 9-2 A comparison of the damping capacity of cast irons and a steel

Material	Relative decrement in amplitude of vibration per cycle
Carbon steel	1–2
Malleable iron	3–6
Ductile iron	3–9
Hypoeutectic gray iron (3.2% C, 2.0% Si)	40
Eutectic gray iron (3.9% C, 0.9% Si)	105
Hypereutectic gray iron (3.7% C, 1.8% Si)	126

Source: C. F. Walton (ed.), "Iron Castings Handbook," p. 155, Iron Founder's Society, Inc., Cleveland, 1971.

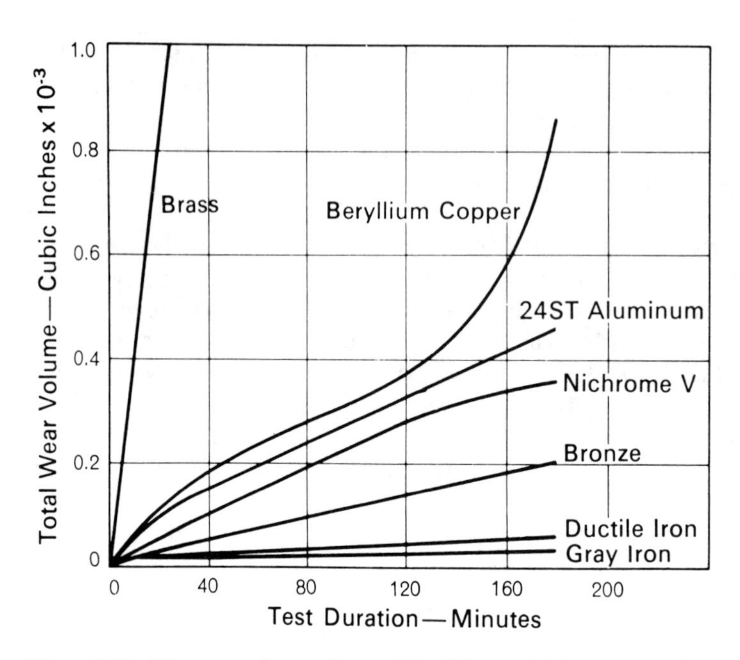

Figure 9-13 The wear of several materials sliding unlubricated against hardened 52100 steel at 5000 ft/min. *(From C. F. Walton (ed.), "Iron Castings Handbook," p. 342, Iron Founder's Society, Inc., Cleveland, Ohio, 1971.)*

Table 9-3 Relative wear rates of 5-in. diameter balls used in the milling of ore

Material	Hardness before testing, R_c	Hardness after testing R_c	Abrasion factor
Martensitic Cr–Mo white cast iron	54	66	89
Martensitic high-Cr white cast iron	53	64	98
Martensitic Cr–Mo steel	49	55	100
Chill-cast Ni–Cr–Mo white cast iron	–	59	107
Pearlitic Cr–Mo steel	38	39	127

Source: C. F. Walton (ed.), "Iron Castings Handbook," p. 337, Iron Founder's Society, Inc., Cleveland, 1971.

APPENDIX

I

SUGGESTED READINGS AND REFERENCES

Review of physical metallurgy

There are several textbooks suitable for review of the basic principles of physical metallurgy. The last two books emphasize specifically physical metallurgy.

van Vlack, L. H.: "Elements of Materials Science and Engineering," Addison-Wesley, Reading, Mass., 1975.
Guy, A. G.: "Introduction to Materials Science," McGraw-Hill, New York, 1976.
Reed-Hill, R. E.: "Physical Metallurgy Principles," 2d ed., van Nostrand, New York, 1973.

General reference on the heat treatment of steels

Sisco, F. T.: "The Alloys of Iron and Carbon," vol. 2, McGraw-Hill, New York, 1937.
Hanson, A., and J. G. Parr: "The Engineers Guide to Steel," Addison-Wesley, Reading, Mass., 1965.
Sacks, G., and K. R. van Horn: "Practical Metallurgy," American Society for Metals, Metals Park, Ohio, 1940.
Brick, R. M., R. B. Gordon, and A. Phillips: "Structure and Properties of Alloys." 3d ed., McGraw-Hill, New York, 1965.
Samans, C. H.: "Engineering Metals and Their Alloys," Macmillan, New York, 1953.
McGannon, H. E. (ed.): "The Making, Shaping and Treating of Steel," 9th ed., United States Steel Corporation, Pittsburgh, 1971.
Hume-Rothery, W.: "The Structure of Alloys of Iron," Pergamon, New York, 1966.
Thelning, K. E.: "Steel and Its Heat Treatment," Butterworths, London, 1975.
"Heat Treatment of Steels," Republic Steel Corporation, Cleveland, 1961.
Speich, G. R., and W. C. Leslie: "The Tempering of Steels," *Met. Trans.*, vol. 3, p. 1043, 1972.
Clark, D. S., and W. R. Varney: "Physical Metallurgy for Engineers," van Nostrand, New York, 1952.
Bullens, D. K.: Steel and Its Heat Treatment, vol. 1, "Principles," 5th ed., Wiley, New York, 1948.
Irvine, K. J.: The Physical Metallurgy of Steel, *J.I.S.I.*, vol. 207, p. 837, 1969.
Duckworth, W. E., and J. D. Baird: Mild Steels, *J.I.S.I.*, vol. 207, p. 854, 1969.
Nutting, J.: Physical Metallurgy of Alloy Steels, *J.I.S.I.*, vol. 207, p. 872, 1969.

Hardenability and heat treatments

McGannon, H. E. (ed.): "The Making, Shaping and Treat of Steel," 9th ed., United States Steel Corporation, Pittsburgh, 1971.

Thelning, H. E.: "Steel and Its Heat Treatment," Butterworths, London, 1975.

Hollomon, J. H., and L. D. Jaffe: "Ferrous Metallurgical Design," Wiley, New York, 1947.

Clark, D. S., and W. R. Varney: "Physical Metallurgy for Engineers," van Nostrand, New York, 1962.

Source Book on Heat Treating, vol. I: "Materials and Processing," American Society for Metals, Metals Park, Ohio, 1975.

Source Book on Heat Treating, vol. II: "Production and Engineering Practices," American Society for Metals, Metals Park, Ohio, 1975.

"Alloying Elements and Their Effects. Hardenability," Republic Steel Corporation, Cleveland, 1973.

Siebert, C. A., D. V. Doan, and D. H. Breen: "The Hardenability of Steels," American Society for Metals, Metals Park, Ohio, 1977.

"Atlas of Isothermal Transformation and Cooling Transformation Diagrams," American Society for Metals, Metals Park, Ohio, 1977.

Lyman, T. (ed.): "Selection of Steel for Hardenability," in Metals Handbook, vol. 1, Properties and Selection of Metals, American Society for Metals, Metal Park, Ohio, 1961.

"U.S.S. Carilloy Steels," United States Steel, Pittsburgh.

"Atlas of Isothermal Transformation Diagrams," United States Steel Corporation, Pittsburgh, 1951.

Supplement to "Atlas of Isothermal Transformation Diagrams," United States Steel Corporation, Pittsburgh, 1953.

Jatezak, C. F.: Determining Hardenability from Composition, *Met. Progr.*, vol. 100, p. 60, September, 1971.

Cias, W. W.: "Phase Transformation Kinetics and Hardenability of Medium-Carbon Alloy Steels," Climax Molybdenum Company, Greenwich, Conn.

Symposium on Hardenability, *Met. Trans.*, vol. 4, p. 2231 ff, October, 1973.

"Quenching and Martempering," American Society for Metals, Metals Park, Ohio, 1964.

"Transformation and Hardenability in Steels," Climax Molybdenum Company, Ann Arbor, Michigan, 1967.

Bain, E. C., and H. W. Paxton: "Alloying Elements in Steels," 2d ed., American Society for Metals, Metals Park, Ohio, 1961.

"Hardenability of Alloy Steels," American Society for Metals, Metals Park, Ohio, 1939.

Grossman, M. A.: "Elements of Hardenability," American Society for Metals, Metals Park, Ohio, 1952.

Schrader, A., and A. Rose: "De Ferri Metallographia," Metallographic Atlas of Iron, Steels and Cast Irons, vol. II, Structure of Steels, Saunders, Philadelphia, 1966.

"Hardenability of Nickel Alloy Steels," International Nickel Company, New York, 1965.

Parker, E. R.: "Interrelations of Compositions, Transformation Kinetics, Morphology, and Mechanical Properties of Alloy Steels," *Met. Trans.*, vol. 8A, p. 1025, 1977.

Doane, D. V., and J. S. Kirkaldy (eds.): "Hardenability Concepts with Applications to Steels," Met. Soc. AIME, Warrendale, Pa., 1978.

Surface treatments

Thelning, K. E.: "Steel and Its Heat Treatment," Butterworths, London, 1975.

"Gas Carburizing," American Society for Metals, Metals Park, Ohio, 1964.

"Furnace Atmospheres and Carbon Control," American Society for Metals, Metals Park, Ohio, 1964.

Bullens, D. K.: Steel and Its Heat Treatment, vol. II, "Tools, Processes, Control," 5th ed., Wiley, New York, 1948.

Dawes, C., and R. J. Cookseg: Surface Treatment of Engineering Components, in "Heat Treatment of Metals, Special Report 95," Iron and Steel Institute, London, 1965.
Metals Handbook, vol. 2, "Heat Treating, Cleaning and Finishing," American Society for Metals, Metals Park, Ohio, 1964.
Source Book on Heat Treating, vol. I, "Materials and Processing," American Society for Metals, Metals Park, Ohio, 1975.
Source Book on Heat Treating, vol. II, "Production and Engineering Practices," American Society for Metals, Metals Park, Ohio, 1975.
"Carburizing and Carbonitriding," American Society for Metals, Metals Park, Ohio, 1977.

Tool steels

Bullens, I. K.: Steel and Its Heat Treatment, vol. III, "Engineering and Special Purpose Steels," 5th ed., Wiley, New York, 1949.
Roberts, G. A., J. C. Hamaker, and A. R. Johnson: "Tool Steels," 3d ed., American Society for Metals, Metals Park, ohio, 1962.
Wilson, R.: "Metallurgy and Heat Treatment of Tool Steels," McGraw-Hill, New York, 1975.
Thelning, K. E.: "Steel and Its Heat Treatment," Butterworths, London, 1975.
Metals Handbook, vol. 2, "Heat Treating, Cleaning and Finishing," American Society for Metals, Metals Park, Ohio, 1964.
Payson, P.: "The Metallurgy of Tool Steels," Wiley, New York, 1962.
Brick, R. M., R. B. Gordon, and A. Phillips: "Structure and Properties of Alloys," 3d ed., McGraw-Hill, New York, 1965.
"Properties and Selection of Tool Materials," American Society for Metals, Metals Park, Ohio, 1975.

Stainless steel

Keating, F. H.: "Chromium-Nickel Austenitic Steels," Butterworths, London, 1956.
Thielsch, H.: Physical and Welding Metallurgy of Chromium Stainless Steels, *Weld. J.*, vol. 30, p. 209s, 1951.
———: Alloying Elements in Chromium-Nickel Stainless Steels, *Weld. J.*, vol. 29, p. 361s, 1950.
———: Physical Metallurgy of Austenitic Stainless Steels, *Weld. J.*, vol. 29, p. 577s, 1950.
Zapfee, C. A.: "Stainless Steels," The American Society for Metals, Metals Park, Ohio, 1949.
"Selection of Stainless Steels," American Society for Metals, Metals Park, Ohio, 1968.
"Source Book on Stainless Steels," American Society for Metals, Metals Park, Ohio, 1976.
Thum, E. E. (ed.): "The Book of Stainless Steels," American Society for Metals, Metals Park, Ohio, 1933.
McGannon, H. E.: "The Making, Shaping and Treating of Steels," United States Steel Corporation, Pittsburgh, 1971.
Brick, R. M., R. B. Gordon, and A. Phillips: "Structure and Properties of Alloys," 3d ed., Mc-Graw Hill, New York, 1965.
Pickering, R. F.: Physical Metallurgy of Stainless Steel Developments, *Int. Met. Rev.*, vol. 21, p. 227, 1976.
Peckner, D., and I. M. Bernstein: "Handbook of Stainless Steels," McGraw-Hill, New York, 1977.

Structural steels

Gray, J. M. (ed.): "Processing and Properties of Low Carbon Steel." The Metallurgical Society, New York, 1973.
"Strong, Tough Structural Steels," Iron and Steel Institute, London, 1967.
"Symposium: Low Alloy, High Strength Steels," The Metallurgy Company, Dusseldorf, 1970.
"Steel-Strengthening Mechanisms," Climax Molybdenum Company, Ann Arbor, Michigan, 1969.

Cast irons

Brick, R. M., R. B. Gordon, and A. Phillips: "Structure and Properties of Alloys," 3d ed., McGraw-Hill, New York, 1965.
A. R. Bailey, and L. E. Samuels: "Foundry Metallography," Metallurgical Services, Surrey, England, 1971.
Hume-Rothery, W.: "The Structure of Alloys of Iron," Pergamon, New York, 1966.
Angus, A. T.: "Cast Iron: Physical and Engineering Properties," Butterworth, London, 1976.

Other useful references

Evans, L. S.: "Selecting Engineering Materials for Chemical and Process Plant," Wiley, New York, 1974.
Hanson, A., and J. G. Parr: "The Engineer's Guide to Steel," Addison-Wesley, Reading, Mass., 1965.
Woldman, N. W., and R. C. Gibbons: "Engineering Alloys," 5th ed., van Nostrand, New York, 1973.

Terminology

Recognition of the exact meaning of words and terms in the heat treatment of steels can be a vexing problem. This is due partly to dual meanings of some terms, and to the shop-origin of many terms. The following references should be consulted for more detailed and extensive definitions.

Simons, E. N.: "A Dictionary of Alloys," Frederick Muller, London, 1969.
———: "A Dictionary of Ferrous Metals," Frederick Muller, London, 1970.
———: "A Dictionary of Metal Heat-Treatment," Frederick Muller, London, 1974.
"Handbook of Terms Commonly Used in the Steels and Non-Ferrous Industries," Iron Age, Chilfton Publishers, Philadelphia, 1954.

APPENDIX

II

ENGLISH/METRIC CONVERSION FACTORS

$1\ N/m^2 = 1$ pascal (Pa)
$= 10^{-3}$ M·Pa
$= 10^{-6}\ N/mm^2$
$= 1.4505 \times 10^{-4}$ lbf/in.^{2} (psi)
$= 1.4505 \times 10^{-7}$ ksi
$= 2.0886 \times 10^{-2}$ lbf/ft^2
$= 7.2520 \times 10^{-8}$ tons/in.^{2}
$= 1.0198 \times 10^{-7}$ kgf/mm^2

1 lbf/in.^{2} $= 6.89417 \times 10^{3}\ N/m^2$
$= 7.03068 \times 10^{-4}$ kgf/mm^2

1 kgf/mm^2 $= 1.4223363 \times 10^{3}$ lbf/in.^{2}

1 ton/in.^{2} $= 1.378929 \times 10^{7}\ N/m^2$

1 Joule $= 1.019716 \times 10^{-1}$ kgf/m
$= 7.3756 \times 10^{-1}$ ft·lbf
$= 2.7777 \times 10^{-4}$ watt·hour
$= 1$ watt·sec
$= 9.48451 \times 10^{-4}$ Btu

1 ft·lbf $= 1.3558$ Joule

1 MPa·m$^{1/2}$ $= 1.09885$ ksi·in.$^{1/2}$
$= 3.1623 \times 10^{-2}$ MN·m$^{-3/2}$

1 ksi·in.$^{1/2}$ $= 9.10038 \times 10^{-1}$ MPa·m$^{1/2}$

1 km $= 10^{3}$ m
$= 10^{5}$ cm
$= 10^{6}$ mm
$= 10^{9}\ \mu m$
$= 10^{12}$ angstroms (A)
$= 3.93700788 \times 10^{4}$ in.
$= 3.2808399 \times 10^{3}$ ft

1 m $= 10^{2}$ cm
$= 10^{3}$ mm
$= 10^{6}\ \mu m$
$= 10^{10}$ A

1 cm $= 10$ mm
$= 10^{4}\ \mu m$
$= 10^{8}$ A

1 μm $= 10^{-6}$ m
$= 10^{-4}$ cm
$= 10^{-3}$ mm
$= 10^{4}$ A

$°C = \frac{5}{9}(°F - 32)$
$K = \frac{5}{9}(°F - 32) + 273.15$
$°F = \frac{9}{5}(°C) + 32$
$K = °C + 273.15$
$°R = °F + 460$

APPENDIX

III

HARDNESS CONVERSIONS FOR STEELS

	Brinell hardness No., 10-mm ball, 3000-kg load			Rockwell hardness No.				Rockwell superficial hardness No., superficial brale penetrator					
Diamond pyramid hardness No.	Standard ball	10-mm ball, 3000-kg load Hultgren ball	Tungsten carbide ball	A-scale, 60-kg load, brale penetrator	B-scale, 100-kg load, 1/16-in. diam. ball	C-scale, 150-kg load, brale penetrator	D-scale, 100-kg load, brale penetrator	15-N scale, 15-kg load	30-N scale, 30-kg load	45-N scale, 45-kg load	Shore scleroscope hardness No.	Tensile strength (approx.), 1000 psi	Diamond pyramid hardness No.
940	...	...	...	85.6	...	68.0	76.9	93.2	84.4	75.4	97	...	940
920	...	...	...	85.3	...	67.5	76.5	93.0	84.0	74.8	96	...	920
900	...	...	...	85.0	...	67.0	76.1	92.9	83.6	74.2	95	...	900
880	...	...	767	84.7	...	66.4	75.7	92.7	83.1	73.6	93	...	880
860	...	...	757	84.4	...	65.9	75.3	92.5	82.7	73.1	92	...	860
840	...	...	745	84.1	...	65.3	74.8	92.3	82.2	72.2	91	...	840
820	...	...	733	83.8	...	64.7	74.3	92.1	81.7	71.8	90	...	820
800	...	...	722	83.4	...	64.0	73.8	91.8	81.1	71.0	88	...	800
780	...	...	710	83.0	...	63.3	73.3	91.5	80.4	70.2	87	...	780
760	...	...	698	82.6	...	62.5	72.6	91.2	79.7	69.4	86	...	760
740	...	...	684	82.2	...	61.8	72.1	91.0	79.1	68.6	84	...	740
720	...	...	670	81.8	...	61.0	71.5	90.7	78.4	67.7	83	...	720
700	...	615	656	81.3	...	60.1	70.8	90.3	77.6	66.7	81	...	700
690	...	610	647	81.1	...	59.7	70.5	90.1	77.2	66.2	...	...	690
680	...	603	638	80.8	...	59.2	70.1	89.8	76.8	65.7	80	329	680
670	...	597	630	80.6	...	58.8	69.8	89.7	76.4	65.3	...	324	670
660	...	590	620	80.3	...	58.3	69.4	89.5	75.9	64.7	79	319	660
650	...	585	611	80.0	...	57.8	69.0	89.2	75.5	64.1	...	314	650
640	...	578	601	79.8	...	57.3	68.7	89.0	75.1	63.5	77	309	640
630	...	571	591	79.5	...	56.8	68.3	88.8	74.6	63.0	...	304	630
620	...	564	582	79.2	...	56.3	67.9	88.5	74.2	62.4	75	299	620
610	...	557	573	78.9	...	55.7	67.5	88.2	73.6	61.7	...	294	610
600	...	550	564	78.6	...	55.2	67.0	88.0	73.2	61.2	74	289	600
590	...	542	554	78.4	...	54.7	66.7	87.8	72.7	60.5	...	284	590
580	...	535	545	78.0	...	54.1	66.2	87.5	72.1	59.9	72	279	580
570	...	527	535	77.8	...	53.6	65.8	87.2	71.7	59.3	...	274	570
560	...	519	525	77.4	...	53.0	65.4	86.9	71.2	58.6	71	269	560
550	505	512	517	77.0	...	52.3	64.8	86.6	70.5	57.8	...	264	550
540	496	503	507	76.7	...	51.7	64.4	86.3	70.0	57.0	69	260	540
530	488	495	497	76.4	...	51.1	63.9	86.0	69.5	56.2	...	254	530
520	480	487	488	76.1	...	50.5	63.5	85.7	69.0	55.6	67	250	520
510	473	479	479	75.7	...	49.8	62.9	85.4	68.3	54.7	...	244	510
500	465	471	471	75.3	...	49.1	62.2	85.0	67.7	53.9	66	240	500
490	456	460	460	74.9	...	48.4	61.6	84.7	67.1	53.1	...	234	490
480	448	452	452	74.5	...	47.7	61.3	84.3	66.4	52.2	64	230	480
470	441	442	442	74.1	...	46.9	60.7	83.9	65.7	51.3	...	224	470
460	433	433	433	73.6	...	46.1	60.1	83.6	64.9	50.4	62	220	460
450	425	425	425	73.3	...	45.3	59.4	83.2	64.3	49.4	...	214	450
440	415	415	415	72.8	...	44.5	58.8	82.8	63.5	48.4	59	210	440
430	405	405	405	72.3	...	43.6	58.2	82.3	62.7	47.4	...	204	430

(From "Metals Handbook," vol. 1, 8th ed., pl. 1234, American Society for Metals, Metals Park, Ohio, 1961.)

Diamond pyramid hardness No.	Brinell hardness No., 10-mm ball, 3000-kg load			Rockwell hardness No.				Rockwell superficial hardness No., superficial brale penetrator			Shore scleroscope hardness No.	Tensile strength (approx.), 1000 psi	Diamond pyramid hardness No.
	Standard ball	10-mm ball, 3000-kg load Hultgren ball	Tungsten carbide ball	A-scale, 60-kg load, brale penetrator	B-scale, 100-kg load, 1/16-in. diam. ball	C-scale, 150-kg load, brale penetrator	D-scale, 100-kg load, brale penetrator	15-N scale, 15-kg load	30-N scale, 30-kg load	45-N scale, 45-kg load			
420	397	397	397	71.8	...	42.7	57.5	81.8	61.9	46.4	57	200	420
410	388	388	388	71.4	...	41.8	56.8	81.4	61.1	45.3	...	195	410
400	379	379	379	70.8	...	40.8	56.0	81.0	60.2	44.1	55	190	400
390	369	369	369	70.3	...	39.8	55.2	80.3	59.3	42.9	...	185	390
380	360	360	360	69.8	(110.0)	38.8	54.4	79.8	58.4	41.7	52	180	380
370	350	350	350	69.2	...	37.7	53.6	79.2	57.4	40.4	...	175	370
360	341	341	341	68.7	(109.0)	36.6	52.8	78.6	56.4	39.1	50	170	360
350	331	331	331	68.1	...	35.5	51.9	78.0	55.4	37.8	...	166	350
340	322	322	322	67.6	(108.0)	34.4	51.1	77.4	54.4	36.5	47	161	340
330	313	313	313	67.0	...	33.3	50.2	76.8	53.6	35.2	...	156	330
320	303	303	303	66.4	(107.0)	32.2	49.4	76.2	52.3	33.9	45	151	320
310	294	294	294	65.8	...	31.0	48.4	75.6	51.3	32.5	...	146	310
300	284	284	284	65.2	(105.5)	29.8	47.5	74.9	50.2	31.1	42	141	300
295	280	280	280	64.8	...	29.2	47.1	74.6	49.7	30.4	...	139	295
290	275	275	275	64.5	(104.5)	28.5	46.5	74.2	49.0	29.5	41	136	290
285	270	270	270	64.2	...	27.8	46.0	73.8	48.4	28.7	...	134	285
280	265	265	265	63.8	(103.5)	27.1	45.3	73.4	47.8	27.9	40	131	280
275	261	261	261	63.5	...	26.4	44.9	73.0	47.2	27.1	...	129	275
270	256	256	256	63.1	(102.0)	25.6	44.3	72.6	46.4	26.2	38	126	270
265	252	252	252	62.7	...	24.8	43.7	72.1	45.7	25.2	...	124	265
260	247	247	247	62.4	(101.0)	24.0	43.1	71.6	45.0	24.3	37	121	260
255	243	243	243	62.0	...	23.1	42.2	71.1	44.2	23.2	...	119	255
250	238	238	238	61.6	99.5	22.2	41.7	70.6	43.4	22.2	36	116	250
245	233	233	233	61.2	...	21.3	41.1	70.1	42.5	21.1	...	114	245
240	228	228	228	60.7	98.1	20.3	40.3	69.6	41.7	19.9	34	111	240
230	219	219	219	...	96.7	(18.0)	...	...	...	...	33	106	230
220	209	209	209	...	95.0	(15.7)	...	...	...	...	32	101	220
210	200	200	200	...	93.4	(13.4)	...	...	...	...	30	97	210
200	190	190	190	...	91.5	(11.0)	...	...	...	...	29	92	200
190	181	181	181	...	89.5	(8.5)	...	...	...	...	28	88	190
180	171	171	171	...	87.1	(6.0)	...	...	...	...	26	84	180
170	162	162	162	...	85.0	(3.0)	...	...	...	...	25	79	170
160	152	152	152	...	81.7	(0.0)	...	...	...	...	24	75	160
150	143	143	143	...	78.7	...	...	...	...	...	22	71	150
140	133	133	133	...	75.0	...	...	...	...	...	21	66	140
130	124	124	124	...	71.2	...	...	...	...	...	20	62	130
120	114	114	114	...	66.7	...	...	...	...	...	...	57	120
110	105	105	105	...	62.3	...	...	...	...	...	...	...	110
100	95	95	95	...	56.2	...	...	...	...	...	...	...	100
95	90	90	90	...	52.0	...	...	...	...	...	...	...	95
90	86	86	86	...	48.0	...	...	...	...	...	...	...	90
85	81	81	81	...	41.0	...	...	...	...	...	...	...	85

INDEX